Sustainable Development Goals Series

The **Sustainable Development Goals Series** is Springer Nature's inaugural cross-imprint book series that addresses and supports the United Nations' seventeen Sustainable Development Goals. The series fosters comprehensive research focused on these global targets and endeavours to address some of society's greatest grand challenges. The SDGs are inherently multidisciplinary, and they bring people working across different fields together and working towards a common goal. In this spirit, the Sustainable Development Goals series is the first at Springer Nature to publish books under both the Springer and Palgrave Macmillan imprints, bringing the strengths of our imprints together.

The Sustainable Development Goals Series is organized into eighteen subseries: one subseries based around each of the seventeen respective Sustainable Development Goals, and an eighteenth subseries, "Connecting the Goals", which serves as a home for volumes addressing multiple goals or studying the SDGs as a whole. Each subseries is guided by an expert Subseries Advisor with years or decades of experience studying and addressing core components of their respective Goal.

The SDG Series has a remit as broad as the SDGs themselves, and contributions are welcome from scientists, academics, policymakers, and researchers working in fields related to any of the seventeen goals. If you are interested in contributing a monograph or curated volume to the series, please contact the Publishers: Zachary Romano [Springer; zachary.romano@springer.com] and Rachael Ballard [Palgrave Macmillan; rachael.ballard@palgrave.com].

Shigeo Fujii · Shinya Funakawa ·
Hirohide Kobayashi · Izuru Saizen ·
Hidenori Harada · Hitoshi Shinjo ·
Hai Trung Huynh · Hai Hoang ·
Duc Thanh Tran · Van Quang Tran ·
An Van Le
Editors

Livelihood and the Environment in Vietnam

Springer

Editors
Shigeo Fujii
Graduate School of Global Environmental Studies
Kyoto University
Kyoto, Japan

Shinya Funakawa
Graduate School of Global Environmental Studies
Kyoto University
Kyoto, Japan

Hirohide Kobayashi
Graduate School of Global Environmental Studies
Kyoto University
Kyoto, Japan

Izuru Saizen
Graduate School of Global Environmental Studies
Kyoto University
Kyoto, Japan

Hidenori Harada
Graduate School of Asian and African Area Studies
Kyoto University
Kyoto, Japan

Hitoshi Shinjo
Graduate School of Global Environmental Studies
Kyoto University
Kyoto, Japan

Hai Trung Huynh
Hanoi University of Science and Technology
Hanoi, Vietnam

Hai Hoang
The University of Danang
Danang, Vietnam

Duc Thanh Tran
Hue University of Agriculture and Forestry
Hue, Vietnam

Van Quang Tran
The University of Danang - University of Science and Technology
Danang, Vietnam

An Van Le
Hue University of Agriculture and Forestry
Hue, Vietnam

ISSN 2523-3084 ISSN 2523-3092 (electronic)
Sustainable Development Goals Series
ISBN 978-981-97-2288-4 ISBN 978-981-97-2289-1 (eBook)
https://doi.org/10.1007/978-981-97-2289-1

This Springer imprint is published by the registered company Springer Nature Singapore Pte Ltd.
The registered company address is: 152 Beach Road, #21-01/04 Gateway East, Singapore 189721, Singapore

Foreword

Tackling environmental challenges requires contributions from diverse disciplines, from a broad range of specializations in sciences (e.g., understanding the dynamics of organisms in their ecosystems, chemistry of hazardous substances, and development of mitigation strategies for greenhouse gas emissions), to humanities (e.g., studying social institutions and human behavior to understand how technological innovations work in the society). Universities are expected to contribute to resolving the current and future environmental challenges by promoting innovation. Typically, academic research is prioritized in universities because of the all-inclusive nature of these institutions; however, regarding environmental issues, efforts contributing to the field are being sought after.

In 2002, the Graduate School of Global Environmental Studies (GSGES) was founded at Kyoto University in response to the above issues. GSGES started with 50 faculty members, 45 of whom were transferred from four other Graduate Schools and one Research Institute in Kyoto University. GSGES has established an interdisciplinary system of education and research collaboration with other organizations in the university. GSGES aims to train not only researchers but also environmental practitioners. In addition, given its focus on global environmental issues, GSGES encourages internationalization by facilitating lectures in English.

GSGES follows the philosophy of field study embodied throughout Kyoto University. The Asian Platform Project, which was launched in 2005 through GSGES resources, provided immense support to researchers and faculty members who, until that time, were often engaged in individual research activities. As part of the project, various activities were realized in Hanoi, Danang, and Hue, which functioned as important sites for research, education, and practice. To achieve the best outcomes, collaborative work was pursued while taking advantage of each researcher's expertise and independence. These collaborations included contributions from local researchers and professionals, primarily from partner universities. Without their contribution, the series of activities of GSGES in Vietnam would not have been possible, and the various international research and educational endeavors would not have been realized. In addition to the local researchers who are authors of this book, many other researchers and experts have contributed to this project. The writer expresses their gratitude toward them.

March 2024

Takeshi Katsumi
Dean, Graduate School of Global Environmental Studies
Kyoto University
Kyoto, Japan

Preface

This book introduces life and the environment in Vietnam and is based on research conducted by the Graduate School for Global Environmental Studies (GSGES), Kyoto University.

The impetus for this publication came from a book, "Exploring Academic Frontiers for a Sustainable Future: Challenges for Japan-ASEAN Research Collaboration—Toward Our Program Goal," which was published in March 2017 as an outcome of the JSPS Program for Advancing Strategic International Networks to Accelerate the Circulation of Talented Researchers (FY2014-2016), led by the Center for Southeast Asian Studies, Kyoto University. As part of the program, young faculty members were invited for long-term transfer to and from Southeast Asian universities, and GSGES played the role of inviting young researchers from Vietnam. In that book, GSGES contributed to Chapter 7, titled "Life and Environment in Vietnam," and 10 collaborative teams with Vietnamese counterparts contributed to a total of 24 pages.

These contents were presented as case studies and did not seem appropriate for the public. In the following year, I planned to publish a book on Vietnam for the public by GSGES and called for other GSGES members to contribute to the book. Most of the manuscripts were ready by the end of FY2017, but due to inadequate editorial work and management by me, the book was not published before March 2021, when I retired from Kyoto University.

In the next fiscal year, after receiving encouragement from my colleagues, the project was re-launched as a GSGES project with a new Editorial Committee consisting of six members—Hitoshi Shinjo, Shinya Funakawa, Hirohide Kobayashi, Izuru Saizen, Hidenori Harada and myself. The tasks of the committee were to review the table of contents and authors as well as collect and publish manuscripts. Committee meetings began on May 11, 2021, and a total of 26 were held until March 26, 2023, when all the manuscripts were ready for submission to a publisher. Thereafter, we communicated with the publisher regarding copywriting and formatting and finally completed the publication. Five key Vietnamese members, namely An Van Le, Hai Trung Huynh, Van Quang Tang, Duc Thanh Tran, and Hai Hoang, were invited as editors to review the manuscripts.

Ms. Saki Honma was of great help in organizing the style, coordinating the index, and communicating to Springer, and Mr. Yosuke Nishida of Springer kindly worked for our publication, patiently waiting for us to submit the manuscript, which was far behind schedule. We would like to express our deepest gratitude to Ms. Honma, Mr. Nishida, as well as the authors, editors, and other members who have helped us in our activities in Vietnam. We also thank Editage for English language editing. Finally, we would like to congratulate our team for the completion of this book.

March 2024

Shigeo Fujii
Chief Editor, Professor Emeritus
Kyoto University
Kyoto, Japan

Contents

Contributors

Misa Aoki Faculty of Human Life and Environment, Nara Women's University, Nara, Japan (Column 9)

Hung Quang Dinh School of Chemistry and Life Sciences, Hanoi University of Science and Technology, Hanoi, Vietnam (Section 3.3, Column 4)

Khac-Uan Do School of Chemistry and Life Sciences, Hanoi University of Science and Technology, Hanoi, Vietnam (Column 11)

Ngoc Thi Nhu Do School of Chemistry and Life Sciences, Hanoi University of Science and Technology, Hanoi, Vietnam (Section 5.5)

Shigeo Fujii Graduate School of Global Environmental Studies, Kyoto University, Kyoto, Japan (Section 1.2, 1.3, 3.1, 3.2, 3.3, 3.4, 3.5, 3.6, 4.7, 5.6, 5.8, Subsection 5.7.2, Column 4)

Shinya Funakawa Graduate School of Global Environmental Studies, Kyoto University, Kyoto, Japan (Section 2.3, 4.1, 4.4, 4.5, 6.1)

Hidenori Harada Graduate School of Asian and African Area Studies, Kyoto University, Kyoto, Japan (Section 4.7, 5.1, 5.8, Subsection 5.7.2, Column 5, 6, 11)

Duc Tan Ho Department of Organization, Administration and Facilities Management, University of Agriculture and Forestry, Hue University, Hue, Vietnam (Section 6.7)

Thong Trung Ho Faculty of Animal Science and Veterinary Medicine,University of Agriculture and Forestry, Hue University, Hue, Vietnam (Section 6.7)

Hai Hoang The University of Danang, Danang, Vietnam (Section 4.9)

Hoa Thi Thai Hoang Faculty of Agronomy, University of Agriculture and Forestry, Hue University, Hue, Vietnam (Section 4.4)

Tin Cong Hoang Faculty of Environmental Science, University of Sciences, Hue University, Hue, Vietnam (Section 4.3)

Akiko Iizuka Center for International Exchange, Utsunomiya University, Utsunomiya, Japan (Section 6.5, 6.6)

Toru Inui Graduate School of Engineering, Osaka University, Suita, Japan (Section 5.9)

Yoshizumi Kajii Graduate School of Global Environmental Studies, Kyoto University, Kyoto, Japan (Section 5.5)

Kinh Thi Kieu University of Science and Education, The University of Danang, Danang, Vietnam (Section 4.9, 6.2)

Hirohide Kobayashi Graduate School of Global Environmental Studies, Kyoto University, Kyoto, Japan (Section 5.1, 5.2, 5.3, 6.8)

Masataka Kuroda Overseas Water Supply Department, Nihon Suido Consultants Co., Ltd., Tokyo, Japan (Section 4.7)

Ha Thi Thu Le Graduate School of Global Environmental Studies, Kyoto University, Kyoto, Japan;
Faculty of Forestry, University of Agriculture and Forestry, Hue University, Hue, Vietnam (Section 2.1, 2.2, 2.6)

Nhan Thi Hien Le Nonprofit Organization FuDO (Fundament Design Organization), Kyoto, Japan (Section 6.3)

Bich-Thuy Ly School of Chemistry and Life Sciences, Hanoi University of Science and Technology, Hanoi, Vietnam (Section 3.2, 5.5, Column 8)

Noriko Nakamura NIJI Commercial and Development Joint Stock Company, Quang Tri, Vietnam (Section 6.3)

Binh Huu Ngo Faculty of Environmental Science, University of Sciences, Hue University, Hue, Vietnam (Section 4.3)

Hong Thi Nguyen Graduate School of Global Environmental Studies, Kyoto University, Kyoto, Japan (Section 6.4)

Lam Ho Nguyen High-Tech Agriculture Division, University of Agriculture and Forestry, Hue University, Hue, Vietnam (Section 4.5)

Lan Chau Nguyen Faculty of Civil Engineering, University of Transport and Communications, Hanoi, Vietnam (Section 5.9)

Lien Pham Hong Nguyen School of Chemistry and Life Sciences, Hanoi University of Science and Technology, Hanoi, Vietnam (Section 3.6, 5.9)

Thai Thi Kim Nguyen Faculty of Environmental Engineering, Hanoi University of Civil Engineering, Hanoi, Vietnam (Section 1.1)

Tung Ngoc Nguyen Faculty of Architecture, University of Sciences, Hue University, Hue, Vietnam (Section 5.2, 5.3, 6.8)

Hung Nguyen Duy SOLEN Environmental Solutions Joint-Stock Company, Hanoi, Vietnam (Section 3.1, 3.4, 3.5)

Yuki Okamoto Technology Integration Section, Solution Management Division, WorldLink & Company Co. Ltd., Kyoto, Japan (Section 4.2)

Giang Huong Pham Graduate School of Global Environmental Studies, Kyoto University, Kyoto, Japan (Section 4.7)

Huu Ty Pham University of Agriculture and Forestry, Hue University, Hue, Vietnam (Section 4.9)

Nguyet-Anh Pham Faculty of Chemistry and Environment, Thuyloi University, Hanoi, Vietnam (Section 5.8, Subsection 5.7.2)

Izuru Saizen Graduate School of Global Environmental Studies, Kyoto University, Kyoto, Japan (Section 4.1, 4.3, 4.8, 5.4, 6.4, Column 11)

Ryo Sakaguchi Graduate School of Global Environmental Studies, Kyoto University, Kyoto, Japan (Section 4.7, Column 5)

Yosuke Sakamoto Graduate School of Global Environmental Studies, Kyoto University, Kyoto, Japan (Section 5.5)

Toshihiko Shine Research Institute for Languages and Cultures of Asia and Africa, Tokyo University of Foreign Studies, Tokyo, Japan (Section 2.4, 2.5, Column 1, 2, 3)

Hitoshi Shinjo Graduate School of Global Environmental Studies, Kyoto University, Kyoto, Japan (Section 2.1, 2.2, 2.7, 4.6)

Jane Singer Department of Global Studies, Kyoto University of Foreign Studies, Kyoto, Japan (Section 4.9, 6.2)

Atsushi Takai Graduate School of Global Environmental Studies, Kyoto University, Kyoto, Japan (Section 5.9)

Shuhei Tanaka Graduate School of Global Environmental Studies, Kyoto University, Kyoto, Japan (Subsection 5.7.1, Column 10)

Ueru Tanaka Faculty of Agriculture, Setsunan University, Neyagawa, Japan (Section 6.5, 6.6, 6.7)

Minori Tokito Graduate School of Agriculture, Kyoto University, Japan (Section 4.8, Column 7)

An Thi Tran Faculty of Management Science, Thu Dau Mot University, Thu Dau Mot, Vietnam (Section 5.4)

Ni Ngoc Khanh Tran Faculty of Environmental Science, University of Sciences, Hue University, Hue, Vietnam (Section 4.3)

Van Quang Tran University of Science and Technology, The University of Danang, Danang, Vietnam (Subsection 5.7.1)

Quynh-Anh Tran-Nguyen The University of Danang - University of Science and Education, Da Nang, Vietnam (Section 5.8)

Minh Phuoc Truong The University of Danang - University of Science and Education, Danang, Vietnam (Section 5.4)

Phuong Hoang Truong Faculty of Architecture, University of Sciences, Hue University, Hue, Vietnam (Section 6.8)

Dieu-Anh Van School of Chemistry and Life Sciences, Hanoi University of Science and Technology, Hanoi, Vietnam (Section 5.6)

Minh Tuan Vu Faculty of Agronomy, University of Agriculture and Forestry, Hue University, Hue, Vietnam (Section 4.6)

Miki Yoshizumi College of Gastronomy Management, Ritsumeikan University, Kyoto, Japan (Section 5.2, 5.3, 6.3)

Abbreviations

ACI	Agrifood Consulting International
APs	Affected People
ARVN	Army of the Republic of Vietnam
AVS	Acid-volatile sulfide
BOD	Biochemical Oxygen Demand
BRI	Bottom Reflectance Index
BS	Base Saturation
BTEX	Benzene, toluene, ethylbenzene, *m,p*-Xylene, and *o*-Xylene
C&D	Construction and Demolition
CDOM	Colored Dissolved Organic Matter
CEC	Cation Exchange Capacity
CENM	Center for Environmental and Natural Resources Monitoring (Vietnam)
COD	Chemical Oxygen Demand
CS	Campus Sustainability
CV	Coefficient of Variation
DII	Depth Invariant Index
DO	Dissolved Oxygen
DONRE	Department of Natural Resource and Environment (Vietnam)
EC	Electrical Conductivity
ECe	Electrical Conductivity of soil saturation paste
ECw	Electrical Conductivity of soil solution
EIA	Environmental Impact Assessment
ESD	Education for Sustainable Development
EURO	European Emission Standards
GDP	Gross Domestic Product
GHGs	Greenhouse Gases
GIS	Geographic Information System
GNSS	Global Navigation Satellite System
GPS	Global Positioning System
GRDP	Gross Regional Domestic Product
GSGES	Graduate School of Global Environmental Studies (Kyoto University)
GSO	General Statistics Office of Vietnam
GWQI	Groundwater Quality Index
HCMC	Ho Chi Minh City

HHs	Households
HTGHs	Hue Traditional Garden Houses
HUAF	University of Agriculture and Forestry, Hue University
HUST	Hanoi University of Science and Technology
JICA	Japan International Cooperation Agency
JPP	JICA Partnership Program
JSPS	Japan Society for the Promotion of Science
JST	Japan Science and Technology Agency
LaL	Land Law
LASV	Liberation Army of South Vietnam
LEP	Law on Environmental Protection
LWR	Law on Water Resources
MEXT	Ministry of Education, Culture, Sports, Science and Technology (Japan)
MH-CEM	Middle and Highland Center for Environmental Monitoring (Vietnam)
MOC	Ministry of Construction (Vietnam)
MONRE	Ministry of Natural Resources and Environment (Vietnam)
MoST	Ministry of Science and Technology
MOU	Memorandum of Understanding
MSW	Municipal Solid Waste
MSWIA	Municipal Solid Waste Incinerator Ash
MSWM	Municipal Solid Waste Management
N-CEM	Northern Center for Environmental Monitoring (Vietnam)
NOx	Nitrogen Oxides
NRTI	Nighttime Radiation Temperature Inversion
OA	Overall Accuracy
OC	Organic Carbon
ODA	Official Development Assistance
PAN	Peroxyacetyl Nitrate
PAVN	People's Army of Vietnam
PFES	Payment for Forest Environmental Services
PM	Particulate Matter
PMD	Prime Minister Decision
RFA	Recycled Fine Aggregate
RON	Research Octane Number
SAR	Sodium Adsorption Ratio
SAV	Submerged Aquatic Vegetation
S-CEM	Southern Center for Environmental Monitoring (Vietnam)
SDGs	Sustainable Development Goals
SS	Suspended Solids
STI	Subsidence Temperature Inversion
SW	Solid Waste
SWM	Solid Waste Management
TKN	Total Kjeldahl Nitrogen
TN	Total Nitrogen
TP	Total Phosphorus
TSP	Total Suspended Particle
TSS	Total Suspended Solids

UDDT	Urine-Diverting Dry Toilet
UD-UED	The University of Danang University of Science and Education
URENCO	Urban Environmental Company
USD	US dollar
USGS	United States Geological Survey
VAST	Vietnam Academy of Science and Technology
VCP	Vietnamese Communist Party
VEA	Vietnam Environment Administration
VietGAP	Vietnamese Good Agricultural Practices
VOCs	Volatile Organic Compounds
WHC	Water Holding Capacity
WHO	World Health Organization
WQI	Water Quality Index
WWTPs	Wastewater Treatment Plants

1 Introduction

Shigeo Fujii and Thai Thi Kim Nguyen

Abstract

This book is a compilation of global environmental research related to the Sustainable Development Goals (SDGs) in Vietnam, which was jointly conducted by faculty members of the Graduate School of Global Environmental Studies (GSGES), Kyoto University, and their Vietnamese counterparts. This chapter introduces the book and presents the status of SDGs in Vietnam in Sect. 1.1. Section 1.2 summarizes the various projects in Vietnam and the activities that have progressed under them, explaining the background and network that facilitated the research presented in this book. Finally, Sect. 1.3 illustrates the structure of the book and introduces its content.

Keywords

SDGs · GSGES · Projects · Education · Research

S. Fujii (✉)
Graduate School of Global Environmental Studies, Kyoto University, Kyoto, Japan
e-mail: fujii.shigeo.6z@kyoto-u.jp

T. T. K. Nguyen
Faculty of Environmental Engineering, Hanoi University of Civil Engineering, Hanoi, Vietnam

1.1 SDGs in Vietnam

SDGs are a novel, universal set of goals and targets that the United Nations Member States are expected to adopt within the framework of their agendas and political policies for the next 15 years. The SDGs were adopted in 2015 by all United Nations Member States as a universal call to action to end poverty, protect the planet, secure everyone, and allow the population to enjoy peace and prosperity by 2030.

Sachs et al. (2023) obtained the Score (present state) of the SDG Index for each country and showed that Vietnam ranked 55th out of 166 countries in the world. Compared with neighboring countries, the rank was lower than Thailand (43rd) but higher than Malaysia (78th), Cambodia (103rd), and Laos (115th). The SDG index has a high-rank correlation coefficient with the economic indicator, GNI per capita (The World Bank (n.d.)) at 0.803, but Vietnam ranked 135th out of 203 countries in GNI per capita, showing more awareness of the SDGs with comparison of the economic level.

Figure 1.1 shows the category of Score (present state) and Trend (to achievement) for each of the 17 Goals. Compared to the global trend, goals 15 (Climate action), 10 (Reduced inequalities), and 7 (Affordable and clean energy) have inferior Trend, while no goals have inferior

S. Fujii et al. (eds.), *Livelihood and the Environment in Vietnam*, Sustainable Development Goals Series,
https://doi.org/10.1007/978-981-97-2289-1_1

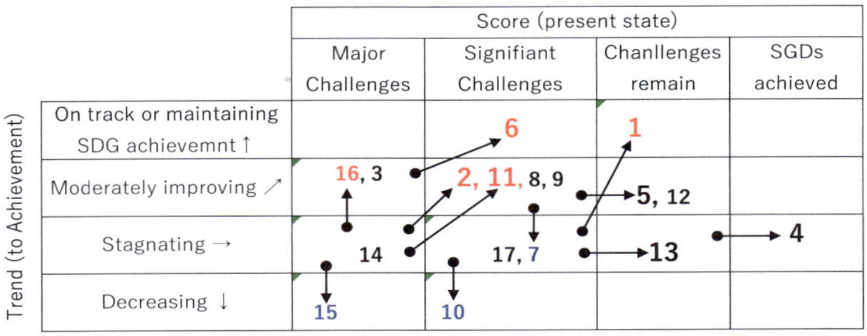

Fig. 1.1 Score and Trend of the SDG Index in Vietnam compared with the world averages, the figure is newly illustrated by the ranks in Sachs et al (2023). Numbers indicate No. of SDGs. Bigger font numbers indicate better state than world average (no worse SDG state). Red indicates better trend than the world averages, while blue indicates lower. World average ●——▶ Vietnam

Score. Three Goals of 1 (No poverty), 2 (Zero hunger), and 11(Sustainable cities and communities) are superior in both Score and Trend.

Vietnam is strongly committed to the implementation of SDGs through the issuance of the Vietnam Sustainable Development Strategy in 2004, the Vietnam Sustainable Development Strategy for the period of 2011–2020, and, most recently, the National Action Plan to implement the 2030 Agenda for Sustainable Development.

On September 25, 2020, the government issued Resolution 136/NQ-CP on Sustainable Development (Government of Vietnam 2020). The resolution sets out 17 SDGs for Vietnam to be achieved by 2030: (1) ending all forms of poverty everywhere; (2) eliminating hunger, ensuring food security, improving nutrition, and promoting sustainable agricultural development; (3) ensuring healthy living and promoting well-being for people of all ages; (4) ensuring quality, equitable, and inclusive education and promoting lifelong learning opportunities for all; (5) achieving gender equality, empowering and creating opportunities for women and girls; (6) ensuring adequate and sustainable management of water resources and sanitation for all; (7) ensuring access to affordable, reliable, and sustainable energy for all; (8) ensuring sustainable, comprehensive, and continuous economic growth, with full employment, productivity, and decent work for all; (9) building resilient infrastructure, promoting inclusive and sustainable industrialization, and fostering innovation; (10) reducing inequality in society; (11) achieving sustainable and resilient urban and rural development, ensuring a safe living and working environment, and reasonably allocating population and labor by region; (12) ensuring sustainable production and consumption; (13) ensuring timely and effective responses to climate change and natural disasters; (14) conserving and sustaining the use of oceans, seas, and marine resources for sustainable development; (15) protecting and developing forests sustainably, conserving biodiversity, developing ecosystem services, combating desertification, preventing degradation, and restoring land resources; (16) promoting a peaceful, democratic, fair, equal, and civilized society for sustainable development, creating access to justice for all, and building effective, accountable, and participatory institutions at all levels; and (17) strengthening implementation practices and promoting global partnerships for sustainable development.

To accomplish these goals, the Vietnamese government has set out tasks and solutions for the synchronous and effective implementation of the National Action Plan for the execution of the 2030 Agenda for Sustainable Development, focusing on perfecting the system of institutions and policies; strengthening information and communication; promoting the role and participation of stakeholders; allocating, mobilizing, and strengthening financial resources; and strengthening international cooperation. The Ministry of Industry and Trade was assigned three tasks: presiding over and organizing the implementation of the National Action Program on sustainable production and consumption for 2021–2030 (Prime Minister 2020); completing policies on sustainable production and consumption; promoting the greening of the distribution system and developing a national supply chain of environmentally friendly products and services; prioritizing products produced and supplied by small- and medium-sized enterprises; and promoting the development of environmental and waste recycling industries.

The Prime Minister issued Decision No. 841/QD-TTg dated July 4, 2023 on the roadmap to implement Vietnam's sustainable development goals until 2030 (Prime Minister 2023). In the Decision setting out a specific roadmap, the lead agency will implement 17 sustainable development goals in Vietnam by 2030 with 117 targets. This decision indicated that by 2030, the multidimensional poverty rate will maintain a decrease of 1–1.5%/year; the proportion of the labor force of working age participating in social insurance is 60%; average income per capita in rural areas increased by more than 2.5–3 times compared to 2020; gross domestic product (GDP) per capita 7,500 USD; the value of exported goods will grow by an average of 5–6%/year in the period 2026–2030.

Another policy related to SDGs was seen in "Water Resources Planning for 2021–2030, with a Vision to 2050" (Prime Minister 2022). Recent progress in SDGs implementation in Vietnam has been reported by the government (Ministry of Planning and Investment 2021).

1.2 GSGES Activities in Vietnam

1.2.1 Establishment of GSGES and Its Characteristics

GSGES is a small, interdisciplinary graduate school with 50 faculty members (1.9% of the total faculty at Kyoto University) specializing in engineering, agriculture, science, economics, humanities, and social sciences.

The admission policy states that "the Graduate School of Global Environmental Studies seeks to secure the current and future health and sustainability of the global environment in two essential ways—first, by training future researchers and practitioners for coming environmental careers in academia, management, policy, advocacy, and project implementation, and second, by establishing sustainability as the basis of an integrated and interdisciplinary field informed by ethics, science, technology, and humanities that can address the environmental challenges confronting the twenty-first century." Accordingly, the school has imparted education with a unique curriculum that is not found in other Japanese universities.

One of the unique features of education at GSGES is that lectures are conducted in English, and students can complete their studies solely in English, even without any knowledge of the Japanese language, which was rather rare in Japan in its founding year (2002). Subsequently, the Japanese Government promoted English education in graduate schools and the number of graduate schools teaching English increased slightly. The program places special emphasis on practical training and interdisciplinary education, with long-term internship programs (three months or longer), field training, and other experiential exercises constituting practical training and four lecture courses on environmental engineering, ecology, environmental ethics and education, and environmental policy and economics constituting interdisciplinary education, all of which are compulsory subjects in the master's degree program.

1.2.2 Development of Various Projects in Vietnam

Based on the above policy and the motto of "think globally, act locally," GSGES is actively promoting overseas activities in collaboration with local researchers and students, with a particular emphasis on Vietnam. Figure 1.2 summarizes the major Vietnamese projects in the form of a chronological table.

The first fully fledged activities of GSGES were initiated through the "Asia Platform" from FY2005 to FY2009 (GSGES 2011), which was implemented with its internal budget. The project established an overseas base in Vietnam for research and educational development and intended to obtain external funds through the accumulation of its achievements. Through this project, several participating faculty members individually received research funds, such as a Grant-in-Aid for Scientific Research, and GSGES itself obtained a Japan International Cooperation Agency (JICA) Grassroots Project, which was implemented in two phases over six years (September 2006–August 2009 (JICA n.d.(a)) and October 2010–September 2013 (JICA n.d.(b)). The base was established at the University of Agriculture and Forestry, Hue University (HUAF). Activities were initially conducted in Hue and its vicinity and subsequently expanded to Danang. The activities of the Asia Platform and JICA Grassroots Projects are described in Sects. 6.5–6.7 of this book.

These activities led to the adoption and application of the "Strategic Environmental Leader Formation Project" (FY2008–2012) under the Special Coordination Funds for Promoting Science and Technology offered by the Japan Science and Technology Agency (JST). JST called for projects to form a center for fostering human resources to exercise leadership in solving environmental problems in developing countries, and GSGES proposed promoting education by establishing overseas bases and activating long-term internships for students there. Vietnam was the major country involved in this project. Specific assistant professors and bases in charge were assigned to the Hanoi University of Science and Technology (HUST) and HUAF, and the dispatch and recruitment of students for enrollment were promoted, further inspiring education and research activities. In addition, with the proposal to educate international students based on the educational program content of this project, GSGES applied and was selected for the Priority Scholarship Program of the Ministry of Education, Culture, Sports, Science, and Technology (MEXT), receiving four scholarships for international doctoral students five times since October 2009. Seven of the twenty scholarship students were from Vietnam, and many were incumbent faculty members who returned to their original positions after obtaining their doctorate degrees, further deepening their educational and research collaboration. Of note, the editor of this book, Tran Thanh Duc, and the

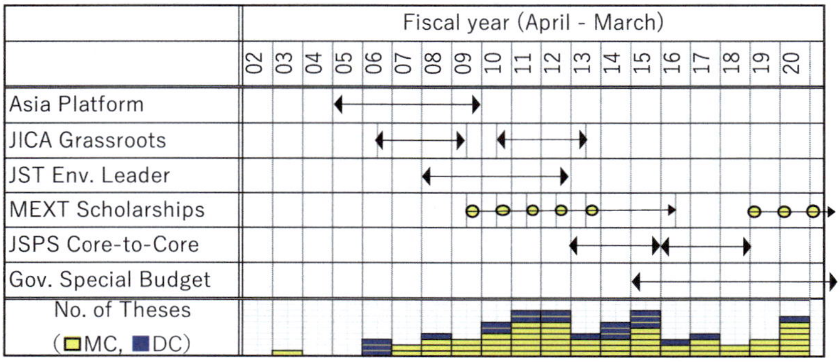

Fig. 1.2 Chronology of important projects in GSGES and theses related to Vietnamese studies

authors Hung Nguyen Duy, Minh Tuan Vu, Tung Ngoc Nguyen, Nguyet-Anh Pham, and Quynh-Anh Tran-Nguyen are GSGES graduates with MEXT scholarships.

The Strategic Environmental Leader Project was completed in 2012; however, the Japan Society for the Promotion of Science (JSPS) Core-to-Core Project (JSPS n.d.) was acquired twice consecutively (FY2013–2015 and FY2016-2018). The Core-to-Core Project aims to establish a world-class or regional core center for cutting-edge research issues that are internationally important or contribute to solving regional problems, in cooperation with overseas research and education institutions. GSGES aimed to build an academic network in Southeast Asian countries to foster the next generation of core young researchers, in which three Vietnamese universities—HUST, Hue University, and Danang University—were expected to work as core institutions. For this project, many researchers and students were invited from Vietnam and dispatched from Kyoto University. Moreover, international symposia were held, more often than annually, to further promote collaborative research.

Furthermore, GSGES directly requested the Japanese government to provide a special budget for the project named "Innovator" (Fujii 2016), and the project was accepted to start in FY2015. Innovators aimed to create new educational programs and research mechanisms in the environmental field using overseas bases, such as those at Hanoi, Hue, and Danang. This project initially had a four-year time limit; however, the time limit has been eliminated since FY2018 and the project has been approved as a core expense.

Table 1.1 lists the Vietnamese organizations that have signed the agreement, along with the year of conclusion. Of these, HUST, Hue University, and Danang University opened overseas bases with video conferencing systems from 2006 to 2008 and have been particularly active in collaborative works.

Table 1.1 Conclusion of the memoranda of understanding (MOU) with Vietnamese organizations and establishment of offices

University	Agreement			Office
	MOU1	MOU2	MOU3	
Hue University	2004*1	2007	2013*3	2006
Vietnam Academy of Science and Technology (VAST)	2004	–	–	–
Hanoi University of Civil Engineering	2005	–	–	–
Danang University	2007*2	2014	2013*2	2007
Hanoi University of Science and Engineering	2008	2009	2013	2008

MOU1: School-level MOU on research exchange
MOU2: University-level MOU on research exchange
MOU3: School-level MOU on student exchange
*1With HUAF
*2With Danang University of Science and Technology
*3with HUAF and Hue University of Science

1.2.3 Education and Research Programs

Utilizing various funds, GSGES has continued to (1) hold international symposia, (2) accept and invite Vietnamese students, (3) dispatch Kyoto University students and researchers to Vietnam, and (4) conduct joint research.

International symposia have been held continuously in Southeast Asia, mainly in Vietnam, since 2009, with twenty-three held before 2021, including eight each in Kyoto and Vietnam (three in Hue, two in Hanoi, one in Danang, one in Hoi An, and one in Can Tho).

The most important aspect of student acceptance is enrollment as a regular student in the master's and doctoral programs of GSGES. From April 2002 to April 2022, the school accepted 467 international students (37% of the total) from 67 countries into its master's and doctoral programs, including 39 students from Vietnam, ranking third after China and Indonesia. This number is quite high at the graduate school level in Japan. Although the number of international students from Vietnam has been increasing rapidly in recent years (Fujii 2021), most are in preparatory education, such as Japanese-language schools. Vietnamese students in GSGES were excellent and received various scholarships, such as MEXT scholarship priority placements, Vietnamese government scholarships, AUN/SEED-Net (the ASEAN University Network, the Southeast Asia Engineering Education Development Network) scholarships, and JICA-JDS (The Project for Human Resource Development Scholarship), among others. After completing their studies, the students returned to Vietnam and played a role in further strengthening research and educational collaboration there.

In addition, educational programs for the preliminary stage of enrollment were conducted at three Vietnamese universities (Hue University, HUST, and Danang University) as designated centers. These include the Special Auditing Student Program (graduate students are invited for six months to attend GSGES lectures and receive research supervision; 13 students before 2022) since 2013 and the two-week Kyoto Study Tour for undergraduate students (including visits to environmental facilities, special lectures, and student exchange meetings; 56 students six times before 2019) since 2009.

Meanwhile, students and faculty members are actively being dispatched to Vietnam. More overseas trips to Vietnam were recorded than to other countries, totaling 154 trips of 1,907 person-days for nearly 200 GSGES members from April 2014 to March 2017 (the second highest number of trips was recorded to Thailand, totaling 117 trips of 457 person-days). The large number of trips was due to international symposia and meetings, and the long total number of days of stay was due to the large number of internships and subsequent dispatches of students to continue research, with major contributions from the three bases at Hanoi, Hue, and Danang.

1.2.4 Research in Vietnam

These activities have led to the production of many theses in Vietnam. As shown in Fig. 1.1, the number of theses was small at the beginning of its establishment, although it gradually increased with the creation of the Asia Platform, averaging over five per year since 2010, totaling 20 doctoral theses and 55 master's theses by FY2020 (both 8% of the total) (Fig. 1.3).

While Hue, Hanoi, and Danang, where overseas offices were established, account for 51, 25, and 21% of the total, respectively, some studies have been conducted in the surrounding provinces and even in southern Vietnam, covering the entire country. Table 1.2 shows the research topics, indicating the diversity of topics across the humanities and sciences. Environmental pollution, livelihoods, and culture were the topics of 39, 27, and 12% of studies, respectively. Many of the studies in this book are based on these dissertations, and the chapters and section numbers of these studies are provided in the table.

In addition, GSGES has provided seed research funds (budgeted at USD 800–4, 800 per project) to young faculty members of the three Vietnamese universities (and faculty members of three other universities in Cambodia, Laos, and the Philippines). The program has been implemented annually since FY2013. By FY2022, 125 researchers had been selected to participate in the program, contributing to the fostering of research by young researchers. Some recipients are the authors of this book and some of their research outcomes are presented.

Many of these studies have contributed significantly to studies on SDGs and their implementation in Vietnam. Other work contributing to the SDGs by Japanese organizations can be observed in the JICA grassroots projects (JICA n.d.(c)) and SATREPS projects (JST (n.d.)).

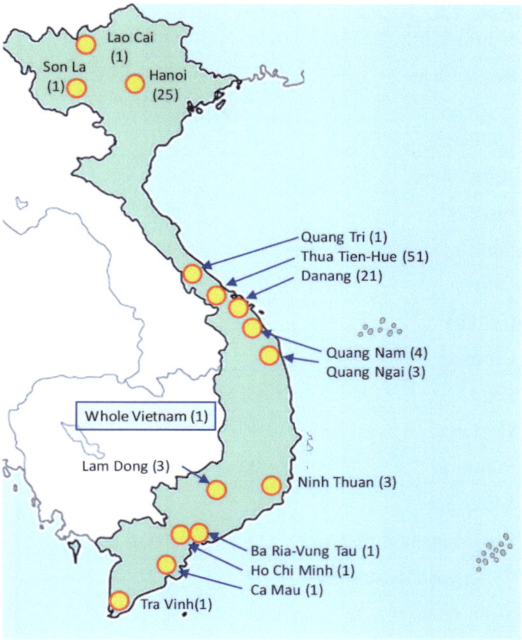

Fig. 1.3 GSGES study sites of master's and PhD doctoral studies in Vietnam (FY2003–2020, %, $n = 75$)

1.3 Structure of This Book

This book is a compilation of the abovementioned research activities in Vietnam conducted by GSGES faculty members and their counterparts. The first chapter is followed by Chaps. 2–6.

Chapter 2 provides an overview of Vietnam's climate, geography, geology, history, races, industries, and relations with Japan, as well as fundamental information for understanding the individual items in the following chapters. Chapter 3 summarizes Vietnam's environmental laws and regulations, highlighting how laws have been developed for air, water, solid waste, and soil, among others, and provides basic information for discussing the environment and pollution in the subsequent chapters.

Chapters 4–6 are the major parts of the book and present many research results. To organize the research results in an easy-to-understand manner, this book divides the research content into two categories: issues in rural areas, as discussed in Chap. 4, and in urban areas, as discussed in Chaps. 4 and 5. Furthermore, Chap. 6 describes the research related to the future of Vietnam. Chapter 4 focuses on the livelihoods of fisheries, agriculture, and forestry and illustrates rural waste and migration due to dam construction, based primarily on research conducted in Hue. Chapter 5 covers environmental pollution and its countermeasures, such as residential, water use, air pollution, wastewater management, waste, and global warming issues, based on research conducted mainly in Hanoi and Danang. Chapter 6, the final chapter, builds on the contents of Chaps. 4 and 5 with a view to the future, including the achievement of SGDs, as well as highlights environmental education, grassroots movements, and ethnic minority issues.

In addition, this book includes columns on unique Vietnamese customs and lifestyles, which are placed in appropriate instances in Chaps. 2–6. These include environmental issues such as novel water purification and sludge treatment processes, customs unique to Vietnam, such as the Chinese New Year and Teachers'

Table 1.2 Research areas of master's and doctoral studies in GSGES (FY2003–2020, %, $N=75$)

Research area	%	Chapter and section
Industries	27	–
Agriculture	7	4.4, 4.5, 4.6
Fishery	9	4.2
Forestry	4	4.8
Others	7	6.4, 6.7
Culture	12	–
Clothing	3	6.6
Food	1	Column
Housing	8	4.9, 5.2, 5.3, 6.8
Pollution	38	5.7
Water use	5	5.7
Environmental pollution	15	5.5, 5.6
Toilets and wastewater	13	5.8
Solid waste	5	4.7, 5.9
Policy and Governance	18	–
Education	3	6.2, 6.3
Disaster prevention	8	None
Climate change	4	5.4
Resident participation	3	6.5
Natural environment	5	4.3

Day, the food culture of ethnic minorities, and bicycle and motorcycle issues, all of which are unique to Vietnam.

Chapters 4–6, the main part of this book, contain 23 sections; excluding Sect. 1.1 (Introduction), they focus on studies or reports indirectly or directly related to the SDGs in Vietnam. Table 1.3 summarizes how the contents of this book relate to each SDG by indicating the corresponding targets. The 17 SDGs include 169 targets, of which studies and research related to 13 goals and 24 targets are presented in the 23 sections of this book. Goal 11 (sustainable cities and communities) has the largest number of sections (six), followed by Goal 2 (zero hunger) with four sections, and Goals 4 (quality education), 6 (clean water and sanitation), and 12 (responsible consumption and production), with three sections each. Vietnam is a developing country with a multi-ethnic population, and its ancient culture and lifestyle are rapidly changing amid rampant modernization and urbanization. The authors are pleased to share with readers how activities are being undertaken to achieve SGDs under these circumstances.

Table 1.3 Relationship of this book's contents with sustainable development goals and their targets

Goal	Chapter and section (target no.)
G01: No poverty	5.4 (1.5)
G02: Zero hunger	4.4 (2.4), 4.5 (2.4), 4.6 (2.3), 6.7 (2.3)
G03: Good health and well-being	5.5 (3.9), 5.6 (3.9)
G04: Quality education	6.2 (4.7), 6.3 (4.a), 6.6 (4.7)
G05: Gender education	6.6 (5.b)
G06: Clean water and sanitation	5.6 (6.3), 5.7 (6.1), 5.8 (6.2, 6.3)
G08: Decent work and economic growth	6.4 (8.9)
G10: Reduced inequalities	4.9 (10.7)
G11: Sustainable cities and communities	5.2 (11.3). 5.3 (11.4), 5.5 (11.6), 5.9 (11.6), 6.5 (11.5), 6.8 (11.c)
G12: Responsible consumption and production	4.7 (12.4, 12.5), 5.9 (12.4, 12.5), 6.4 (12.b)
G13: Climate action	5.4 (13.1)
G14: Life below water	4.2 (14.4), 4.3 (14.2)
G15: Life on land	4.8 (15.2)

References

Fujii S (2016) A new challenge by GSGES, Kyoto University—Activities after EML/HSE projects and launching of a renewed comprehensive project. Environ Sanitary Eng Res 30(2):12–18 (in Japanese)

Fujii S (2021) International contribution in water environment by universities. In: Japan Society of Water Environment (eds) Encyclopedia of water environment. Asakura Publishing (in Japanese), pp 504–507

Graduate School of Global Environmental Studies (GSGES) (2011) Project Report of GSGES Asian Platform

Government of Vietnam (2020) Resolution 136/NQ-CP dated September 25, 2020 on Sustainable Development

JICA (Japan International cooperation Agency) (n.d.(a)) 2005 accepted project of Partner Type (Citizen Participation) of Grassroots. Accessed 28 March 2023 from https://www.jica.go.jp/partner/kusanone/partner/ku57pq00000x9tkg-att/vie_05_p.pdf (in Japanese)

JICA (Japan International Cooperation Agency) (n.d.(b)) 2009 accepted project of Partner Type (Citizen Participation) of Grassroots. Accessed 28 March 2023 from https://www.jica.go.jp/partner/kusanone/partner/ku57pq00000x9tkg-att/vie_11_p.pdf (in Japanese)

JICA (Japan International cooperation Agency) (n.d.(c)) Grassroots technical cooperation projects. Accessed 10 Nov 2023 from https://www.jica.go.jp/activities/schemes/partner/kusanone/index.html (in Japanese)

JSPS (Japan Society for the Promotion of Science) (n.d.) Core-to-Core Program. Accessed 28 March 2023 from https://www.jsps.go.jp/english/e-c2c/completed_b.html

JST (Japan Society for the Promotion of Science) (n.d.) SATRREPS (Science and Technology Research Partnership for Sustainable Development). Accessed 10 Nov 2023 from https://www.jst.go.jp/global/english/index.html

Ministry of Planning and Investment (MPI, Vietnam) (2021) National Report 2020 Progress of Five-Year-Implementation of Sustainable Development Goals

Prime Minister of Vietnam (2020) Decision No. 889/QD-TTg, issued on June 24, 2020

Prime Minister of Vietnam (2022) Decision No. 1622/QD-TTg dated December 27, 2022

Prime Minister of Vietnam (2023) Decision No. 841/QD-TTg dated July 4, 2023

Sachs JD, Lafortune G, Fuller G, Drumm E (2023) Sustainable Development Report (2023)–Implementing the SDG Stimulus Includes the SDG Index and Dashboards-, Dublin University Press. Accessed 19 Jan 2024 from https://s3.amazonaws.com/sustainabledevelopment.report/2023/sustainable-development-report-2023.pdf

The World Bank (n.d.) GNI per capita, Global Economic Prospects 2024. Accessed 19 Jan 2024 from https://data.worldbank.org/indicator/NY.GNP.PCAP.CD

Overview of Vietnam

2

Hitoshi Shinjo, Ha Thi Thu Le, Shinya Funakawa and Toshihiko Shine

Abstract

Vietnam is situated along the eastern coast of the Indochina Peninsula in Southeast Asia. It appears as an S-shaped strip of land on the map, extending for 1,650 km from the north to south. Its population was 97.6 million people in 2020, with a population density of 295 people/km^2. Its diverse topography includes hills, mountains, plains, coastlines, and continental shelves, reflecting a long history of geology and topography formation in a monsoon, humid climate, and extreme weather. Based on the climate, geology, topography, and soils, the ecological conditions are categorized into three major regions; northern mountainous area, southern central plateau, and alluvial plains. Its history has been shaped by China's occupation, independent dynasties, French colonization, Vietnam War, and socialist market economy. Its ethnic groups are divided into 54 groups, of which one majority (Kinh) occupies 85%. The share of services in GDP in 2020 was overwhelming (42%) compared with industry (approximately 34%) and agriculture, forestry, and fishing (15%). Cooperation between Vietnam and Japan in a wide range of fields such as politics, science and technology, culture, and human exchange is being undergone. As columns, Hue city, *Tết*, and Teacher's day are introduced.

Keywords

Culture · Economy · Ethnicity · History · Topography

2.1 Basic Information

Vietnam is situated along the eastern coast of the Indochina Peninsula in Southeast Asia with a land area of 331,344 km^2 (GSO 2022). The Vietnam mainland spans across longitudes 102°8′E to 109°27′E and latitudes 8°27′N to 23°23′N (Fig. 2.1). It appears as an S-shaped strip of land on the map, extending for 1650 km from the north to south, and is approximately 50 km wide at its narrowest point. The country has a long land border of 4,550 km, bordering China to the West, Laos and Cambodia in the West, the East Sea of the Pacific Ocean in the

H. Shinjo (✉) · H. T. T. Le · S. Funakawa
Graduate School of Global Environmental Studies, Kyoto University, Kyoto, Japan
e-mail: shinjo.hitoshi.4x@kyoto-u.ac.jp

H. T. T. Le
Faculty of Forestry, University of Agriculture and Forestry, Hue University, Hue, Vietnam

T. Shine
Research Institute for Languages and Cultures of Asia and Africa, Tokyo University of Foreign Studies, Tokyo, Japan

S. Fujii et al. (eds.), *Livelihood and the Environment in Vietnam*, Sustainable Development Goals Series,
https://doi.org/10.1007/978-981-97-2289-1_2

Fig. 2.1 Administrative map of Vietnam

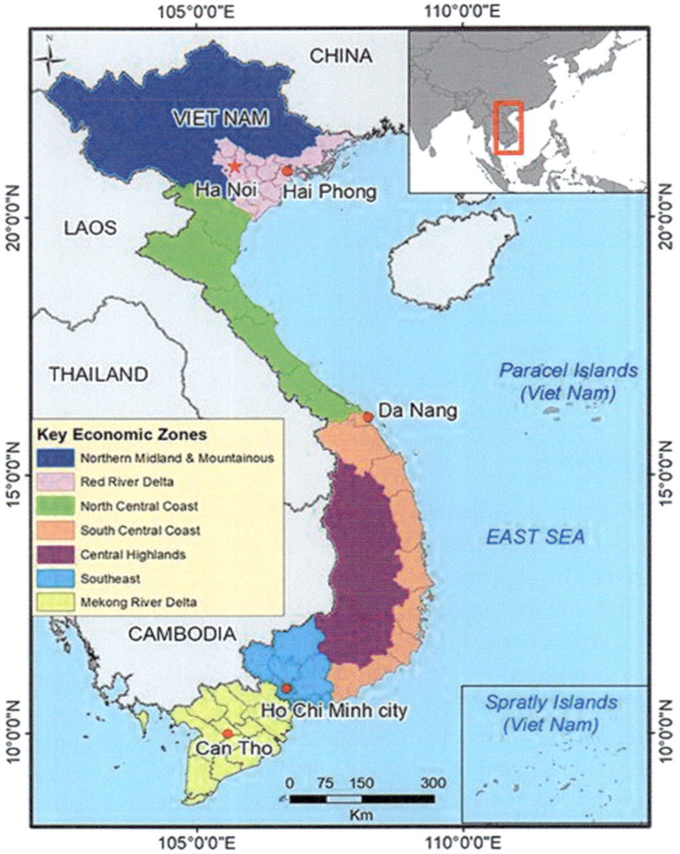

East, and the Gulf of Thailand in the South. In 2020, forestry land accounted for 46% (154,049 km^2) of the total area of the country, followed by agricultural production land with 117,184 km^2 (35%). The other agricultural land such as land for fishing, salt production occupied 8,603 km^2 (2.6%). Non-agricultural land, including homesteads, government offices, public works, religious sites, cemeteries, etc., covered 39,310 km^2 (12%). The remaining 12,198 km^2 (3.7%) was classified as unused land (GSO 2022). The territory of Vietnam is divided into seven administrative regions: Northern midlands and mountain areas, Red river delta, North-Central and Central coastal areas, Central highlands, South East, and Mekong river delta (GSO 2021) (Fig. 2.1). There are 63 provincial administrative units, comprising five municipalities (Hanoi, Ho Chi Minh city, Hai Phong, Danang, and Can Tho) and 58 provinces (JMLIT 2015). Hanoi is the capital of Vietnam.

In 2020, the population of Vietnam was estimated at 97.6 million people (GSO 2021), with a population density of 295 people/km^2, among the highest in the world. Ho Chi Minh city (formerly Saigon) is the most populous city with 9,227,600 people, and Hanoi ranks second with approximately 8,246,600 people (GSO 2021). The life expectancy in 2020 was 76.4 and 71.1 years for females and males, respectively (GSO 2021). Approximately 36.82% of the total population in 2020 resided in cities, while 63.18% in the rural areas (GSO 2021). In 2020, the GDP per capita was 2,786 USD, categorized as lower and middle income (World Bank 2020).

The total working population in 2020 aged 15 years and above in the economic sectors was 53.6 million; with 17.7 million among them in the agriculture, forestry, and fishing sector; 16.5 million in the industry and construction sector; and 19.4 million in the service sector (GSO 2021). In 2020, the literacy rate in adults aged 15 and above was approximately 95.4% (Statista 2020). However, only 24.1% of the labor force aged 15 years and above was trained with diplomas and certificates, with 39.7% among them in urban areas and 16.3% in rural areas (GSO 2021). In 2020, the COVID-19 pandemic increased the unemployment rate in at working age to 2.48% (GSO 2021).

Vietnam is a multi-ethnic nation with 54 ethnic groups. The Vietnamese (Kinh) constitute 85% of the country's population (CIA Factbook 2024) and mainly dwell in the Red river delta, Central coastal delta, Mekong river delta, and major cities. The other 53 ethnic minority groups, with a total population above exceeding 8 million, are scattered over the highlands that cover two-thirds of the country from the north to the south (Government of Vietnam 2021). Vietnamese is the official language of the Vietnamese (Kinh) people, as well as the first or second language for the other ethnic groups. The Vietnamese or *Quốc Ngữ* script was devised by Portuguese missionaries in the mid-seventeenth century, who modified the Roman alphabet with accents to suit specific Vietnamese consonants, vowels, and tones. Although English has been increasingly accepted as a second language, a little influence of the French language remains. Other spoken languages include Chinese, Khmer, and mountainous dialects. The majority of Vietnamese (86%) do not follow any organized religion. Catholicism (6.1%) is the most popular religion in Vietnam, followed by Buddhism (5.8%) and Protestantism (1%) with other minor faiths (0.8%) such as Caodaism, Hoahaoism and Islam (CIA Factbook 2024).

2.2 Geography

Vietnam's diverse topography includes hills, mountains, plains, coastlines, and continental shelves, reflecting a long history of geology and topography formation in a monsoon, humid climate, and extreme weather (Embassy of the Socialist Republic of Vietnam in the United States of America 2019). The nation is partitioned into the highlands and Red river delta in the north, and the Truong Son range (Central mountains, or the Chaîne Annamitique, occasionally referred to simply as the Chaîne), the coastal lowlands, and the Mekong river delta in the south (Fig. 2.2). The terrain is lower in the northwest–southeast direction, which is evident by the flow direction of large rivers.

Three-quarters of Vietnam's territory is covered by low mountains and hilly regions (Embassy of the Socialist Republic of Vietnam in the United States of America 2019). Areas lower than 1,000 m above sea level constitute 85% of the territory compared with only 1% of mountainous areas (over 2,000 m above sea level). The hills and mountain ranges form a 1,400 km-long large arc from the northwest to the southeast. The topographical structure is quite diverse owing to the rejuvenation of the neo-tectonic movement, creating a clear gradation by altitude, descending from the northwest to the southeast. The highest mountain ranges are located in the west and northwest, including Mt. Fansipan (3,143 m), which is the highest mountain in the Indochinese Peninsula. Further east, the mountain ranges decrease in height and frequently end with coastal lowland strips. From Hai Van pass (the border between Thua Thien Hue province and Danang city) to the south, the terrain is simpler. Although there are no long limestone mountains, large blocks of granite occasionally protrude into peaks; the rest of the terrain contains consecutive plateaus forming the Central highlands, with their eastern edge raised to the Trường Sơn range, which is the main mountain range of Indochina and the longest in Vietnam. The mountain range, which is approximately 1,100 km long and constitutes the boundary between Vietnam and Laos, extends parallel to the coast in a generally gentle curve in the northwest–southeast direction, with a considerably rugged eastern ridge leaving narrow coastal deltas (Vietnam Times 2021).

Deltas occupy only 25% of the land area and are separated by hills and mountains into

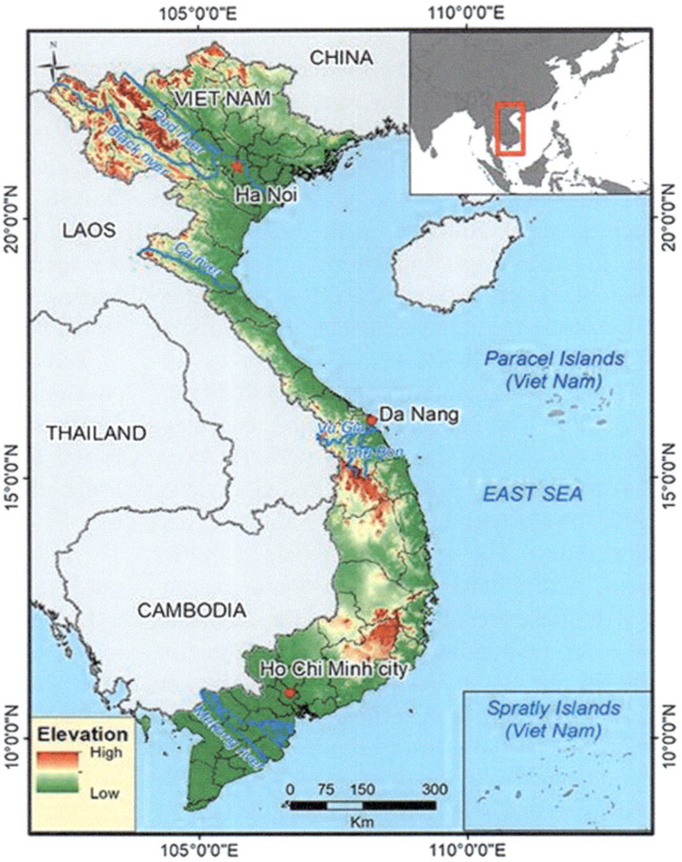

Fig. 2.2 Geography of Vietnam (Only major rivers are shown)

several regions. At the two ends of the country are large and fertile deltas: the Northern delta (Red River basin with an area of 16,700 km^2) and the Southern Delta (Mekong River basin with an area of 40,000 km^2) (Fig. 2.2). Lying between these two large deltas is a chain of narrow deltas, distributed along the central coast, including the delta in the Ma river basin (Thanh Hoa) to Phan Thiet with a total area of 15,000 km^2. Vietnam has a coastline of 3,260 km, from Mong Cai in the north to Ha Tien in the southwest. The part of the East Sea under Vietnam's sovereignty extends to the east and southeast, surrounded by a continental shelf, large and small islands, and archipelagos (Embassy of the Socialist Republic of Vietnam in the United States of America 2019).

Vietnam has nearly 3,500 rivers, including 109 main rivers (Fig. 2.2). There are 16 river basins nationwide with a total catchment area larger than 2,500 km^2. The river basins cover a total area of over 1,167,000 km^2 across the country. Owing to the influence of topographic factors, the river basins in the north (Northern midland and mountainous, and Red river delta) have a lower surface, are fan-shaped, and flow in the northwest–southeast direction. In addition, the hydrological regime of the rivers is also influenced by climatic factors, with the rainy season coming early, leading to abundant water in the rainy season. Slow drainage tends to prolong the flood because of the fan-shaped rivers that flow into a few main streams, leading to stagnant river water, and causing the floodwater to rise quickly and descend slowly (Ministry of Natural Resources and Environment 2012). The topography of the central region (North-Central coast, South-Central coast) tends to slope from

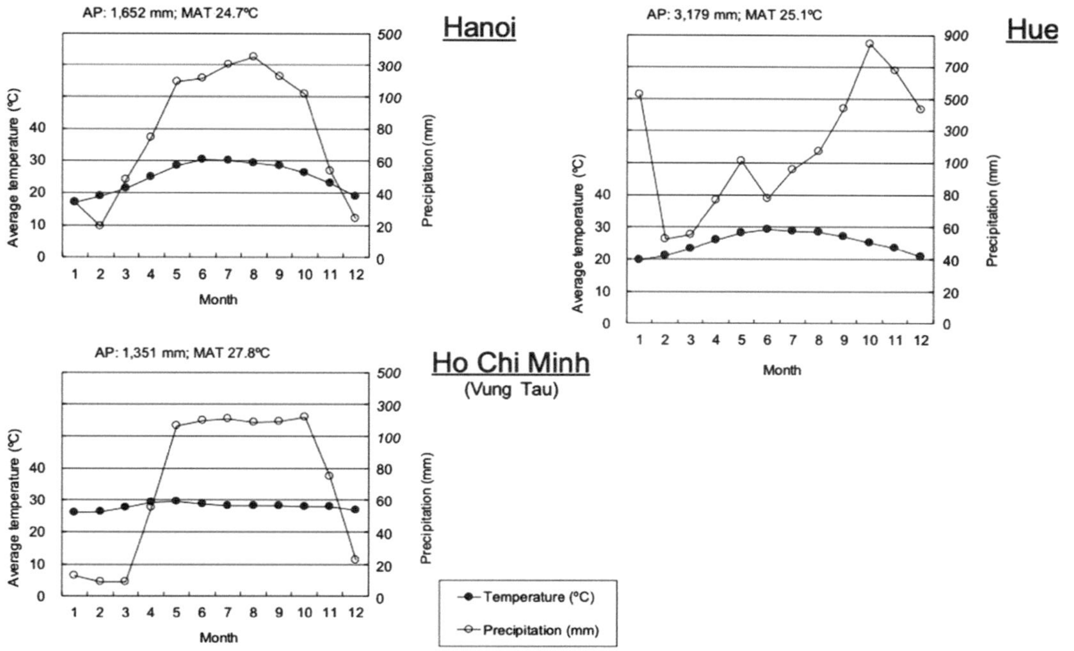

Fig. 2.3 Climate diagram of different regions of Vietnam

the west to east. The Central region has a high density of rivers and streams, divided into several small river basins, frequently with narrow river beds, steep slopes, and small catchment areas. The river flow is generally fast, with a large flow in the rainy season, and frequently causes flooding in the downstream areas. In the Central highlands, rivers run virtually from east to west and the yearly flow distribution is seasonal. The flood season starts from August to November with the flow volume accounting for approximately 70% of the total for the entire year. The southeast region includes the Dong Nai river basin and a system of other small river basins located in the coastal area. The surface flow in rivers in this region is divided into two distinct seasons, with the flood season usually 1–2 months later than the rainy and dry seasons. The flood season lasts for 6 months, from June to November. Saltwater intrusion is a concern for the downstream area of the Dong Nai–Saigon river basin. The level of saline intrusion is influenced by the upstream flow rate and gradually increases at the end of the flood season, reaching the highest level at the end of the dry season (Ministry of Natural Resources and Environment 2012). The Mekong delta terrain slopes in two directions: from the north to south and from the west to east. The hydrological and hydraulic regimes in the Mekong delta are very complicated; accordingly, the quality of the water environment also varies by region. The mangrove distribution and saltwater intrusion process in the Mekong delta are influenced by the irregular semi-diurnal tide regime of the East sea, which has affected approximately 1.4–1.5 million hectares of land (Ministry of Natural Resources and Environment 2012). This high density of the river networks provides about 20% of the energy consumed in Vietnam through hydropower plants (Our World in Data 2023). The construction of the hydropower plants has brought several social impacts. As one example, some hydropower plants have caused the migration of ethnic minorities, who had lived in the catchments (Sect. 4.9). Approximately 70,000 households experienced displacement in 2020, causing uncertainty and social upheaval (Pham et al. 2023). Another example is the establishment of Payment for Forest Environmental Services (PFES), which was legislated in 2011 in Vietnam. In this scheme, hydropower plants are defined as the users of forest ecosystem services and are required to contribute to the forest conservation by monetary payment (Le et al. 2023).

Vietnam is located at the Southeastern tip of the Eurasian continent and stretches over 15° latitude in the tropical Northern hemisphere, closer to the Tropic of Cancer than the Equator, and has a tropical monsoon climate. Due to the diverse topography and elongated shape of the country, and the manifest influence of the East Sea, its climate is very heterogeneous across regions (Ministry of Natural Resources and Environment 2010). The East sea plays an important role in the climate conditions in Vietnam owing to the influence of the spatial and temporal heterogeneity of the atmospheric circulation, which causes different weather patterns in the coastal areas of Vietnam (Nguyễn Ngọc Khánh 2009). The country receives an extremely large amount of solar radiation (1,400–3,000 h/year sunshine) and the annual mean air temperature ranges from 13 °C to 28 °C, which tends to rise gradually from the north to south. There is significant annual rainfall in all regions and torrential in a few, ranging from 600 to 5,000 mm, and is typically measured between 1,400 and 2,400 mm, with 80–90% rainfall during the rainy season. There are approximately 60–200 rainy days per year (Ministry of Natural Resources and Environment 2010).

Along the stretching territory, Vietnam's climate is categorized into three regions: the North, Central, and South. The North (Northern midland and mountainous, and Red river delta) is hot and humid and experiences heavy rainfall in summers owing to the Southwest monsoon (April–September); there is little rain and extreme cold in winters because of the influence of the Northeast monsoon (October–March). The average monthly precipitation in the North ranges between 200 and 300 mm, with the highest and lowest rainfall observed in July–August and January, respectively (JICA 2000). The Central region has a shorter and less cold winter than the North, with the rainfall concentrated in the last months of the year, and an extremely hot and dry summer influenced by the Southwest monsoon. The North-Central coast regions (from Thanh Hoa to Thua Thien Hue) have a transitional climate because of their location between the northern and southern climate zones separated by the Hai Van pass. The rainy season coincides with the Northeast monsoon (September–December) and the dry season is affected by the Southeast monsoon. The average annual precipitation ranges from 1,500 to 3,000 mm, of which approximately 80% is in the rainy season. The climate in the South-Central region (South-Central coast and Central highlands) is also transitional owing to the influence of the northern and southern monsoon climate. In southern Vietnam (Southeast region and Mekong river delta), the climate is hot all year round with rainy (May–October) and dry (November–April) seasons. The average annual rainfall in the south ranges from 1,900 to 2,600 mm, 90% of which is observed during the rainy season (JICA 2000).

2.3 Climate, Geology, and Soils

The production infrastructure that supports primary production in Vietnam can be understood by combining the geological structure, climate, and soils. Refer FAO-UNESCO (1979) and Kyuma (1972) for details on geology, climate, and soils.

2.3.1 Climate of Vietnam

Vietnam is long and narrow from north to south, with mountainous areas over 2,000 m high; therefore, the climatic conditions vary greatly depending on the region. Annual precipitation and average temperatures are approximately 1,650 mm and 24.7 °C in Hanoi, 3,200 mm and 25.1 °C in Hue, and 1,900 mm and 27.8 °C in Ho Chi Minh, respectively (Fig. 2.3). Temperature differs significantly depending on

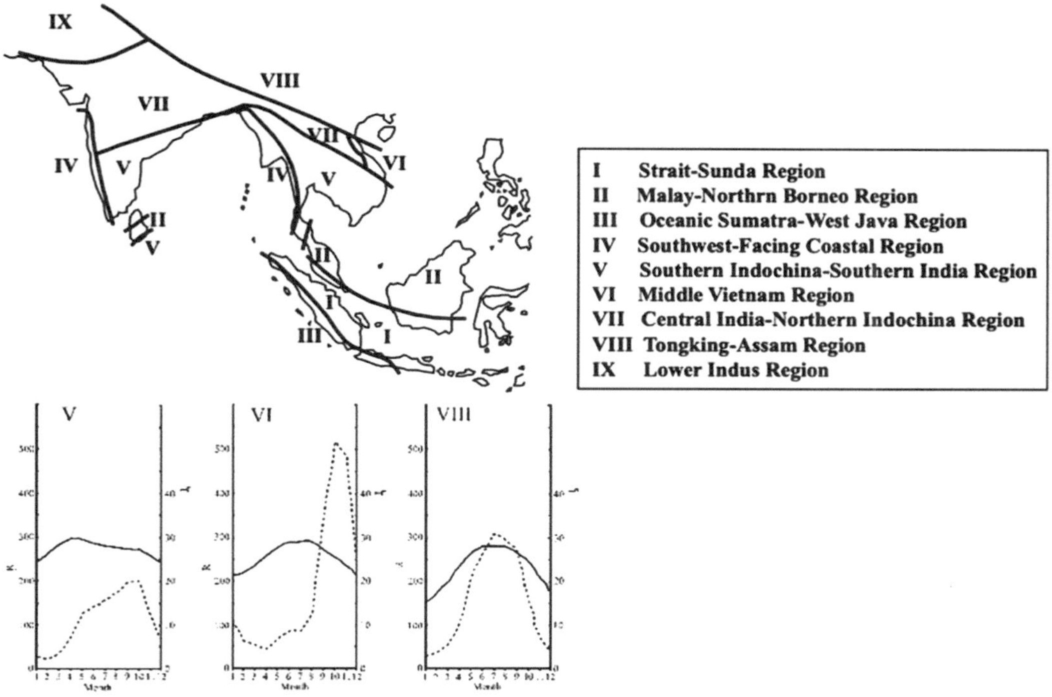

Fig. 2.4 Numerical classification of the climates of South and Southeast Asia (Created based on Kyuma 1972)

latitude and altitude (0 to over 2,000 m). On the other hand, the annual precipitation is frequently less than 2,000 mm, and according to Köppen's climate classification, the northern region belongs to the warm winter temperate climate (Cwa), central to the tropical rainforest and monsoon climates (Af/Am), and south to the savanna climate (Aw). According to the climate classification of Kyuma (1972), which applies the idea of Thornthwaite that emphasizes water balance, the climate of Vietnam is classified into Group VIII (subtropical monsoon climate), VI (tropical monsoon climate with moderate temperature fluctuation), and V (tropical monsoon climate with small temperature fluctuation) (Fig. 2.4). Hue (Group VI), a city in central Vietnam where a considerable part of the research presented in this book was conducted, can be a fairly peculiar area where the impact of the typhoon invasion from September to November is huge in the annual precipitation pattern.

2.3.2 Geological Structure of Indochina Peninsula and Vietnam

The basic geological structure of the Indochina Peninsula comprises three cratons (Sudo et al. 1996). The first is the Indochina craton (Kon Tum craton), which expands from eastern Thailand, southern Laos to central and south Vietnam, and occupies most of Cambodia. Next is the Southern China craton, which comprises South China from the south of Changjiang River to northeastern Vietnam. The third, the Shan-Thai craton, stretches from northeastern Myanmar/Shan Plateau, western Thailand, to Malaya. Several terrains fill the periphery of these cratons. The exposure of the bedrock is limited, and in a large area, the bedrock is covered with Mesozoic and Paleozoic sedimentary rocks as well as igneous rocks from the Archean to the Cenozoic eras.

Fig. 2.5 Landscape and soil profiles in the South-Central plateau (**a**), North-Central highland (**b**), and lowlands (**c**) of Vietnam

In the Indochina craton, the oldest Precambrian bedrock in Vietnam is partially exposed on the Kon Tum Plateau in Central and South Vietnam and the Song Coi river basin in the north, together with igneous rocks derived from igneous activity from the end of the Tertiary to the Quaternary era. The basalts of Vietnam are scattered, and red and infertile Ferralsols (Oxisols) are distributed on a relatively flat and stable topographical surface (Fig. 2.5a). On the other hand, in northeastern Thailand and the Cambodian plains on the same Indochina craton, coarse-grained sedimentary rocks cover the basement rocks, with a wide distribution of sandy soils (Arenosols) generated from them.

Most of the mountainous areas of Vietnam other than those described above are thickly covered with sedimentary rocks of the Mesozoic and Paleozoic eras. These areas form the Vietnam Transverse Mountain Belt, whose topography is generally steep owing to the orogeny at the craton boundary associated with the collision and compression of the Indian plate with the Eurasian plate since the Eocene (Himalayan orogeny). The soil repeatedly erodes and forms, and with the contribution of intermediate to acidic igneous rocks, the distribution of acidic soils such as Acrisols, Alisols (Ultisols), and their related Cambisols (Inceptisols) is predominant under warm and humid climatic conditions (Fig. 2.5b).

Large rivers in the Indochina region flow mainly along the fold zone, and a vast delta is formed downstream where a large amount of sediment is deposited. As for Vietnam, these are the Red River in the north and the Mekong River in the south. Large-scale deltas produce alluvial soils with various particle sizes and mainly coarse-textured alluvial soils from the lower reaches of steep rivers that flow directly from mountainous areas to the sea. It is a paddy rice cultivation area (Fig. 2.5c).

2.3.3 Summary of the Production Infrastructure of Vietnam

Based on the climate, geology, topography, and soils mentioned above, the ecological conditions for the production infrastructure of Vietnam are roughly summarized as follows.

(1) **Northern mountainous area** A large part of this area is the Cwa type, where the annual average temperature is below 25 °C. The steep topography and acidic soils derived from sedimentary rocks are widely distributed. These conditions are similar to those in northern Indochina and the southernmost part of China, such as northern Laos and northern Myanmar. The region is populated by widely distributed ethnic minorities on the mountain slopes whose main livelihood is slash and burn agriculture. It is essential from the social and economic aspects to establish sustainable primary production and livelihoods in this region, which is still economically disadvantaged owing to Indochina's unstable social and economic conditions in the latter half of the twentieth century.

(2) **Southern Central Plateau** Except for high altitude areas, it belongs to the Aw-type climate, where the annual average temperature exceeds 25 °C. The region is centered on the Central and Southern Kon Tum plateau, which has relatively flat plateaus with an altitude of approximately 1,000 m. The widely distributed Ferralsols are originally oligotrophic; however, their excellent physical properties and fertilizer response enable it to be an acceptable area for developing intensive modern agriculture combined with a humid and warm climate. Currently, it is a significant producer of commercial crops such as coffee.

(3) **Alluvial plains** These are large lowland areas distributed in the offshore regions of the Song Cai (Song Hong) river in the north, Mekong river in the south, and large and small rivers along the coast. They are currently the main production areas for paddy rice. Still, their productivity is variable because of climatic, water-use, and soil conditions in terms of soil texture and exhibit considerable regional specificity. The main issues are the establishment and modernization of lowland agriculture centered on paddy rice that suits each region's characteristics.

2.4 History

In writing the following article, the author referred to Ban Mai (2008), Doãn Quốc Sỹ (1970), Willbanks (2012) and Yoshii (2012).

2.4.1 Summary of Vietnam's Historical Eras

Lyrics of the song "*Gia tài của mẹ*" (母之家財, Mother's property) (*Trịnh Công Sơn* 鄭公山, 1965), which is still popular in Vietnam, says;

> 一千年奴隷賊唐 *Một nghìn năm nô lệ giặc Tàu* (One thousand years ruled by the Chinese),
>
> 一百年都護賊西 *Một trăm năm đô hộ giặc Tây* (One hundred years enslaved by the French),
>
> 二十年内戦層日 *Hai mươi năm nội chiến từng ngày* (And twenty years of civil war every day).

China's occupation of Vietnam spanned a thousand years (43–939). Then, Vietnam experienced saw the rule of 11 independent feudal dynasties for another thousand years (939–1847, *Ngô-Đinh-Lê* 呉丁黎, *Lý-Trần-Hồ* 李陳胡, *Lê-Mạc-Trịnh* 黎莫鄭, and *Tây Sơn-Nguyễn* 西山, 阮). Since the mid-*Nguyễn* era to 1945, the French occupied Vietnam for a hundred years (1848–1945). Subsequently, the country witnessed civil war for 20 years (1945–1965, although, according to the current view, the period is considered to be from 1955 to 1975).

2.4.2 Vietnam War

For about 20 years, the armed forces of the Democratic Republic of Vietnam (both the People's Army of Vietnam (PAVN) and its allied Liberation Army of South Vietnam (LASV)) fought against the Army of the Republic of Vietnam (ARVN) and US military. The most famous battles are the Tet Offensive (*Tổng Công Kích Tết Mậu Thân* 戊申節総攻撃, 1968) that temporarily occupied parts of Hue (化), Danang (沱瀼), Saigon (西貢/柴棍), and the Ho Chi Minh Campaign (*Chiến dịch Hồ Chí Minh* 胡志明戦役, 1975) that aimed to occupy the Saigon

(Saigon was codenamed Ho Chi Minh city by PAVN and LASV, and renamed actually after 1975). Three years after the withdrawal of the US military, the Vietnam war once ended with the liberation of Saigon on April 30, 1975. However, after 1975, Vietnam still faced various challenges, including its resistance to China's invasion, its interference to stop the Cambodian genocides, and its patience against the US embargo.

2.4.3 Post-war

Immediately before the collapse of the Soviet Union, the Communist Party of Vietnam adopted the so-called *Đổi Mới* policy at the end of 1986, which focused on economic rehabilitation as the first priority, cooperating with Western countries, and discontinuing the Soviet model. Before *Đổi Mới*, there were several trials and errors on socio-economic reforms implemented by local young communist generations who followed the Chinese model (although their bilateral relations were not friendly because of the Cambodian issue). The Communist Party elders accepted the achievements of the young leaders. After the PAVN withdrew from Cambodia in 1991, diplomatic relations between China, Japan, the United States, and Vietnam have normalized gradually. By 1995, Vietnam fully returned to the international community after joining the Association of South-East Asian Nations (ASEAN). Since then, Vietnam is receiving support from the World Bank and other donors and is developing a "socialist market economy;" moreover, environmental restoration, reforestation, poverty reduction, industrialization, and large-scale infrastructure construction are in progress.

2.5 Ethnicity

2.5.1 Ethnicity and Affirmative Actions in Vietnam

The great communist leader V. I. Lenin wrote a famous pamphlet titled "The Right of Nations to Self-Determination" during February–May, 1914. As a communist country, Vietnam also followed this ideology and established affirmative actions and incentives for 53 minority nations or ethnic minorities inside Vietnam. Affirmative actions and incentives for ethnic minorities are measures to achieve socio-economic equality. This measure is not always specified in the basic laws and regulations, except for the "Decree on the Ethnic minority's work," that is, Decree 05/2011/ND-CP in Vietnam. It is frequently documented in relative detail in enforcement regulations. Revolutionary meritorious people (revolutionary bereaved families, war-dead bereaved families) are also provided preferential treatment along with ethnic minorities. If a person is from an ethnic minority as well as a revolutionary meritorious person, the preferential points are doubled (cf.: Decision 72/HDBT year 1990). This type of point-plus system has been repeatedly criticized by ethnic minority education specialists (teachers/professors who are minorities) for spoiling ethnic minority students, especially, in higher education. This is because the graduation grades of even lazy ethnic minority students are frequently higher than those of hard-working majority Kinh students. Vietnamese minority intellectuals frequently use the term "We Vietnamese people" (*Dân tộc Việt Nam ta* 我等越南民族). They do not prefer the term "Kinh" (京人), because it differentiates demarcates between the ethnic majority (Kinh) and minorities (Non-Kinh). The myth that they were once the same drives them to use this title "We Vietnamese people."

2.5.2 Vietnamese (the Kinh) Ethnic Identity and Its Relation with the Chinese (the Hoa) in Their Common Myth

According to Vietnamese mythology described in "*Đại Việt Sử Ký*" (大越史記) (edited by *Lê Văn Hưu* 黎文休, 1272 in the *Trần* 陳 era), *Lộc Tục* (禄續), the first Vietnamese king *Hùng Vương* I (雄王一世), called *Kinh Dương Vương* (涇陽王)), was the fifth-generation descendant of the legendary Chinese emperor *Shén-nóng/*

Thần Nông (神農), who was the inventor of the harrow/hoe (耒耨＝鋤鍬) and bull-plow (牛犁). After the demise of *Lộc Tục*'s father (Chinese emperor *Dì Míng/Đế Minh* 帝明), *Lộc Tục* was enthroned at *Phong Châu* (峰州) (Phu Tho province 富寿省). Thus, the year of the demise of *Đế Minh* is the year of the enthronement of *Lộc Tục*, also called the year of independence, the Independence Day of Vietnam from China (the 10th of the 3rd month of Sino-Vietnamese Lunisolar calendar in 2879 BC, "The anniversary of the *Hùng Vương*'s ancestor" called "*Giỗ Tổ Hùng Vương*" (雄王祖忌) in Vietnamese), and the beginning of the idiom "5,000 years of the land" (*đất nước năm nghìn năm* 國土五千年) of Vietnam. This corresponds to the "4.2 kaclimate event" (cooling/drying) era.

2.5.3 Highlands and Lowlands

Subsequently, *Sùng Lãm* (崇纜), son of *Lộc Tục* and grandson of *Đế Minh* (帝明), was enthroned (*Hùng Vương* II 雄王二世, called *Lạc Long Quân* 貉龍君); he married *Đế Minh*'s great-grand-daughter (daughter of Chinese emperor *Đế Lai* 帝來), Princess *Âu Cơ* (嫗姬). A hundred eggs were laid between the two, and various ethnic groups called *Bách Việt* (百越) were born. When the two were separated, **50** children left for the highlands with their mother, Princess *Âu Cơ*. And the other **50** children who remained with their father, *Sùng Lãm*, in the lowlands became the ancestors of Vietnamese and other lowland people. The above is a myth, not a historical fact. However, it still matches a little with reality. According to Decision No. 121 of the General Statistics Office of Vietnam (GSO) on "Vietnam's ethnic classification" (1979), the number of highlands ethnic groups remains the same as *Đại Việt Sử Ký* (**50** ethnic groups in the highlands), with four ethnic groups in the lowlands (the Kinh, Hoa, Khmers, and Chams). As of 2021, the 63 provinces and cities of Vietnam's local administrative divisions comprise 58 provinces and five centrally controlled cities, among which 13 provinces are located in the highlands (two in the Northeast region: Cao Bang (高平省) and Lang Son (諒山省) provinces, six in the northwestern region, and five in the Central Highlands), and **50** in the lowlands, that is, same as *Đại Việt Sử Ký*. The implications of this alignment with the Fig. 50 are mysterious.

2.6 Industry

Until the mid-nineteenth century and the beginning of French domination, the Vietnamese economy was characterized by village-based subsistence agriculture (Cima 1987). Nevertheless, the French colonialists deliberately developed the regions differently, designating the south for agricultural production and the north for industrial development (Cima 1987).

In the north, while wet rice remained the principal subsistence crop for locals, the French introduced agriculture to the north with products such as coffee, tea, cotton, and tobacco. The French also developed several extractive industries, such as coal, iron, and non-ferrous metal mining for export because the north is rich in natural mineral resources. In the south, agricultural development focuses on rice cultivation and rubber plantation, the main export commodities. Industry in the south largely comprises food processing plants and factories manufacturing consumer goods (Cima 1987).

In 1954, the north and south were not only split politically, but also espoused different economic ideologies: the north followed communism, and the south adhered to capitalism (Cima 1987). All private enterprises and private ownership were prohibited under the communist regime in the north. Meanwhile, south Vietnam opened its free-market economy to conduct extensive trade with other anti-communist or non-communist countries, such as the US, Canada, France, West Germany, Japan, and Thailand. In that context, socialist industrialization was promoted in the north to support efforts to reunify the country after the US intervention in the south in 1964 (Anh et al. 2014). In the north, the communist regime's First Five-Year Plan (1961–65) placed priority on heavy industry. Notably, virtually the entire

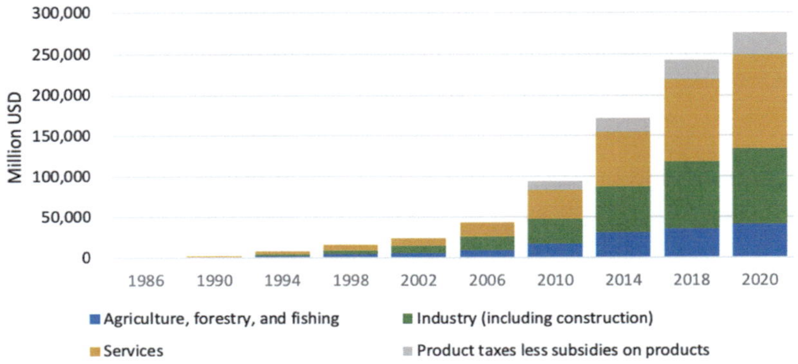

Fig. 2.6 Gross domestic product (GDP) across economic sectors from 1986 to 2020 (GSO 2001, 2011, 2021)

industrial development during this period was attained with foreign assistance from Vietnam's socialist allies (China, the Soviet Union, and other Eastern Bloc nations) (Anh et al. 2014). Nonetheless, the plan was interrupted in 1964 when the US began airstrikes in the north, halting the advancement of heavy industry and shifting the remaining economic activity to light industry and agriculture (Anh et al. 2014; Cima 1987). During 1965–1972, the US army bombings damaged all six industrial cities as well as most of the provincial and district towns in the north. All power stations, railway lines, roads, bridges, as well as sea and inland ports were seriously damaged or destroyed, postponing transportation by road and energy supplies. As a result, the distribution of raw materials and consumables was severely influenced and delayed large-scale construction work (Anh et al. 2014).

In the post-war period (1976–1985), the unified country planned to boost the domestic economy. Although the government's Second Five-Year Plan (1976–1980) set extraordinarily high targets for the annual growth rates of industry, agriculture, and national income, and aspired to unify the north and south Vietnam, the plan could not attain its goals: the economy was still dominated by small-scale production, low labor productivity, joblessness, raw material and technology shortage, and scarcity of food and consumer goods (Cima 1987). The Third Five-Year Plan (1981–1985) set slightly modest targets, which were a compromise between the ideological and pragmatic factions, and the government highlighted the development of agriculture and industry (Cima 1987).

By 1985–1986, Vietnam was on the verge of a socio-economic crisis. The government initiated a comprehensive reform, called *Đổi Mới* (renovation) in December 1986. *Đổi Mới* was intended to alter the economy from a centrally planned one to a form of market socialism officially termed "Socialist-oriented market economy." *Đổi Mới* combined state intervention (mainly at the planning stage) and free-market rules and incentives (Anh et al. 2014; Meditz 2005). The program abolished agricultural collectives, eliminated price controls on agricultural goods, and permitted farmers to trade their goods in the marketplace (Meditz 2005). It encouraged private commerce in the production of consumer goods and foreign investment from foreign-owned enterprises (Anh et al. 2014; Meditz 2005). In the late 1990s, the results of business and agricultural reforms initiated in the *Đổi Mới* era were visible (Meditz 2005). Within 35 years (1986–2000), Vietnam's economic sectors developed rapidly, notably the industry and services (Fig. 2.6). In this period, the average GDP per year increased by 6.5%, with agriculture, forestry, and fisheries increased by 3.7%, industry and construction by 9.1%, and services by 6.7%. Compared with the general growth rate of the world economy and the rapid decline in the centrally planned economies of Eastern Europe and the Soviet Union in their transition to market economies, the growth rates of the Vietnamese economy were a remarkable achievement (Nguyễn Thị Hương 2020). There was a dramatic increase in the GDP at the current prices of services from just 9 million USD to around 114,741 million USD between 1986 and 2020. Similarly, the contribution of industry (including construction) to the GDP increased significantly from only 8 million USD in 1986 to 92,961 million USD in 2020. The contribution to GDP from agriculture, forestry, and fishing also experienced a gradual growth from approximately 10 million USD to approximately 40,943 million USD in 2020. In 35 years, the structure of economic sectors has changed remarkably. Agriculture, forestry, and fishing had the largest percentage share in GDP compared with industry and services in 1990 and earlier (GSO 2001, 2011, 2021) because agricultural production was considered the backbone of Vietnam's development strategy for economic revival after the country's reunification in 1975 (Cima 1987). Nevertheless, the GDP now focuses greatly on the development of industry (manufacturing) and services. In particular, in 1986, the share of agriculture, forestry, and fishing in the GDP was dominant at 38%, followed by service at approximately 33%, and then industry at 29% (GSO 2001). In contrast, the share of services in GDP was at the top (37%) in 2010, industry ranked second at 32%, and agriculture, forestry, and service sector contributed only 19% in the same period (GSO 2011). Similarly, in 2020, the share of services in GDP was overwhelming (42%) compared with industry (approximately 34%) and agriculture, forestry, and fishing (15%) (GSO 2021).

In 2001, the Vietnamese Communist Party (VCP) approved a 10-year economic plan that heightened the role of the private sector while reasserting the importance of the state sector in the economy (Meditz 2005). The country gained World Trade Organization (WTO) membership in 2007 along with an excellent opportunity to boost the country's economy by trading with major partners such as the US and China (Anh et al. 2014).

The agriculture, forestry, and fisheries sector has proven its role as a pillar of the economy by ensuring the supply of food and essential goods. In 2020, the country's cereal production reached 47.3 million tons, down 1.9% from 2019, mainly because of a decrease in cultivated area. The area under perennial crops was 3,616,300 hectares in 2020, 1.8% higher than in 2019, out of which the area under fruit trees was 1,135,200 hectares, an increase of 6.4% (GSO 2021). Cattle and poultry production were quite stable in 2020. Buffalo population in 2020 was 2.3 million head, down 2.3% from 2019; other cattle population was 6.2 million head, up 2.8%; pig population was 22 million head, up 12%; and poultry population was 512.7 million head, up 6.6% (GSO 2021). As regards forestry, the timber production for the entire year was estimated at 17,169,700 m^3, an increase of 5.2% over the previous year (GSO 2021). Fisheries production in 2020 was estimated at 8,497,200 tons, an increase of 2.7% compared with 2019. Aquaculture production was 4,633,500 tons, up 3.1% from 2019, while fisheries production in 2020 was 3,863,700 tons, up 2.3% from the previous year (GSO 2021).

Industrial production was concentrated in a few key sectors such as mining and quarrying, manufacturing, electricity, gas and steam supply, as well as water supply, sewerage, and waste disposal. The growth rate of the industrial production index in 2020 compared with that in 2019 for the entire industry was 3.3%. Among them, mining and quarrying declined by 7.5% in 2020 compared with that in 2019, while manufacturing, electricity, gas and steam supply, as well as water supply, sewerage, and waste management increased by 4.8%, 3.1%, and 4.2%, respectively (GSO 2021).

The services sector covers a wide range of economic activities, with a few of the most important subsectors such as wholesale and retail trade, services and tourism, accommodation, food, and beverages. Total retail sales of consumer goods and services in 2020 were estimated at 4,976.5 trillion VND or 218.0 billion USD,[1] up 1.7% year-on-year. Retail sales of goods reached 3,944.9 trillion VND or 172.8

[1] 1 USD = 22,830 VND (April 3, 2022).

billion USD, accounting for 79% of the service sector, up 6.8% from 2019; accommodation and catering services reached 493.3 trillion VND or 21.6 billion USD, accounting for constituting 9.9% (down 17% from 2019); services and tourism reached 538.3 trillion VND or 23.6 billion USD, accounting for 11% (down 11% from the previous year) (GSO 2021).

2.7 Relations with Japan

During World War II, after co-occupying the French-colonized Vietnam, Vietnam was solely under Japanese rule at the end of the war. Later, during the Vietnam War, Japan recognized and provided post-war compensation to only South Vietnam; however, the assistance was subsequently renamed as the Economic Cooperation Fund and Official Development Assistance (ODA) and continued. It should be remembered that Okinawa (1952–1972), the southernmost region of Japan, which was under US administration during the Vietnam War, was a sortie base for its military. After the withdrawal of US troops from Vietnam in March 1973, negotiations to normalize diplomatic relations between Japan and North Vietnam (Democratic Republic of Vietnam) began, and diplomatic relations were established in September 1973. After the unification of North and South Vietnam in July 1976, Japan has maintained diplomatic relations with the Socialist Republic of Vietnam, the successor to the Democratic Republic of Vietnam. After the unification of Vietnam, exchanges between the two countries became sparse owing to the influx of refugees and the Cambodia conflict, and in October 1979, aid to Vietnam was suspended. Under the *Đổi Mới* policy, which was adopted in December 1986, economic assistance was resumed in November 1992 owing to the market economy, open-door policies, and the end of the Cambodia conflict in 1991. Using this opportunity, Japanese companies started expanding earnestly in Vietnam. In April 2009, the "Japan–Vietnam Joint Statement on Strategic Partnerships for Peace and Prosperity in Asia" was announced, and cooperation in not only economic aspects but also a wide range of fields such as politics, science and technology, culture, and human exchange is expected. In fact, Japan is Vietnam's largest donor with extremely active investments.

With regard to human exchange, Japan actively accepts Vietnamese students with scholarships provided by the Japanese and Vietnamese governments, as well as the Japan International Cooperation Agency (JICA). It also accepts several trainees, who learn and virtually comprise the essential labor force in several sectors of Japan, including agriculture and industry. As of June 2020, the number of Vietnamese living in Japan has exceeded 420,000 (Ministry of Foreign Affairs of Japan 2022). One of the outstanding achievements in the field of education is the establishment of Vietnam Japan University. The Japan–Vietnam Joint Statement issued when the Prime Minister of Japan visited Vietnam in 2010 included a proposal to establish a high-quality university in Vietnam with Japanese cooperation. This concept was reincorporated in the Japan–Vietnam Joint Statement when the President of Vietnam visited Japan in March 2014. In response to this, the Vietnamese government decided to establish Vietnam Japan University as the seventh member university of Vietnam National University Hanoi, which was established in July 2014, and the first enrollment was achieved in the first semester (master course) in September 2016 (Vietnam Japan University 2022).

Column 1

***Huế* (Hue)**

In writing the following article, the author referred to Nguyễn Thanh chủ biên (2005).

Thua Thien Hue province (承天-化省) is located in the North-Central coastal area of Vietnam, at 16° north latitude, almost in the center of Vietnam. It borders Quang Tri province (廣治省) to the north, Quang Nam province (廣南省) and Danang Central Direct-Controlled Municipality (沱瀼城舖) to the south, Laos to the west, and the South China Sea (東海, East Sea) to the east. The *Trường Sơn* Range (長山山脈) with fertile Basalt soil forms the border with

Fig. 2.7 ERSDAC satellite image of Hue 10 years after the Deluge in 1996 (October 30, 2007)

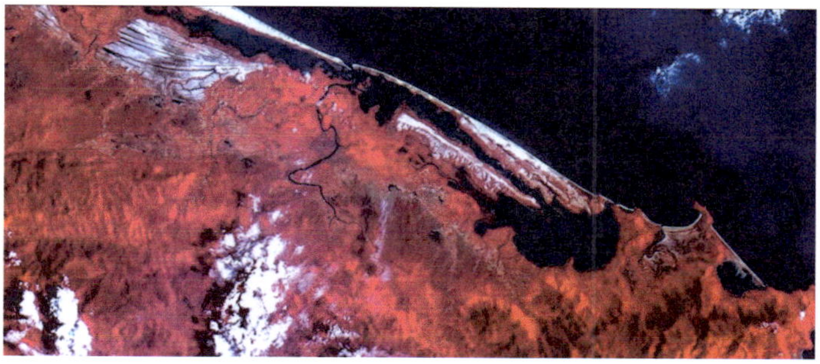

Laos. This province has a population of approximately 1 million, of which 350,000 live in the Hue city (化城舖). The area is approximately 5,000 Km^2, including 2,000 Km^2 of forests and 220 Km^2 of landwater (lagoons) such as *Tam Giang* (三江) and *Cầu Hai* (鬟海). The region has a tropical monsoon climate, with an average annual temperature of 25 °C, humidity of 84%, and rainfall of 2,750 mm.

In 1306, *Huế* (Hue) was handed over from Champa (Cham) to *Đại Việt* (the *Trần* Dynasty (大越陳朝)) as a betrothal gift of marriage between the Cham king *Chế Mân* (制旻) and *Trần* princess *Huyền Trân Công Chúa* (玄珍公主). Before 1306, the indigenous people here were Austronesian, such as the Chams who left inscriptions in old Cham and Sanskrit languages, and Austroasiatic speakers. In the *Nguyễn* era ((阮朝), nineteenth century), several ethnonyms were written in Sino-Vietnamese script including the *Tà Ôi* (斜隈人) and *Vân Kiều* (圓橋人). They were the successors of *Sa Huỳnh* (沙黃) ironware culture and produced their own iron farming tools along with elaborate textiles and constructed huge wooden community houses called Guol. The Vietnamese (Kinh) language is also an Austroasiatic language. Certain numbers and animal names are similar between Vietnamese and *Tà Ôi*, *Bru-Vân Kiều*, and Katu languages. For example, the numbers 1, 2, 3, and 4 are named *mui*, *bar*, *pei*, and *pon* in *Tà Ôi*, *Bru-Vân Kiều*, and *Katu*, while they are called *một*, *hai*, *ba*, and *bốn* in Vietnamese. Dogs, cats, and birds are called *aso*, *meo*, and *sim* in *Tà Ôi*, *Bru-Vân Kiều*, and *Katu* while *chó*, *mèo*, and *chim* in Vietnamese.

Hue was earlier known as *Thuận Hóa* (順化), the capital of the *Nguyễn* Dynasty (阮朝) from 1636 to 1945. The *Sông Bồ* (蒲江) and *Sông Hương* (香江) rivers as well as several lagoons moisten the vast alluvial plains of Hue and result in abundant agricultural production. However, Hue is also a typhoon-prone area, and the abundant water occasionally poses a flood risk. The *Nguyễn* Dynasty's architectural heritage in central Hue and its suburbs were severely destroyed by war. The rich basalt soil forests in the hill areas provided a hideout for the Liberation army in war time; therefore, they were severely polluted because of the spraying of Agent Orange by the US military (1964–1973). Since 1975, the Hue Monument Conservation Center (HMCC) has been preserving the Hue heritages. The government/municipality, experts from International NGOs, and Hue University of Agriculture and Forestry (HUAF) are working to support environmental restoration and livelihood reconstruction for both ethnic majority and minority residents in areas contaminated with Agent Orange or damaged by natural disasters like the Deluge happened in 1996 (Figs. 2.7 and 2.8).

Column 2

Tết

In writing the following article, the author referred to Chu Húc Cường (2013), Chu Quang Trứ (2000), Chu Quang Trứ, Nguyễn Bá Vân (1984), Kỳ Dương Nhật Linh (2021), Lê Trung Vũ (1996), and Trương Anh Thuận (2021).

The traditional New Year, called *Tết Nguyên Đán* (元旦節), or *Tết* (Tet) in short in Vietnamese, is the largest nationwide festival. However, this occasion has various meanings. Here I will try to explain the three main meanings.

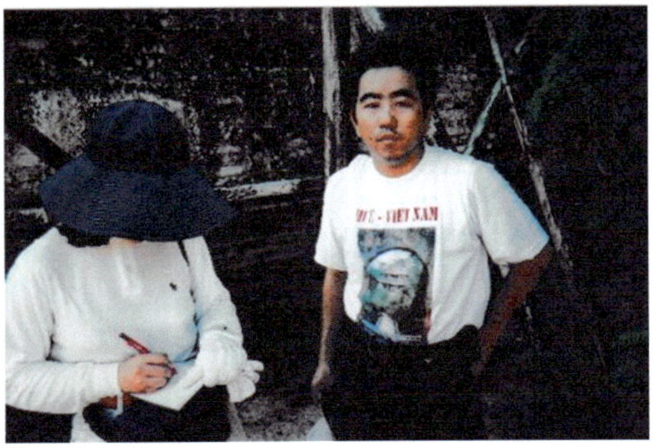

Fig. 2.8 The author at HMCC's restoration work of the *Minh Mạng* Mausoleum (明命帝陵) in the suburbs of Hue approximately in 1996, the year the Deluge happened

(1) Independency and loyalty. Although Vietnam, China, and Korea have been following the Christian era for greater than 50 years since the 1940–1950s, the Vietnamese still celebrate the New Year according to their own luni-solar calendar called *Hiệp Kỷ Lịch* (阮朝協紀曆). This is because celebrating the traditional New Year is a "pledge of allegiance" to the nation's independence. In Sino-Vietnamese vocabulary, this is called *Phụng Chính Sóc* (奉正朔), which means following the right New Year Day. Before 907, under the *Táng* Dynasty (唐朝)'s rule, Vietnam could not have its own calendar and followed the Chinese calendar. After re-independence from the *Míng* Dynasty (明朝) in 1428, although Vietnam still diplomatically followed the Chinese calendar similar to the Christian era presently, it domestically created its own calendar. Different from Japan, Vietnam's luni-solar calendar is still an alternative national calendar authorized by the government. The calendar is dissimilar to the Chinese one (it occasionally differs by year). Celebrating the traditional New Year is a testament to the Vietnamese peoples' loyalty to the nation.

(2) Gratitude to the spirits of ancestors. Even those people who work all year long take a break from work for two weeks before and after the traditional New Year. However, they do not play or rest at home. On this occasion, all the family members meet, discuss, and decide the annual schedule for ancestral rituals such as the anniversary of passing (*giỗ*).

(3) Implementation of political and moral correctness that is in accord with universal law or god's law. According to the people's belief on stove deities "*zào-jūn*" (*táo quân* 灶君) in Vietnam and China, the 23rd night of the 12th month is considered their night of Ascension. Although similar to the Japanese belief on *sān-shī* deities (*sanshi* 三尸, deities from Taoism), the responses are quite different between Japan and Vietnam.

In Japan, on the night of *gēng-shēn* (*kōshin*, 庚申) six times a year, the *sān-shī*, which lurk in the human body, escape and ascend to heaven, report each person's 60-day behavior to the "*yù-huáng*" (*gyokkō*, 玉皇, heavenly god), and decide award and punishment. Japanese who fear punishment hold a feast, drink good liquor, sing, and dance, trying to prevent the *sān-shī* in their bodies from ascending and stop them from reporting to "*yù-huáng*." The *sān-shī* are terrifying, evil deities, who allow "*yù-huáng*" to punish people.

On the other hand, in Vietnam, on the 23rd night of the 12th month, the stove deities called "*zào-jūn*" (*táo quân* 灶君) escape from the three stones of the stove in each family/household, ascend to heaven, report the family's behavior for one year to the "*yù-huáng*" (*ngọc hoàng* 玉皇, heavenly god), and decide award and punishment. Therefore, Vietnamese must return home before that date. They return home, greet their neighbors, visit pagodas, clean their ancestors' graves, clean the shrines/chapels dedicated to their ancestors, line up offerings in front of the altar for ancestors, and steam the New Year's rice cakes. They discuss and set an annual schedule for ancestral rituals. They exhibit their behavior as exemplary villagers and family members to the "*zào-jūn*" (*táo quân*) just before ascension. After the 23rd, people spend days without watchmen/heavenly spies. On New Year' Eve, the "*zào-jūn*" (*táo quân*) return to each family after finishing the report. The Vietnamese have delicious dinner with their families and enjoy conversations on politically and morally correct topics. The returning "*zào-jūn*" (*táo*

Fig. 2.9 *Đông Hồ* (東湖, suburb of Bac Ninh province 北寧省 nearby Hanoi city河内城舖) popular print for New Year: "Marriage of Mouse": On the 23rd night of the 12th month in the luni-solar calendar, when the stove deities (*táo quân* 灶君) ascend to heaven, mice organize a wedding march. People should not create noise or disturbance. Females must not work and must spend time with their children so that they do not interfere with the mouse wedding out of curiosity

quân) will be satisfied with the actions of serious and nice people. The "*zào-jūn*" (*táo quân*) will surely report the virtuous deeds of his family to the "*yù-huáng*" (*ngọc hoàng*) next year and bring awards. Vietnam's "*zào-jūn*" (*táo quân*) are deities of good fortune, unlike *sān-shī* (*sanshi*) in Japan. On New Year's Day, people install a New Year print called "*nián-huà*" (*niên họa* 年画) of auspicious subjects at home. Although the *sān-shī* (*sanshi*) and "*zào-jūn*" (*táo quân*) have almost similar functions in belief, the characters are actually opposite, along with the people's responses to them. These are a few of the motivations for Vietnamese to cherish the *Tết*, the traditional New Year, and to always return home for the occasion (Figs. 2.9 and 2.10).

Fig. 2.10 *Làng Sình* (廊呈, suburb of Hue city 化城舖) popular print for Ritual of the three stove deities (*táo quân* 灶君): On the 23rd night of the 12th month in the luni-solar calendar, when the stove deities ascend to heaven, people burn these prints and pray to the heavenly god, *Ngọc Hoàng* (玉皇) for their satisfactory reports

Column 3

Vietnam Teacher's Day

In writing the following article, the author referred to Hứa Văn Ân (2000).

November 20 is observed as Vietnam Teacher's Day (*Ngày Nhà giáo Việt Nam*). It is not a holiday legally but is a special celebrating day. Traditionally, educational anniversaries in Vietnam, China, Korea, and Japan were observed on Confucius's birthday and demise anniversary which are called *Thích Điện* (釋奠) (the 28th of the 9th month and the 11th of the 4th month according to the Luni-solar calendar).

And people like *Chu Văn An* (朱文安, 1292–1370), *Lương Thế Vinh* (梁世榮, 1441–1496), *Lê Quý Đôn* (黎貴惇, 1726–1784), and Madame *Nguyễn Nhược Thị* (阮若氏夫人, 1830–1909) have been traditionally respected as educators and scientists by the Vietnamese (Kinh). Even today, various streets and schools in Vietnam are named after them. In 1954, the World Teachers Charter included content such as "*Respect for pedagogical and occupational freedom*" and "*Encouragement of teacher's creativity in textbook selection*," which was adopted by a joint committee of the World Federation of Free Teacher's Unions (FISE) and other two associations that supported the international communist movement. At the FISE Warsaw Convention in August 1957, it was resolved that November 20, 1958 would be observed as the "Charter's Anniversary" and was celebrated in FISE member countries including Vietnam. This is the origin of "Vietnam Teacher's Day." In Vietnam, World Teacher's Day is also a celebration day, although not as large as the Vietnam Teacher's Day. Both days were celebration days in originally socialist countries. However, as UNESCO officially designated October 5, 1994 as "World Teacher's Day," it has also been celebrated in Japan unofficially.

The Council of Ministers Decision No. 167-HDBT dated September 28, 1982 defines "Vietnam Teacher's Day" (it was signed by General *Võ Nguyên Giáp*) as follows:

> Article 1. Every year from now on, the November 20th is the Vietnam Teacher's Day.
>
> Article 2. To make this event meaningful, from October every year, each municipality must review the labor situation and activities of local teachers, and raise awareness of teacher's honor and its responsibility in society.
>
> Article 3. Each municipality can visit teachers and organize intimate discussions with teachers, and take this opportunity to award and reward teachers who achieve results. Vietnam Teacher's Day events must avoid flashy forms that bother students and their parents.
>
> Article 4. Each school can coordinate its learning and education plans so that teachers can take a break and attend school and community events in the November 20th.

Fig. 2.11 Author receiving a gift from a student on November 20, 1994 at Hue University

To realize the above-mentioned articles, on the Vietnam Teacher's Day, the heads of government/municipality meet the presidents/principals, commendation events are organized for excellent teachers by the municipality/school, and multi-generational alumni associations recognize and felicitate the teachers. The day is celebrated as a thank-you party to teachers. Although it is a weekday, teachers can take a break and attend school and community events to meet their old students (cf.: above-mentioned Article 4, Decision No. 167-HDBT, 1982). If you happen to be in Vietnam on November 20, I recommend you to write postcards to your nostalgic teachers in Japan on occasion of the Vietnam Teacher's Day (Fig. 2.11)!

References

Ban Mai (2008) Trịnh Công Sơn—vết chân dã tràng, Lao động, Hà Nội. Accessed 19 Oct 2022 from http://www.vietnamvanhien.org/trinhcongsonvcdt.pdf

CIA Factbook (2024) Vietnam - The World Factbook. Accessed 30 Apr 2024 from https://www.cia.gov/the-world-factbook/countries/vietnam/

Cima RJ (ed) (1987) Vietnam: A country study. Federal Research Division, Library of Congress. Accessed 19 Oct 2022 from http://memory.loc.gov/master/frd/frdcstdy/vi/vietnamcountryst00cima_0/vietnamcountryst00cima_0.pdf

Chu Húc Cường (2013) Tín ngưỡng thờ táo quân ở Việt Nam từ góc độ khảo sát thư tịch học. In: Văn hóa thờ

Nữ Thần (Mẫu) ở Việt Nam và Châu Á bản sắc và giá trị. Thế giới, Hà Nội, pp 308–325. Accessed 19 Oct 2022 from https://repository.vnu.edu.vn/bitstream/VNU_123/88513/1/VAN%20HOA%20THO%20NU%20THAN%20MAU%20O%20VIET%20NAM%20VA%20CHAU%20A%20BAN%20SAC%20VA%20GIA%20TRI_p306-323.pdf

Chu Quang Trứ (2000) Đông Hồ's Folk Paintings. Đông Hồ—Bắc Ninh

Chu Quang Trứ, Nguyễn Bá Vân (1984) Tranh dân gian Việt Nam. Văn Hóa, Hà Nội

Doãn Quốc Sỹ (1970) Thần thoại Việt Nam-Trung Quốc. Sáng Tạo, Sài Gòn. Accessed 19 Oct 2022 from http://tusachtiengviet.com/images/file/NFjR1d3h0wgQANct/than-thoai.pdf

Embassy of the Socialist Republic of Vietnam in the United States of America (2019) Vietnam, geography. Accessed 3 Nov 3 2022 from https://vietnamembassy-usa.org/vietnam/geography

FAO-UNESCO (1979) Soil map of the world: Volume IX. Southeast Asia. UNESCO, Paris

Government of Vietnam (2021) Ethnic groups in Vietnam. Accessed 18 Oct 2022 from https://vietnam.gov.vn/ethnic-groups-in-viet-nam

GSO (General Statistics Office of Vietnam) (2001) Statistical yearbook of Vietnam 2000. Statistical Publishing House, Hanoi. Accessed 18 Oct 2022 from https://www.gso.gov.vn/wp-content/uploads/2020/02/Ni%C3%AAn-gi%C3%A1m-th%E1%BB%91ng-k%C3%AA-2000-%C4%91%C3%A3-n%C3%A9n.pdf

GSO (General Statistics Office of Vietnam) (2011) Statistical yearbook of Vietnam 2010. Statistical Publishing House, Hanoi. Accessed 18 Oct 2022 from https://www.gso.gov.vn/wp-content/uploads/2019/10/NGTK-2010-pdf.pdf

GSO (General Statistics Office of Vietnam) (2021) Statistical yearbook of Vietnam 2020. Statistical Publishing House, Hanoi. Accessed 18 Oct 2022 from https://www.gso.gov.vn/wp-content/uploads/2021/07/Sach-NGTK-2020Ban-quyen.pdf

GSO (General Statistics Office of Vietnam) (2022) Statistical yearbook of Vietnam 2021. Statistical Publishing House, Hanoi. Accessed 30 Apr 2024 from https://www.gso.gov.vn/wp-content/uploads/2022/08/Sach-Nien-giam-TK-2021-1.pdf

Hứa Văn Ân (2000) Truyền thống tôn sư trọng đạo: Chào mừng ngày nhà giáo Việt Nam. Trẻ, Thành Phố Hồ Chí Minh

JICA (Japan International Cooperation Agency) (2000) The Study on the national transport development strategy in the Socialist Republic of Vietnam (VITRANSS). Environment Technical Report No. 11. Accessed 19 Oct 2022 from https://openjicareport.jica.go.jp/pdf/11596897_02.pdf

JMLIT (Japan Ministry of Land, Infrastructure, Transport and Tourism) (2015) Vietnam. Accessed 19 Oct 2022 from https://www.mlit.go.jp/kokudokeikaku/international/spw/general/vietnam/index_e.html

Kyuma K (1972) Numerical classification of the climate of South and Southeast Asia. Southeast Asian Stud 9(4):502–521. https://doi.org/10.20495/tak.9.4_502

Kỳ Dương Nhật Linh (2021) Changing painting career in Sình Village in the current context. J Sci Ind Dev Thua Thien-Hue Province 2(156):41–55. Accessed 19 Oct 2022 from https://vjol.info.vn/index.php/ncpt-hue/article/view/50327 (in Vietnamese)

Lê Trung Vũ (1996) Tết cổ truyền người Việt. Văn Hóa, Hà Nội

Le TTH, Nguyen THM, Le TH, Shinjo H (2023) Implementation of payment for forest environmental services and its influence on the livelihood of ethnic minorities in Thua Thien Hue Province, Vietnam. Trees, Forests People 13:1000404. https://doi.org/10.1016/j.tfp.2023.100404

Meditz SW (2005) Country profile: Vietnam. Federal Research Division, Library of Congress. Accessed 19 Oct 2022 from https://tile.loc.gov/storage-services/master/frd/copr/Vietnam.pdf

Ministry of Foreign Affairs of Japan (2022) Basic data on Socialist Republic of Viet Nam. Accessed 19 Oct 19 2022 from https://www.mofa.go.jp/mofaj/area/vietnam/data.html (in Japanese)

Ministry of Natural Resources and Environment (2010) Vietnam's second national communication to the United Nations framework convention on climate change. Accessed 3 Nov 2022 from https://unfccc.int/resource/docs/natc/vnmnc02.pdf

Ministry of Natural Resources and Environment (2012) State of the Environment Report

Nguyễn Ngọc Khánh (2009) Typical traits of the east sea as regards geography and resources. J Res Dev 4(75):3–22. Accessed 19 Oct 2022 from https://vjol.info.vn/index.php/ncpt-hue/article/view/4143/3930 (in Vietnamese)

Nguyễn Thanh chủ biên (2005) Địa chỉ Thừa Thiên-Huế. Thuận Hóa, Huế

Nguyễn Thị Hương (2020) Những dấu ấn quan trọng về kinh tế – xã hội trong hành trình 75 năm thành lập và phát triển đất nước qua số liệu thống kê. General Statistics Office. Accessed 19 Oct 2022 from https://www.gso.gov.vn/su-kien/2020/09/23410/

Nguyen Thi Tue Anh, Luu Minh Duc, Trinh Duc Chieu (2014) The evolution of Vietnamese industry. WIDER Working Paper 2014/076. UNU-WIDER. Accessed 19 Oct 2022 from https://www.wider.unu.edu/sites/default/files/wp2014-076.pdf

Our World in Data (2023) Vietnam: energy country profile. Accessed 30 Nov 2023 from https://ourworldindata.org/energy/country/vietnam#energy-mix

Pham Huu Ty, Baas Laura, Tran Thi Phuong, Nguyen Quang Tan (2023) Impoverishment persistence in hydropower dam-induced resettled communities: a sociological investigation on livelihood and food security in Vietnam. Soc Sci 12(4):222. https://doi.org/10.3390/socsci12040222

Statista (2020) Total literacy rate of adults aged 15 and above in Vietnam from 2011 to 2020. Accessed

19 Oct 2022 from https://www.statista.com/statistics/1008115/vietnam-literacy-rate-of-adults/

Sudo S, Kamiya M, Hirano H (1996) Mineral resources of Indochina: (1) overview of Indochina's geology and mineral resources. Geol News 502:42–47 (in Japanese)

Trương Anh Thuận (2021) The Hiệp Kỷ calendar in the Nguyễn dynasty. Rev Soc Sci Ho Chi Minh City 1(269):54–64. Accessed 19 Oct 2022 from https://vjol.info.vn/index.php/khxh/article/view/53986 (in Vietnamese)

United Nations (2020) Vietnam population. Accessed 19 Oct 2022 from https://www.worldometers.info/world-population/vietnam-population/

United Nations (2021) Demographic Yearbook—2021. Accessed 9 Dec 2022 from https://unstats.un.org/unsd/demographic-social/products/dyb/dyb_2021/

Vietnam Times (2021) What is the longest mountain range in Vietnam? Accessed 19 Oct 2022 from https://vietnamtimes.org.vn/what-is-the-longest-mountain-range-in-vietnam-33929.html

Vietnam Japan University (2022). History. Accessed 19 Oct 2022 from https://vju.ac.vn/about-us/su-kien-quan-trong-ste5.html

Willbanks JH (2012) Vietnam: the course of a conflict. Army University Press, Kansas. Accessed 19 Oct 2022 from https://www.armyupress.army.mil/Portals/7/combat-studies-institute/csi-books/Vietnam-Willbanks.pdf

World Bank (2020) GDP per capita (Current US$)—Vietnam. Accessed 19 Oct 2022, from https://data.worldbank.org/indicator/NY.GDP.PCAP.CD?locations=VN

World Data (2021) Vietnam. Accessed 19 Oct 2022, from https://www.worlddata.info/asia/vietnam/index.php

Yoshii M (2012) Study on anti-nuclear songs of Japan in 2011: compared with anti-war songs of Vietnam in 1960s. In: Bulletin of Center for International Education and Research, vol 7. Mie University, pp 55–65. Accessed 19 Oct 2022 from https://www.crc.mie-u.ac.jp/seeds/pdf/20100112-094154_kenkyu1.pdf

3 Regulations and Policies on Environment

Shigeo Fujii, Hung Nguyen Duy, Bich-Thuy Ly, Hung Quang Dinh and Lien Pham Hong Nguyen

Abstract

Vietnamese laws and policies may greatly influence environmental issues. This chapter introduces legal and administrative systems related to the environment. Section 3.1 provides an overview of the systems, and Sects. 3.2, 3.3, and 3.4 describe the laws and policies related to air, water, and soil environments, respectively. Section 3.5 explains the environmental monitoring, and the last section, Sect. 3.6, summarizes solid waste management.

Keyword

Regulations policies · Law on Environmental Protection (LEP) · Environmental standards · Emission standards

S. Fujii (✉)
Graduate School of Global Environmental Studies, Kyoto University, Kyoto, Japan
e-mail: fujii.shigeo.6z@kyoto-u.jp

H. Nguyen Duy
SOLEN Environmental Solutions Joint-Stock Company, Hanoi, Vietnam

B.-T. Ly · H. Q. Dinh · L. P. H. Nguyen
School of Chemistry and Life Sciences, Hanoi University of Science and Technology, Hanoi, Vietnam

3.1 Legislation System and Administration

In Vietnam, the environmental legislation system is relatively new. However, all legislative systems in Vietnam have the same structure and components that include: constitution, laws, and under-law documents such as decrees, circulars, and decisions as shown in Fig. 3.1.

3.1.1 Constitution

The constitution is the primary law with the highest legal validity in the legal system of Vietnam. The term "environment" first appeared in the constitution in 1980, which described "state agencies, enterprises, cooperatives, people's armed forces units and citizens all have the obligation to implement policies to protect, renovate and regenerate natural resources, and protect and improve living environment" under Article 36 (only once). Further, "environment" appeared three times in two articles in 1992 and nine times in five articles in 2013. Several articles in the current constitution (amended in 2013) use "environment", such as Article 43 ("everyone has the right to live in a healthy environment and has an obligation to protect the environment"); Article 50 ("the Socialist Republic of Vietnam builds an independent and self-reliant economy, ……, ………, ……., and

S. Fujii et al. (eds.), *Livelihood and the Environment in Vietnam*, Sustainable Development Goals Series,
https://doi.org/10.1007/978-981-97-2289-1_3

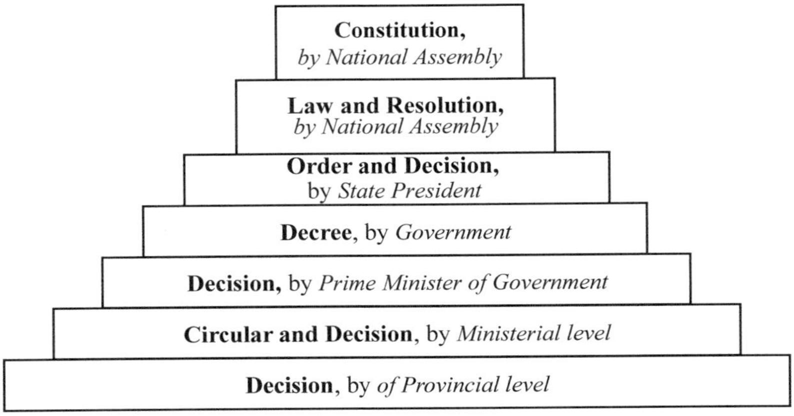

Fig. 3.1 General legislation system in Vietnam

realizes progress and social equity, environmental protection, industrialization and modernization of the country)"; and Article 61 ("the state has policies on protection to environment; manage and use effectively and sustainably natural resources; nature conservation; biodiversity; proactively prevent and combat natural disasters and respond to climate change").

3.1.2 Law on Environmental Protection

The Law on Environmental Protection (LEP) was first issued in 1993 by the Vietnam National Assembly (LEP1993). LEP was amended three times in 2005 (LEP2005), 2014 (LEP2014), and 2020 (currently effective, LEP2020) as shown in Table 3.1. LEP gives Vietnamese basic concepts and policies on the environment, and it also describes the fundamentals of different environmental issues, such as water, air, and solid waste. The first LEP gave emission/discharge standards in 1993. Subsequently, they included solid waste management in 2005, an action plan for air pollution in 2014, and environmental management plans by provinces in 2020, thereby increasing the number of chapters and articles. In addition to LEP, there are other laws mentioning environmental protection missions, such as the Law on Biodiversity (2008), the Law on Water Resources (LWA, 2012), and the Law on Natural Resources and Environment of Sea and Islands (2015).

3.1.3 Under-Law Documents

Several under-law documents provide details regarding the regulations and requirements of LEP. These documents are classified into decrees and decisions issued by the central government, circulars and decisions issued by ministries, and decisions issued by the local government, as shown in Fig. 3.1. For example, LEP is a law that states the establishment of "environmental standards", but it does not define specific indices such as BOD on water environment. The details of the other contents are determined in other under-law documents such as orders, decrees, and circulars.

Those legal documents are often described in Vietnam, such as "45/2022/ND-CP" (decree on environmental protection for wastewater standards). The first number (45) shows the code number of the under-law documents in the year that is denoted by the second number (2022), whereas the first abbreviation (ND) expresses the brevity code (initial letters) of the regulation type in Vietnamese, and the second one (CP) does that of the responsible organization. Some

Table 3.1 Summary of law on environmental protection in Vietnam

Issued year/ Enforced year	# Chapters / # Articles	Major new contents
1993/1993	7/55	Emission/discharge standards; Environmental impact assessment (EIA); Prevention, Resistance, and remediation of environmental degradation, pollution, and incidents; State management in environmental protection; Complement and handling of violations; International relationship in environmental protection
2005/2006	15/136	Strategic environmental assessment; Environmental protection Commitment; Conservation and reasonable usage of natural resources; Environmental protection in activities of manufacturing, Business, and services; Protection of the environments of seas, river water, and other water bodies; Waste management
2014/2015	20/170	Master plan in environmental protection; Environmental protection plan; Environmental protection in mining and using of natural resources; Responses to climate change; Protection of island environment; Protection of air and soil environments; Treatment of contamination; National technical regulations
2020/2022	16/171	Environmental protection of natural heritages; National environmental protection strategy; Environmental protection contents in regional planning and provincial planning; Tools of economic and policy in environmental protection; Education and communication in environmental protection

decisions are expressed without the issued years, such as 1622/QD-TTg.

On the other hand, national technical regulations are expressed such as "QCVN 03:2008/BTNMT" (national technical standards for soil heavy metals). The first abbreviation (QCVN) shows the code of the regulation type, and the second one (BTNMT) expresses that of the responsible organization, while the first number (03) expresses the code number of the under-law document issued in the second number's year (2008). A pattern of "QCVN 01—MT:2015/BTNMT" is also used recently.

Table 3.2 shows some brevity codes and their meanings in regulations.

Table 3.2 Brevity codes and their meaning in regulations

Abbreviation: original Vietnamese	English translation (Meaning)
TT: *Thông tư*	Circular
ND: *Nghị Định*	Decree
QD: *Quyet Dinh*	Decision
NQ: *Nghi Quyet*	Resolution
TTg: *Thu Tuong*	Prime Minister
CP: *Chinh Phu*	Government
BTNMT: *Bộ Tài Nguyên và Môi Trường*	Ministry of Natural Resource and Environment
MT: *Môi trường*	Environment
TCVN: *Tiêu chuẩn Việt Nam*	Vietnam National Standard[a]
QCVN: *Quy chuẩn Việt Nam*	Vietnam National Technical Regulation[a]

[a]English translation in TCVN/QCVN documents

3.1.4 Environmental Standards and Emission/Discharge Standards

LEP1993, the first LEP, declared that the government agencies have to evaluate environmental status to report. Subsequently, environmental standards were established to highlight desirable conditions to protect human health and conserve the living environments for air, water, soil, noise, light, and radiation. The items and materials for which the standards were set in different environments, along with the desired values (concentrations) of these items are given in various circulars issued at the ministerial level. These environmental standards are guidelines to improve the environment, and no penalty is born even if a value exceeds its standard.

To achieve these standards, the government determines emission/discharge standards for waste air, wastewater, and solid waste. Emission/discharge standards were declared in LEP1993, and the items and standard values of the items were provided in various circulars issued at the ministerial level. Violence of the standard may result in a penalty to the person or organization in charge of its discharge as regulated by Decree 45/2022/ND-CP on Penalties for administrative environmental protection offenses.

3.1.5 Administration

The prime minister of Vietnam is at the top position in environmental management in the general organizational structure and directly manages the ministries. At the ministerial level, the Ministry of Natural Resources and Environment (MONRE) is in charge of environmental protection and natural resources. MONRE has several state management units, public non-business units, and state-owned enterprises including some divisions functioning in air-related issues: the Department of Climate Change, Department of Environmental Pollution Control, and Vietnam Meteorological and Hydrological Administration (MONRE n.p.).

At the provincial level, provincial people's committee is at the top, and the Department of Natural Resources and Environment (DONRE) conducts the environmental management activities under the committee. DONRE is administratively controlled by MONRE.

3.2 Air Pollution

3.2.1 Policy and Legal System on Air Quality Management

There is no specific clean air act in Vietnam, but LEP covers most of the legal systems and policies related to air quality. Other laws, such as the Law on Transport, Law of Construction, and Law on Environmental Protection Tax, also partly include legal requirements of air quality. The emission standards of the industrial sections were required in LEP1993. The requirement of an environmental impact assessment (EIA) also appeared in this LEP. Further, LEP was revised in 2014 to include new requirements: setting up a national action plan to reduce air pollution until 2020, with a vision up to the year 2030, requiring significant sources of air pollution to install continuous emission measurement systems, and issuing a guideline for the emission inventory of such significant sources. LEP revised in 2020 requires each province to set up and implement its air quality management plan. The requirement for the best available technology first appeared in the LEP2014. In addition, economic tools such as incentives were introduced in LEP2014. However, until now, its application has been limited. The fees for air pollution emissions must be set up in LEP2020 based on the polluter pays principle.

In addition to the requirements in LEP, there are several regulations at the provincial level to improve air quality. For example, Hanoi banned beehive coal in 2020. Quang Ninh issued provincial environmental standards, including ambient air standards that are tighter than the national air quality standards. Currently, as required by LEP2020, all provinces have prepared their own clean air management plans that include an action list to meet their air quality targets in each period.

Table 3.3 Change of national ambient air quality standards (g/m^3)

Code of standards	Issued year	TSP –1 h	PM_{10} –24h	$PM_{2.5}$ –24 h	NO_2 –1 h	O_3 –8 h	CO –8 h	SO_2 –24 h	Pb –24 h
TCVN 5937:1995	1995	300	–	–	400	60[a]	5,000	300	5
TCVN 5937:2005	2005	300	150	–	200	120	10,000	125	1.5
QCVN 05: 2009/BTNMT	2009	300	150	–	200	120	10,000	125	1.5
QCVN 05: 2013/BTNMT	2013	300	150	50	200	120	10,000	125	1.5
QCVN 05: 2023/BTNMT[c]	2023	300	100	50 / 45[d]	200	120	10,000	125	1.5
WHO Guideline	2005	–	50	25	200	100	10,000	20	–[b]
	2021	–	45	15	–	100	4,000	40	

Note Not all average periods are presented
–Not regulated, 60[a]: Not regulated, but O_3-24 h (24 h average) was given in 60
–[b]No fixed value was given, but the WHO Guideline for Clinical Management of Exposure recommended that lead levels should keep a blood lead concentration to be less than 5 μ/dL
–[c]: Hazardous parameters are included in this standard version but are not presented here
–[d]: Valid from January 1, 2026

The most effective tool to control air pollution is emission standards, and the standards target two key categories of industrial sectors and transportation. Section 3.2.2 summarized these standards as well as ambient air quality standards.

3.2.2 National Standards on Air Pollution

3.2.2.1 Ambient Air Quality Standards

Vietnam has applied nationwide ambient standards to key air pollutants, including total suspended particles (TSP), particulate matter $(PM)_{10}$, $PM_{2.5}$, NO_2, O_3, SO_2, CO, and Pb. QCVN 05:2013/BTNMT gives the latest standards, and Table 3.3 shows the values with the past standards and the WHO (World Health Organization) guidelines (WHO 2021). The standards were first established in 1995 and further strengthened by including more parameters and tightening the limited levels in 2005, 2009, and 2013. However, the standards except for NO_2 and CO are not as strict as those by both WHO guidelines.

3.2.2.2 Air Pollution Emission Standards for Industrial Fields

The first emission standards in Vietnam were issued in 1995, regulating the emission levels of general and specific industry sources. Therefore, new emission standards for general and six major industries were issued between 2009 and 2013. Tables 3.4 and 3.5 illustrate those for general industrial sources (QCVN 19: 2009/BTNMT) and thermal power plants (QCVN 22: 2009/BTNMT), respectively. The limit of PM for coal power plants is 400 mg/Nm^3, which is much higher than those in other countries in the region. China has a limit of 10 and 5 mg/Nm^3 for key regions (Wang et al. 2020), whereas India has set a limit of 100, 50, 30 mg/Nm^3 for plants installed before 2004, between 2004 and 2016, after 2017 (Ganguly et al. 2022), respectively. There are also other national technical regulations (QCVNs) of emission standards on (1) chemical fertilizer manufacturing industry, (2) thermal power industry, (3) cement manufacturing industry, (4) industrial waste incinerators, (5) refining and petrochemical industry, and (6) steel industry.

Table 3.4 Major emission standards for general industry (mg/Nm^3)

Issued	Implemented	PM	NOx (as NO_2)	CO	SO_2
2009	2009	400	1,000	1,000	1,500
	2015	200	850	1,000	500

The required limit values of each plant are multiplied by capacity factor (1, 0.85, 0.7) and area factor (0.6, 0.8, 1.0, 1.2, 1.4)

Table 3.5 Emission standards for thermal power plants (mg/Nm[3])

Issued	Duration	Fuel type	PM	NOx	CO*	SO_2
2009	2009–2014	–	400	1,000	1,000	1,500
2009	After 2015	Coal	200	650[a]/1,000[b]	1,000	500
		Oil	150	600	1,000	500
		Gas	50	250	1,000	300

The required limit values of each plant the values are multiplied by the capacity factor (1, 0.85, 0.7) and the area factor (0.6, 0.8, 1.0, 1.2, 1.4)
[a]coal with volatiles>10%
[b]coal with volatiles ≦10%

3.2.2.3 Transportation Standards

Emission control for transportation is based on two different regulations: standards for vehicle emissions and standards for fuels. Currently, standards were issued for new cars, new motorcycles (motorcycles and mopeds), and on-road cars, but not for on-road motorcycles. Since 2007, new motorcycles and cars are required to meet level 2 (equivalent to the European emission standards (EURO 2)). Recently, stricter standards and regulations have been proposed and implemented for new vehicles. Since 2017, new motorcycles must meet EURO3, and new cars using gasoline and other fuels, except diesel, must meet EURO4. Diesel vehicles are required to follow EURO4 since 2018. Moreover, there are plans to have more stringent standards for new cars and motorcycles in the future; however, on-road cars are only required to meet EURO2. The Emission standards for on-road cars in Vietnam are much looser than those for new cars. Table 3.6 presents the emission standards for passenger cars and motorcycles.

In 2013, Vietnam banned gasoline RON83[1] and gasoline RON90 (the numbers here denote the octane rating of gasoline). In 2014, Vietnam started introducing 5% ethanol-blended gasoline (E5 RON92), which became widely used in Vietnam by 2018 (Ly et al. 2020). Currently, three main types of gasoline are used: gasoline RON92, RON95, and E5 RON92. Before 2007, the benzene limitation was 5%. Since 2007, all types of gasoline have a benzene limitation of 2.5%. Furthermore, standards have been issued for fuels, including gasoline, diesel, and biofuel, which correspond to EURO2, EURO3, and EURO4, respectively. Currently, a regulation including EURO5 for fuels has been proposed and will soon be applied. Fuels that meet Euro5 or higher have sulfur contents of less than 10 ppm.

[1] The Research Octane Number (RON) is the most common octane rating, a standard measure of a fuel's ability to withstand compression in an internal combustion engine without detonating. The value following RON shows the percentage of anti-knocking tendency, ranging from 0 to 100.

3.3 Water Environment

3.3.1 Background on Laws and Policies Related to Water Environment

More than 300 regulations have been used to establish water policies at the national and sub-national levels. A summary of water policies and their development over time is presented in Table 3.7.

In addition to LEP, the Law on Water Resources (LWR) is important for the aquatic environment. The first LWR was approved in 1998 (enforced in 1999) and was revised in 2012 (enforced in 2013), and again in 2021 (enforced in 2023). LWR provides a framework, including (1) basic surveys of water resources; (2) preparation of strategies and master plans; (3) responsibilities to protect water resources; (4) rights and obligations to exploit and use water resources; (5) the prevention, control, and

Table 3.6 Emission standards (g/km) for new passenger cars and new motorcycles

Category		Size[$]	Stage	Implemented year	CO	HC+NOx	NMHC	HC	NOx	PM
Passenger cars	Positive Ignition (Gasoline)	All	EURO2	2007–	2.20	0.50	–	–	–	–
			EURO4	2017–	1.00	–	–	0.10	0.08	–
	Compression Ignition (Diesel)	All	EURO2	2007–	1.00	0.70	–	–	–	0.08
			EURO4	2018–	0.50	0.30	–	–	0.25	0.025
Motorcycles	2-wheeled vehicles	<150	EURO2	2007–	5.50	–	–	1.20	0.30	–
			EURO3	2017–	2.00	–	–	0.80	0.15	–
		≥150	EURO2	2007–	5.50	–	–	1.00	0.30	–
			EURO3	2017–	2.00	–	–	0.30	0.15	–

Note There are different standards for other vehicles, such as commercial cars, mopeds, and used vehicles

Table 3.7 Key water policies related to water resources in Vietnam

Year of Issue	Legal document
1993[a]	Law on Environmental Protection (LEP1993)
1998[a]	Law on Water Resources (LWA)
2005[a]	Law on Environmental Protection, revised (LEP2005)
2006[$]	National strategy on water resources to 2020 Establishment of Environmental Policies
2012[a]	Law on Water Resources (LWA), revised Approving the Strategy on sustainable exploitation and use of natural resources and protection of the marine environment until 2020, with a vision to 2030
2014[a]	Law on Environmental Protection, revised (LEP2014)
2014[$]	Regulating registration for groundwater extraction, form of application, permit, contents of scheme and report in dossier for issue, extension, modification, re-issue of water resource permit (Circular 27/2014/TT-BTNMT)
2015[$]	Regulating technique for water resources planning (Circular 42/2015/TT-BTNMT)
2016[$]	Regulating environmental protection fee for wastewater (Decree 154/2016/ND-CP)
2017[$]	Regulating the method for calculating and charges for granting water entitlements (Decree 82/2017/ND-CP)
2020[b]	Law on Environmental Protection, revised (LEP2020) Fully and comprehensively mention the management, use, quality improvement, and sustainable exploitation of natural resources, solving the requirements of current practice on water environment protection
2021[$]	Regulations on calculation method and rate of charge for granting water right Regulations on supervision of exploitation and use of water resources
2023[$]	Revision of Law on Water Resources

[$]Enforced in the same year
[a]Enforced in the next year
[b]Enforced two years later
TT, BTNMT, ND, CP: See Table 3.2

mitigation of harmful effects caused by water; (6) water-related financial issues; (7) international relations in water-related financial issues; (8) international relations with respect to water resource management; (9) state management of water resources; (10) inspections and dispute settlement; and (11) implementation provisions.

Since LWR was issued and amended, many important laws and related legislation have been developed to meet the requirements of the country's development and its international integration in different issues on environmental protection, particularly water resource protection. The most recent important legal documents are LEP2020, Decision 1622/QD-TTg in 2022, and Decree 02/2023/ND-CP in 2023, which indicate the master plan for the water resources in the period of 2021–2030, with a vision to 2050.

As regulated under LEPs, the wastewater discharge activities of industries and services are licensed by an authorized government office and are regulated by the water resources law, the order of exploitation and protection of irrigation work, and the decrees on drainage, sewage and wastewater management, and waste and materials management. Wastewater must be collected, treated, reused, or transferred to functional units suitable for reuse or treatment according to environmental technical standards before being discharged into the environment. Industrial zones are obliged to have wastewater treatment systems to treat all generated wastewater with online monitoring systems. Companies outside industrial zones must also have wastewater treatment systems. For craft villages (See the column of "Craft Village"), it is necessary to expand their industrial cluster of craft villages to bring production households out of residential areas and introduce new technologies to treat wastewater from craft villages to reduce pollution from the source of production.

Apart from the legal groundwork for managing the water sector, as mentioned above, various secondary regulations have also been developed to protect water resources in different areas, as well as to deal with violations against laws and ordinances. Important secondary regulations include two circulars. Circular 27/2014/TT-BTNMT regulates registration for groundwater extraction, type of groundwater usage, permission, and contents of scheme as well as reporting issues, extension, modification, and re-issue of water resource permission. Circular 42/2015/TT-BTNMT regulates techniques for water resource planning.

In addition, the government has economic instruments for charging organizations and individuals that use water and water surface areas, discharge wastewater into water sources, and violate water-related regulations. These include (1) fees for domestic clean waters, (2) fees for industrial water uses, (3) fees for water use from irrigation works, (4) fees for wastewater discharge and environmental protection, and (5) fines on acts of polluting the water environment.

The issuance of permits for water resource exploitation, extraction and utilization, and wastewater discharge into water sources is further regulated by Decree 149/2004/ND-CP (Government of Vietnam 2004) and Circular 02/2005/TT-BTNMT. Accordingly, exploration, exploitation, and inventory and the use of water resources must comply with the plans approved by competent state agencies. In addition, permits for exploitation, use of water resources, and discharge of wastewater into water sources must stipulate environmental protection conditions. The exploitation and use of water resources, as well as the discharge of wastewater into water sources, must comply with the contents of environmental water resource protection permission granted by competent state agencies. Organizations and individuals must comply with environmental water resource protection requirements during the inventory, exploration, exploitation, and use of water sources. State agencies must control waste from industrial processes, business, service, construction, transport activities, exploitation of riverbed minerals, and garbage from households living near rivers.

Wastewater discharge permission is issued initially for a duration of ten years from the date of approval and periodically renewed to extend the permit for up to five years. The renewal of wastewater discharge permission must comply with the provisions in Article 5 of Decree149/2004/ND-CP. In particular, the

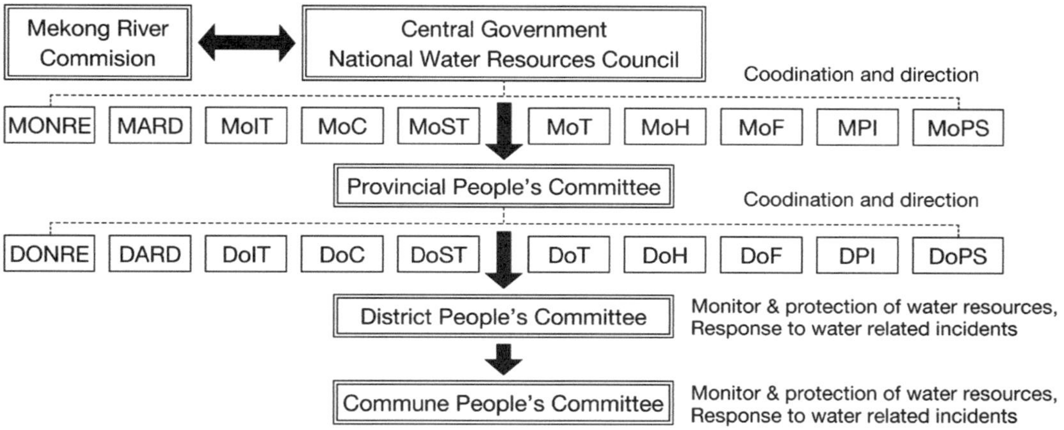

Fig. 3.2 Framework for water environment management in Vietnam. Modified from a Figure 2030 Water Resources Group (2017)

(Abbreviations (Original words): Roles: M/Mo/MO (Ministry of), D/Do/DO (Department of) NRE (NaturalResources and Environment): water resource management, water volume & quality monitoring; ARD (Agriculture and Rural Development): irrigation work management, flood control, rural water supply & drainage, water for agriculture; IT (Industry and Trade): hydropower; water for industry; C (Construction): urban water supply and drainage; ST (Science and Technology): water quality standards, water technology innovation license; T (Transportation): water way transport; H (Heath): water quality for drinking; F (Finance): state budget for water projects; PI (Planning and Investment): overall planning and investment; PS(Public Security): pollution investigation and monitoring.)

licensing of wastewater discharge into water sources must be based on the following foundations: (1) National Technical Specifications on the Environment; (2) Assimilative capacity of water sources; and (3) Sanitary protection zones defined by competent state agencies.

3.3.2 Organizational Structure of Vietnam's Water Management

Water resource management in Vietnam is organized at four administrative levels: nation, provincial/city, district, and commune. The framework for water environment management is shown in Fig. 3.2. In general, ministries, ministerial-level agencies, and other agencies attached to the government are responsible for water resource management at the national level and for the water management in large river basins, reservoirs, industrial zones, and factories. People's Committees at province/city and district levels and their attached agencies/divisions are responsible for water resource management within localities, water management in medium-sized industrial zones and enterprises, and the response to water-related incidents. Clean water supplies, irrigation and drainage companies, water professional associations, and groups of water users are also stakeholders in water and wastewater management.

3.3.3 Environmental Water Quality Standards, and Other Standards

Since the protection of water resources has become an important issue owing to the rapid development and industrialization of Vietnam, the government first issued a set of environmental water quality standards for surface water, groundwater, and sea water in 1995 and continued to revise them over the years. The standards include approximately 40 parameters (water quality indices) with several categories for each water type. For example, Vietnam's surface water quality standards (QCVN 08-MT:2023/BTNMT) are shown in Table 3.8, which compares these standards with those that are applied in rivers in Japan.

To perform these standards and serve as guides for environmental impact assessment, the government established several regulations (technical regulation). The Technical Regulation on Marine Water Quality is to assess and control the quality of sea water, serving the purposes of sports, aquatic entertainment, aquaculture, marine environmental protection, and other purposes for coastal waters, sea near the shore, and sea far from the shore. Wastewater quality standards are used to assess and control the quality of domestic wastewater, livestock effluent, and industrial wastewater. Table 3.9 shows some of these standards for major water quality indices.

In addition, the national technical process has been developed and revised by MONRE to control wastewater discharge from 10 specific industries: (1) natural rubber processing industry, (2) seafood processing industry, (3) pulp and paper mills, (4) textile industry, (5) solid waste landfill sites, (6) health care wastewater, (7) petroleum terminals and stations, (8) offshore oil and gas facilities, (9) steel industry, and (10) bioethanol processing.

Table 3.8 Major values of surface water environment quality standards with comparison of river water environment quality standards in Japan

	Class	pH (–)	BOD (mg/L)	COD (mg/L)	SS (mg/L)	DO (mg/L)	Coliform$ (MPN/100 mL)
Vietnam (Surface water, QCVN 08-MT:2023 / BTNMT)	A	6.5–8.5	≦4	≦10	≦5	≧6	≦1,000
	B		≦6	≦15	≦15	≧5	≦5,000
	C	6.0–8.5	≦10	≦20	≦15	≧4	≦7,500
	D		>10	>20	>15	≧2	>7,500
Japan (River water, Environment Agency Notification)	AA	6.6–8.5	≦1	–	≦25	≧7.5	≦50
	A		≦2		≦25	≧7.5	≦1,000
	B		≦3		≦25	≧5	≦5,000
	C		≦5		≦50	≧5	–
	D	6.0–8.5	≦8		≦100	≧2	–
	E		≦10		–	≧2	–

A—Use for domestic water supply purposes (after applying conventional treatment), conservation of aquatic plants, and for other purposes such as type B, C, and D
B—For the purpose of industrial and agricultural production, but must apply appropriate processing technology or the purpose of use as type C and D
C—For industrial production purpose but must apply appropriate processing technology or purposes as type D
D—Waterway transport and other purposes with low quality of water requirements
AA—Nature Conservation, Use for water supply class 1, including the purpose of A and below
A—Swimming, Use for water supply class 2, Fishery class 1, including the purpose of B and below
B—Use for water supply class 3, Fishery class 2, including the purpose of C and below
C—Use for Fishery class 3, industrial water class 1, including the purpose of D and below
D—Use for industrial water class 2, Agricultural water, including the purpose of E
E—Environmental preservation and use for Industrial water class 3
Coliform$: fecal coliform (Vietnam), coliform group (Japan)

Table 3.9 Major values of wastewater quality standards (with a comparison of Japanese standards)

	Category	Class	pH (–)	BOD(mg/L)	COD (mg/L)	SS (mg/L)	Coliform$ (MPN/100 mL)
Vietnam	Domestic wastewater	A	6–9	≦30	≦75	≦50	≦5,000
		B (1–4)	6–9	≦60–100	≦175–250	≦125–200	≦10,000
	Effluent of livestock	A	6–9	≦40	≦100	≦50	≦3,000
		B	5.5–9	≦100	≦300	≦150	≦5,000
	Industrial wastewater	A	6–9	≦30	≦75	≦50	≦3,000
		B	5.5–9	≦50	≦150	≦100	≦5,000
Japan	All business enterprise		5.8–8.6*1	≦120*3	≦120*3	≦150*3	≦3,000*3
			5.0–9.0*2	≦160*4	≦160*4	≦200*4	

Coliform$: fecal coliform (Vietnam), coliform group (Japan)
*1Discharge to areas other than the sea
*2Discharge to the sea
*3Daily average
*4Maximum

3.4 Soil Environment

The present form of government (The Socialist Republic of Vietnam) came into existence in 1945, but no laws related to the soil environment were issued for a long time. There were only some under-law documents regulating the right of management and usage of land property. Later, two important laws related to the soil environment were issued, namely the Land Law (LaL) in 1987 and LEP1993. The term "environmental protection" was mentioned in the LaL (1987) before the first LEP was issued.

LaL was revised in 1993, 2003, and 2013, and Table 3.10 compares each revision, including the first one (1987). The number of chapters and articles was increased from the first to the fourth versions, adding more detailed regulations. One of the most important aspects is that soil use is restricted by land classification, thereby preventing adverse effects on the natural environment caused by unreasonable activities on certain lands from the standpoint of soil environment conservation. For example, agricultural land is not allowed to be used for industrial activities.

LEP has been dealing with environmental protection since its first release in 1993, but there was no specific description on the protection of the soil environment in the first version, but only on general descriptions. The second version, LEP2005, was still in general compared to air and water environment protection. Two years later the first standards on soil environment protection were released namely "National technical regulation on the allowance limits of heavy metals in the soils" (QCVN 03:2008/BTNMT). The third version, LEP2014, included one section for soil environmental protection. The fourth version has a section namely "Sect. 3.3. Soil Environment Protection (Chap. II)" with five articles. Table 3.11 shows a summary of the heavy metal standards for soil environment protection mentioned above.

Soon after, the regulation was amended and QCVN 03-MT:2015/BTNMT was released. In 2023, "National technical regulation on Soil quality" (QCVN 03:2023/BTNMT) was issued, by merging standards for heavy metals, pesticide residues and dioxins. Soil environment protection was also generally mentioned in some other laws, including law on forest protection and development, mineral law, and law on biodiversity.

Table 3.10 Summary of land law in Vietnam

Year of Issue/ Enact	# of Chapters / Articles	New contents added/ updated	Land Use Classification					
			A	F	RR	UR	O	N
1987/1987	6/57	Mechanism of management and usage of land property	○	○	○[*1]		○	○
1993/1993	7/89	+ Right and responsibility of people using land	○	○	○	○	○	○
2003/2004	7/146	+ Right and responsibility of the government as the land owner	○	○[*2]				○
2013/2014	14/212	+ Land ownership, right and responsibility of the government	○	○[*2]				○
2024/2025	16/260	+Land price framework; rent payments, use term for housing projects, fund for commercial housing projects,	○	○[*2]				○

A: agricultural, F: forestry, RR: rural residential, UR: urban residential, O: others, N:not-in-use
[*1]Residential
[*2]Non-Agricultural

Table 3.11 Soil standards for allowance limits of heavy metals (mg/kg dry soil)

	As	Cd	Cu	Pb	Zn	Cr
	08[*1]/15[*2]	08[*1]/15[*2]	08[*1]/15[*2]	08[*1]/15[*2]	08[*1]/15[*2]	08[*1]/15[*2]
Agricultural	12/15	2/1.5	50/100	70/70	200/200	–/150
Forestry	12/20	2/3	70/150	100/100	200/200	–/200
Residential	12/15	5/2	70/100	120/70	200/200	–/200
Industrial	12/25	10/10	100/300	300/300	300/300	–/250
Commercial[*3]	12/20	5/5	100/200	200/200	300/300	–/250

08[*1]: QCVN 03:2008/BTNMT, 15[*2]: QCVN 03-MT:2015/BTNMT
Commercial[*3]: Commercial (08[*1]), Commercial and Service (15[*2]).
The land use classifications are not the same in both standards

3.5 Government Environmental Monitoring System

3.5.1 Environmental Monitoring Organization

The environmental monitoring mission in Vietnam is currently managed by the Vietnam Environment Administration (VEA) through three centers: the Northern Center for Environmental Monitoring (N-CEM), the Middle and Highland Center for Environmental Monitoring (MH-CEM), and the Southern Center for Environmental Monitoring (S-CEM).

N-CEM was an extension of the "Database Office" of the Environment Department, which was set as the first center for environmental monitoring activities in 1996 under the former Minister of Science, Technology, and Environment. Due to the establishment of MONRE, the office was renamed three times and finally became N-CEM in 2018 when MH-CEM and S-CEM were established. Currently, N-CEM oversees the northern 28 administration units (provinces and cities), whereas MH-CEM oversees 14 units and S-CEM oversees the remaining 21 units. On behalf of VEA, N-CEM gathers, stores, and manages environmental monitoring data and information from all centers.

Table 3.12 List of environmental monitoring stations, sites, and works

Fields of observation			Monitoring stations$1	Monitoring sites$2	Monitoring works$3
Hydrometeorology			571/208/213/179	846/2748/500/373	0/0/0/0
Water resources	Surface water		15/41/41/0	–	–
		$	80/33/8/17	–	–
	Ground water		44/27/12/15	375/192/211/0	735/411/411/0
		$	–	412/(NA)/197/60	805/(NA)/305/89
Environment			23/16/25/29	571/200/259/97	0/0/0/0
Sea resources and environment			4/31/34/23	0/0/0/0	0/0/0/0
Satellite positioning and geo-dynamic			79/59/27/38	0/0/0/0	0/0/0/0
Remote sensing			1/1/1/1	0/0/0/0	0/0/0/0

Numbers: (current*)/(2016–2020)/(2021–2025)/(2026–2030) * as of January 12, 2016
All data are based on "Decision 90/QD-TTg" except those in the row$, which are updated date in "Decision 432/QD-TTg" issued in 2021
Monitoring Stations$1: Stations which have automatic monitoring instruments
Monitoring Sites$2: Positioned sites for regular manual sampling activities
Monitoring Works$3: Installed offices to support the monitoring activities

Each of the 63 administrate units also has its own local center for environmental and natural resources monitoring (CENM), which is directly managed by its local Department of Natural Resources and Environment (DONRE).

3.5.2 Environmental Monitoring Network

Environmental monitoring plays a key role in environmental management and protection. LEP1993 clearly defined the contents of state management of environmental protection including the organization, development, and management of environmental monitoring systems at the national level, and LEP2005 had a chapter on regulations on environmental monitoring and information activities. Subsequently, the role of environmental monitoring activities became increasingly important, and a modern network of environmental monitoring was expected to be established for a more positive contribution to environmental protection in the future.

LEP2014 included a chapter on regulating environmental monitoring, and a Decision of 90/QD-TTg was issued in 2016 to approve the national monitoring network for natural resources and the environment for a period between 2016 and 2025, with a view of 2030. As a result, many monitoring stations, sites, and works were established as shown in Table 3.12. In 2021, a new decision was made to revise these numbers, listed in Table 3.12.

3.6 Municipal Solid Waste Management (MSWM)

3.6.1 Regulation on Municipal Solid Waste Management

LEP1993 did not regulate the components of solid waste management (SWM), but LEP2005 included various articles on SWM in general and municipal solid waste management in particular. Subsequently, Decree 59/2007/ND-CP was issued, providing regulations on the activities, rights, and obligations of the entities engaged in SWM. LEP2014, which superseded LEP2005, determined waste management for all stages of generation, reduction, sorting, collection, transportation, reuse, recycling, and destruction. Decree 38/2015/ND-CP was issued to regulate waste and scrap management. Subsequently,

Decree 40/2019/ND-CP was issued for amendments and supplements. After the implementation of LEP2014, technical standards and regulations were issued for MSW, including national technical regulations on MSW incinerators and the technical infrastructure for solid waste treatment and toilets.

The LEP was subsequently modified in 2020. It contains many new regulations, strongly reforming all stages of domestic solid waste management, from classification at the source, storage, and transfer to collection, transportation, and treatment. The law has a separate section on domestic solid waste management, requesting that domestic waste owners be responsible for reducing and classifying waste rights from sources. Regarding solid waste treatment facilities, when investing in or setting up new operations, modern and environmentally friendly technologies must be prioritized to ensure a gradual reduction in the proportion of domestic solid waste treated by direct landfills. Decree No. 08/2022/ND-CP, which was issued recently, details several articles on LEP2020.

Regarding Institutional Responsibility on MSWM, MONRE is responsible for the unified state management of environmental protection. The Ministry is developing (1) technical guidelines; (2) management guidelines for the classification, storage, gathering, transportation, preliminary processing, reuse, recycling, and treatment of MSW; and (3) recovery of energy from MSW. Furthermore, MONRE organizes, manages, and inspects environmental protection activities associated with MSWM. At the local level, the Provincial People's Committee is responsible for (1) organizing the management of SW in the province; (2) issuing specific regulations on MSW, mechanisms, and preferential policies; (3) providing support to encourage the collection, transportation, and investment of SW treatment facilities suitable for the socio-economic development of the locality; and (4) organizing the formulation and approval of SWM planning according to the authority. After the government issued Resolution 09/2019/NQ-CP, which identified MONRE as the unified state management agency for SW, some localities assigned DONRE as a specialized agency to assist.

3.6.2 Strategies on MSWM

The SWM strategy for cities and industrial parks was implemented from 1999 to 2020. This was probably the first SWM policy implemented in Vietnam. However, this strategy was limited to cities and industrial parks. Several policies and strategies for SWM have been issued since Decree 59/2007/ND-CP, as listed in Table 3.13.

Prime Minister Decision (PMD) #3 in Table 3.13 (issued in 2009) set a comprehensive national strategy for integrated SWM until 2025, with a vision for 2050. This included specific targets for the collection and treatment of different types of SW in urban and rural areas, in 2015, 2020, and 2025. The target items related to MSWM and the other strategies (PMD #6 and PMD #12) are listed in Table 3.14.

PMD #6 was approved in 2012 to provide the National Strategy on Environment Protection for 2020 with visions for 2030, particularly complementing PMD #3 for SWM by strengthening the implementation of solid waste collection and treatment facilities. It provides orientation for measures to protect the environment and solutions to increase the percentage of collected, recycled, and reused solid waste and reduce the production and use of difficult-to-disintegrate bags and packing.

PMD #12 was approved in 2018 as a revision of PMD #3. According to this revision, 85% of hazardous waste generated in households and individuals must be collected, transported, and treated to meet the requirements of environmental protection; 100% of eco-friendly plastic bags should be used at shopping centers and supermarkets for domestic purposes, replacing hard-decomposition plastic bags; and 80% of daily-life solid waste generated in rural areas should be collected, stored, transported, self-treated, and centralized. These are the basic objectives of the strategy until 2025.

Table 3.13 Summary of strategies and plannings/program on management of municipal solid waste over the last 15 years

#	Year of issue	Law or strategies
1	2007	Decree on Solid Waste Management (Decree 59/2007/ND-CP)
2	2008	PMD: Approve the plan on construction of solid waste treatment facilities in three key economic regions (northern, central, and southern) up to 2020
3	2009	PMD: National strategy for the management of solid waste by 2025 and the orientation toward 2050
4	2010	PMD: Approve the plan on building solid waste treatment facilities in the Mekong River delta key economic region through 2020
5	2011	PMD: A program for investment in solid waste treatment, run during 2011–2020
6	2012	PMD: National strategy on environment protection until 2020, with visions for 2030
7	2013	PMD: A project on enhancing the control of environmental pollution due to the use of non-biodegradable plastic bags by 2020
8	2014	PMD: A master plan on solid waste disposal of Hanoi capital until 2030, with a vision for 2050
9	2014	Revision of law on environment protection (LEP2014)
10	2015	Decree on the management of waste and scraps (Decree 38/2015/ND-CP)
11	2016	PMD: Solid waste management planning in northern key economic region by 2030
12	2018	PMD: Revision of the national strategy on integrated management of solid waste until 2025, with a vision for 2050
13	2020	Revision of law on environmental protection (LEP2020)
14	2022	Decree detailing some articles on LEP2020 (Decree 08/2022/NĐ-CP)

PMD: Prime Minister Decision

3.6.3 Plannings on MSWM

PMD #2 (2008) established a plan for the construction of solid waste treatment centers in three key economic regions—northern, central Vietnam, and southern Vietnam—up to 2020. The plan mentioned the construction of two, three, and six SW treatment facilities with 240–260, 180, and 100 ha in the northern, central, and southern key economic regions, respectively. MSW treatment technologies include sanitary landfills, composting, and recycling. Incineration of hazardous MSW was planned. The total investment cost was 9,683 billion Vietnamese dongs (approximately 598 million USD).

PMD #4 (2010) approved a plan for the construction of five solid waste treatment facilities in the Mekong Delta region by 2020, according to Decree #1 (2007). The plan covers an area of 16,617 km^2 with a forecasted MSW of 4,200 tons/day by 2015 and 7,550 tons/day by 2020. These solid waste treatment facilities were designed to treat municipal solid waste and other solid wastes by sanitary landfilling, composting, recycling, and incineration.

PMD #11 (2016), on SWM planning in the northern region, aimed to improve the integrated management of solid waste to prevent and minimize the generation of waste at the source, enhance reuse and recycling, and minimize the volume of waste to be landfilled. Zoning for solid waste treatment was proposed in the Master Plan.

There were also three river basin SWM plans: the Cau River Basin (2013), the Dong Nai River (2015), and the Nhue-Day River (2015). Regarding waste management planning, 59 provinces and 63 cities prepared and approved SWM planning in the area. In conclusion, regional treatment zones for SWM are currently limited to the country; the rest are provincial, district, and commune treatment zones.

Table 3.14 Percentages of selected targets in three national strategies

	PMD #3			PMD #6		PMD #12
	2015	**2020**	**2025**	**2015**	**2020**	**2025**
Urban solid waste collection rate	85%	90%	100%	90%	95%	90%[a]
Rural solid waste collection rate	40%	70%		60%	75%	80%[a]
Amount of urban MSW treatment by reuse/recycle/energy recovery/composting	60%	85%	90%			
Amount of solid waste treatment by reuse/recycle/energy recovery/composting				55%	85%	–
Cities having their own recycling facilities, carrying out waste separation at source	50%	80%	100%			
Handling of severe polluted landfilling with appropriate treatment listed in Decision 64/2003/QD-TTg	100%					
Reduction of Severe Environmental Pollution site				20% of 2010	50% of 2015	
Number of engineered and sanitary landfill sites				50%	90%	
Direct landfilling of urban MSW						<30%
Number of closed landfill sites for urban MSW that have land reformed and reused						90–95%
Output waste to be disposed by landfill required for newly invested MSW treatment facility						<20%
Spontaneous open dumps in rural areas to be treated						90–95%
Reduction of plastic bags at shopping center and supermarket	40%	65%	85%			
Using eco-friendly plastic bags instead of hardly biodegradable plastic bags at shopping centers and supermarkets for daily living purposes						100%
Hazardous waste from households and individuals to be collected and treated						85%

[a]With appropriate treatment

MONRE issued the State Environmental Report 2019 specifically on MSWM, reviewing some of its difficulties and constraints. Despite incentives, policy, mechanisms, and support for investment capital and sales, the guide for implementation is still lacking and not timely. The number of solid waste treatment projects that receive concessional loans is small. Coordination among relevant ministries, branches, and localities has not yet been implemented in a unified and effective manner. Investment capital and the cost of collection, transportation transfer, and the MSW process mostly come from the state budget and ODA, but they have not yet responded to meet actual needs. Most imported SW treatment technologies are not suitable for the characteristics of SW in Vietnam, and domestic treatment equipment and technologies have not been synchronized; therefore, they cannot be disseminated and replicated. Meanwhile, the state has not been given a well-defined orientation toward technology application, and selection criteria for suitable equipment and technology have not been issued.

3.6.4 Other Solid Waste Management

According to Decree 08/2022/ND-CP, industrial waste is generated by production, business, and service activities. They are classified as hazardous waste, must-be-controlled industrial waste, or ordinary industrial solid waste. The National Assembly assigned MONRE to promulgate a list of hazardous waste, industrial waste that must be controlled, and ordinary industrial solid waste. The classification of must-be-controlled industrial waste is hazardous waste or ordinary industrial solid waste, according to environmental technical regulations on hazardous waste thresholds. When the industrial waste that must be controlled is not classified, it is managed as hazardous waste. In cases where there are no technical regulations for certain hazardous properties and components, the national standards on the environmental protection of one of the countries belonging to the industrialized countries will apply. Circular No. 02/2022/TT-BTNMT (MONRE) details the regulations for ordinary industrial solid and hazardous waste management.

Circular No. 08/2017/TT-BXD regulates construction solid waste management. This circular provides detailed regulations on the classification, collection, reuse, recycling, and treatment of construction solid waste, as prescribed in Decree No. 38/2015/ND-CP on waste and scrap management. New points and regulations of LEP2020 include environmental protection by industry and field. Environmental protection in the field of construction is regulated by Article 64 of the law. Provincial People's Committees regulate the collection, transportation, and treatment of construction solid waste and plan the locations for dumping waste from construction activities, sludge from septic tanks, and sludge from drainage systems. The Minister of Construction is responsible for developing standards and technical regulations for the design of solid waste collection systems in accordance with the classification of solid waste at the source of commercial centers combined with apartments, households, apartments combined with offices, and high-rise building complexes with mixed functions.

Column 4

Craft Village (Làng thủ công)

Vietnam has a unique form of industry known as a craft village ("Làng thủ công" in Vietnamese). A craft village is a village, hamlet, or community, usually in a rural area, that specializes in the production of a paricular product. The products include not only traditional crafts (processed foods, textiles, and furniture), but also new goods and recycled materials (waste paper, waste plastics, metals, batteries, and used vehicles recycling). According to MONREI (2021) there are 4,575 craft villages in Vietnam.

While these villages play an important role in the rural economy and help preserve traditional crafts and cultural heritage, they may also contribute to environmental degradation, especially in recycling villages. These villages specialize in the collection, sorting, and processing of recyclable materials, such as paper, plastics, and metals, which are then transformed into new products or raw materials. Many craft villages operate with traditional techniques and equipment, small-scale operations. In some places, the working environment is not adequately maintained, and the emissions of gases, effluent and wastes are not well managed.

To address these issues, the Vietnamese government has implemented policies and initiatives to promote cleaner and more sustainable practices in craft villages. For example, the government provides funding and support for the adoption of cleaner technologies, and for the development of wastewater treatment facilities. By implementing these measures, craft villages can continue to play a role in the rural economy while also minimizing their environmental impact.

References

2030 Water Resources Group (2017) Vietnam: Hydro-Economic Framework for Accessing Water Sector Challenges. Accessed 24 August 2023 from https://2030wrg.org/wp-content/uploads/2017/08/Vietnam-Hydro-Economic-Framework.pdf

Ganguly D, Dey S, Arora LK et al (2022) Study to assess the compliance of thermal power plants in India to new SO_2 emission norms (2015) and lay out phased plan for FGD implementation. Accessed 28 March 2023 from https://cea.nic.in/wp-content/uploads/tprm/2022/07/Report_on_Emission_norms.pdf

Government of Vietnam (2004) Decree 149/2004/ND-CP dated on July 27, 2004 regulating the licensing of water resources exploitation, abstraction and utilization and waste water discharge into water sources

Ly BT, Kajii Y, Nguyen TYL, Shoji K, Van DA, Do TNN, Nghiem TD, Sakamoto Y (2020) Characteristics of roadside volatile organic compounds in an urban area dominated by gasoline vehicles, a case study in Hanoi. Chemosphere 254:126749. https://doi.org/10.1016/j.chemosphere.2020.126749

MONRE (n.d.) Organizational Structure of the Ministry of Natural Resources and Environment. Accessed 31 March 2023 from https://monre.gov.vn/portal/pages/cocautochuc.aspx (in Vietnamese)

Wang G, Deng JG, Zhang Y, Zhang Q, Duan L, Hao JM et al (2020) Air pollutant emissions from coal-fired power plants in China over the past two decades. Sci Total Environ 741:140326. https://doi.org/10.1016/j.scitotenv.2020.140326

WHO (World Health Organization) (2021) What are the WHO air quality guidelines? Accessed 23 March 2023 from https://www.who.int/news-room/feature-stories/detail/what-are-the-who-air-quality-guidelines

Livelihood and Its Surrounding Environments in Rural Areas

4

Izuru Saizen, Shinya Funakawa, Yuki Okamoto, Tin Cong Hoang, Binh Huu Ngo, Ni Ngoc Khanh Tran, Hoa Thi Thai Hoang, Lam Ho Nguyen, Minh Tuan Vu, Hitoshi Shinjo, Giang Huong Pham, Ryo Sakaguchi, Masataka Kuroda, Hidenori Harada, Shigeo Fujii, Minori Tokito, Jane Singer, Hai Hoang, Huu Ty Pham and Kinh Thi Kieu

Abstract

This chapter focuses on livelihood and the surrounding environment in Central Vietnam. Based on detailed field surveys, valuable insights are provided, alongside proposals to improve the current conditions. The main topics are (1) transitions of fishery aquaculture practices toward the natural environment challenges of brackish ecosystems. (2) Spatial and temporal changes in submerged aquatic vegetation ecosystems. This is assessed by monitoring seagrass beds using satellite data, and interviewing household members. (3) Cropping systems in lowland sandy soil, which is normally not suitable for agriculture, and linkages between cropping patterns and soil properties, identified by through sampling soil in several coastal districts. (4) Rice cropping in saline soils and the challenges to increasing rice yield and income of farmers. (5) The characteristics of home gardens, such as mixed multi-layer cropping and the possibility of vanilla production to improve livelihood. (6) Historical changes in waste and wastewater management methods, and flows in nitrogen and phosphorus, evaluating the impacts of nutrient management on the environment. (7) Livelihood in an ethnic minority called Cotu, which was closely associated with natural resources, and is now moving drastically toward mono-culture, specifically to acacia forestry. (8) Challenges of dam-displaced communities.

Keywords

Primary industry · Ethnic minority · Cropping system · Vulnerability · Rural sustainability

The original version of the chapter has been revised: Author's name Thuc Dinh Do has been removed. A correction to this chapter can be found at https://doi.org/10.1007/978-981-97-2289-1_7

I. Saizen (✉) · S. Funakawa · H. Shinjo · G. H. Pham · R. Sakaguchi · S. Fujii
Graduate School of Global Environmental Studies, Kyoto University, Kyoto, Japan
e-mail: saizen.izuru.4n@kyoto-u.ac.jp

Y. Okamoto
Technology Integration Section, Solution Management Division, WorldLink & Company Co. Ltd., Kyoto, Japan

T. C. Hoang · B. H. Ngo · N. N. K. Tran
Faculty of Environmental Science, University of Sciences, Hue University, Hue, Vietnam

H. T. T. Hoang · M. T. Vu
Faculty of Agronomy, University of Agriculture and Forestry, Hue University, Hue, Vietnam

L. H. Nguyen
High-Tech Agriculture Division, University of Agriculture and Forestry, Hue University, Hue, Vietnam

S. Fujii et al. (eds.), *Livelihood and the Environment in Vietnam*, Sustainable Development Goals Series, https://doi.org/10.1007/978-981-97-2289-1_4

4.1 Introduction

4.1.1 Overview of This Chapter

Vietnam can be roughly divided into three regions, consisting of Northern, Southern, and Central Vietnam. Ho Chi Minh city and Hanoi as mega cities are well known in the world. Ho Chi Minh city is located in Southern Vietnam and has the role as the economic center, while Hanoi is located in Northern Vietnam and is the capital as the center of politics. Two mega cities have been leading rapid economic growth in two regions as well as the entire country especially after Vietnam War. Central Vietnam, however, has not been well developed economically compared to two regions, and the percentage of primary industry is high. In addition, poverty rate is relatively high so we often witness active interaction between humans and nature, and livelihoods in many living people are closely related to and depend on natural resources. Accordingly, their livelihoods are vulnerable to environmental problems induced by not only climate changes but also the anthropogenic activities.

In this chapter mainly focuses on livelihood and its surrounding environments in Central Vietnam. Based on detailed field surveys, valuable insights are detected and proposals to improve current condition are mentioned in this chapter. Figure 4.1 shows the study areas in this chapter, and most study areas other than Sect. 4.7 can be found in Central Vietnam. Section 4.1 introduces the summary of this chapter, natural environment, and the primary industry and its changes, which deliver readers basic and comprehensive understanding of Central Vietnam. Section 4.2 refers transitions of fishery's aquaculture practices to natural environment challenges of brackish ecosystem and proposes the way to sustain the fishery products at Tam Giang lagoon. Section 4.3 is about spatial and temporal changes of submerged aquatic vegetation ecosystems at An Hoa lagoon. By monitoring seagrass beds by using satellite data and interviews with the representatives of households, a baseline for conservation of seagrass ecosystems is described. Section 4.4 shows the cropping systems in lowland sandy soil, which is normally not suitable for agriculture, and discusses linkages between cropping patterns and soil properties through sampling soil at several coastal districts, in the Thua Thien Hue province. Section 4.5 treats rice cropping in saline soils and its challenges. Through field surveys at the Quang Phuoc commune, feasible solutions are mentioned to increase rice yield and income of farmers. Section 4.6 describes the characteristics of home garden, such as mixed multi-layer cropping, at Hong Ha commune and discusses the possibility of vanilla production to improve livelihood. Section 4.7 shows the historical changes on waste and wastewater management manners and those flows in nitrogen and phosphorus at Trai village in suburban Hanoi, and then evaluates the impacts of nutrient management on the environment. Section 4.8 introduces livelihood in an ethnic minority called Cotu at Doi village. Their livelihood used to be closely related to natural resources but is changing drastically moving to mono-culture, especially to acacia forestry. Section 4.9 introduces challenges of dam-displaced communities. The field surveys are conducted at five villages widely as shown in Fig. 4.1 and reveal the current difficult condition they are facing.

Originally, people living in rural areas of Vietnam had lived a self-sufficient lifestyle

M. Kuroda
Overseas Water Supply Department, Nihon Suido Consultants Co., Ltd., Tokyo, Japan

H. Harada
Graduate School of Asian and African Area Studies, Kyoto University, Kyoto, Japan

M. Tokito
Graduate School of Agriculture, Kyoto University, Kyoto, Japan

J. Singer
Department of Global Studies, Kyoto University of Foreign Studies, Kyoto, Japan

H. Hoang
The University of Danang, Danang, Vietnam

H. T. Pham
University of Agriculture and Forestry, Hue University, Hue, Vietnam

K. T. Kieu
University of Science and Education, The University of Danang, Danang, Vietnam

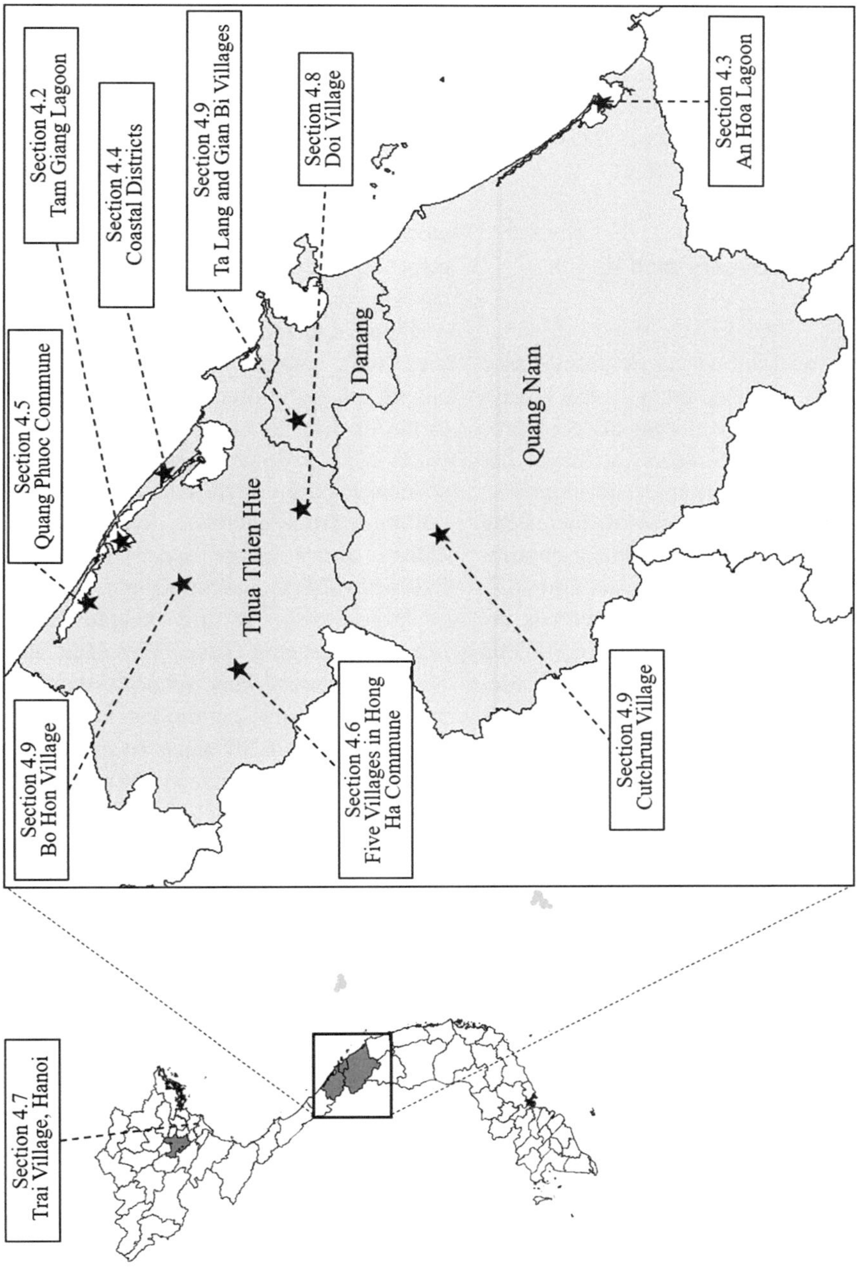

Fig. 4.1 Study area locations in this chapter

through the effective use of natural resources in a sustainable way. With the encroachment of the monetary economy in recent years, this lifestyle has changed drastically. As a result, natural resources have been overused and the surrounding environment has deteriorated. This chapter provides readers with a detailed explanation of the social impact brought about by these changes in rural areas, and conveys this to readers in an easy-to-understand manner.

4.1.2 Primary Industry and Its Changes

Vietnam is a well-known agricultural-based country, where the primary industry has always been the most prominent among all economic sectors. Over the last 30 years, Vietnam has achieved significant economic development, with the annual rate of GDP growth increasing from around 6.5–7.0%. Agriculture contributed 40% of GDP in 1990 compared to 20% in 2020; however, this production sector is the main source of income for around 16,800,000 households. In other words, income per household had been quite low at the beginning of this period. In the early 1990s, the Vietnamese government launched a series of policies designed to enhance economic development. Agricultural reforms were a key element of this strategy. The introduction of Land Law in 1992 allowed the private sector and farmer households to arrange land allocation and ownership more easily. Although Vietnam experienced regular food shortages before 1990, the Vietnamese agricultural sector has produced enough in the time since for both domestic consumption and exportation. Agricultural products such as rice, coffee, black pepper, rubber, and seafood are now exported worldwide.

Although Vietnam has achieved strong development in rural and agricultural settings, the country still faces many challenges. Vietnam possesses 54 ethnic groups, of which 53 are minorities living in remote mountainous areas. As these groups still practice traditional slash-and-burn farming, their quality of life remains behind in terms of development (JICA 2013; Inoue et al. 2021).

In Central Vietnam, farmers in the agricultural industry have low incomes compared to North and South Vietnam. In an attempt to stabilize farm management, in addition to paddy rice the central government encourages the cultivation of commercial crops such as acacia and rubber in mountainous areas. In flatlands and coastal areas, fish and shrimp farming are encouraged (Figs. 4.2 and 4.3). Raising livestock has also been encouraged in a further effort to alleviate poverty, supported in many cases by the international aid program.

Flood damage is a serious problem in this area, with countermeasures being a vital consideration. Agricultural sectors in particular are prone to occasional flood damage. In terms of water quality, domestic wastewater and factory wastewater are not sufficiently regulated. Water pollution from chemical fertilizers and pesticides is also a serious issue, which directly and indirectly affects livelihoods in Central Vietnam.

Target areas in this chapter are mainly located in Danang, Thua Thien Hue, and Quang Nam provinces. This section summarizes the primary industry conditions in the previous decade (2011–2020) using secondary statistical data provided by the General Statistics Office of Vietnam, which can help to provide readers with a comprehensive background to this chapter.

The development of planted paddy areas is shown in Table 4.1, while development of paddy production is shown in Table 4.2. From 2011 to 2020, the total size of paddy areas was almost stable or decreased slightly. However, paddy production has increased from 751,400 to 802,600 tons. Thus, the paddy production rate has increased significantly over the past decade. Paddy production in Central Vietnam is a vital source of food production, and has been well maintained throughout Vietnam's period of rapid economic growth.

The development of aquaculture areas from 2011 to 2020 is shown in Table 4.3. While the total size of aquaculture areas in Danang decreased from 6 km^2 in 2011 to 2 km^2 in 2020, Thua Thien-Hue and Quang Nam both saw an

Fig. 4.2 Paddy rice fields in a mountainous area in Central Vietnam

Fig. 4.3 Aquaculture practice in the lagoon in Central Vietnam

increase from 58 km^2 and 48 km^2 in 2011 to 76 km^2 and 54 km^2 in 2020, respectively. Thus, the total size of aquaculture areas within the study areas increased from 112 km^2 in 2011 to 132 km^2 in 2020. Tam Giang lagoon in Thua Thien-Hue and An Hoa lagoon in Quang Nam are

Table 4.1 Development of planted paddy area in Thua Thien-Hue, Danang, and Quang Nam from 2011 to 2020 (thousand ha)

	2011	2012	2013	2014	2015	2016	2017	2018	2019	2020
Thua Thien-Hue	299.1	299	284.9	317.1	319.3	324.3	327.4	334.4	326.7	320.2
Danang	34.4	35.3	31.3	33.9	31.6	32.2	31.4	31.9	30.8	30.4
Quang Nam	417.9	447.3	440.3	466.9	461.2	441.7	461.1	462.6	452.3	452.0
Total	751.4	781.6	756.5	817.9	812.1	798.2	819.9	828.9	809.8	802.6

Table 4.2 Transition of production of paddy production in Thua Thien-Hue, Danang, and Quang Nam from 2011 to 2020 (thousand tons)

	2011	2012	2013	2014	2015	2016	2017	2018	2019	2020
Thua Thien-Hue	53.5	53.8	53.7	53.7	54.4	54.5	54.9	54.7	54.8	54.3
Danang	6.5	5.9	5.4	5.5	5.4	5.4	5.2	5.2	5.1	4.8
Quang Nam	87.7	88.6	87.9	87.4	88.5	86.7	86.8	86.2	84.9	83.5
Total	147.7	148.3	147	146.6	148.3	146.6	146.9	146.1	144.8	142.6

Table 4.3 Transition of aquaculture areas in Thua Thien-Hue, Danang, and Quang Nam from 2011 to 2020 (km^2)

	2011	2012	2013	2014	2015	2016	2017	2018	2019	2020
Thua Thien-Hue	58.0	62.0	72.0	75.0	72.0	71.0	71.0	74.0	75.0	76.0
Danang	6.0	5.0	5.0	5.0	5.0	4.0	5.0	5.0	5.0	2.0
Quang Nam	48.0	48.0	47.0	49.0	50.0	51.0	52.0	55.0	55.0	54.0
Total	112.0	115.0	124.0	129.0	127.0	126.0	128.0	134.0	135.0	132.0

known as fertile fishing grounds, highly suitable for aquaculture. As a result, aquaculture areas have been developed in both regions during this period.

The development of fishery production from 2011 to 2020 is presented in Table 4.4. Fishery production increased in all areas from 153,518 tons in 2011 to 219,512 tons in 2020. During this decade, this figure increased by 43%, meaning that fishery was one of the key primary industries in these three regions.

The development of the concentrated planted forest areas is shown in Table 4.5. The size of this area has increased constantly in Thua Thien-Hue and Quang Nam, though it has decreased in Danang. In total, the concentrated planted forest areas have increased from 168.0 km^2 in 2011 to 275.0 km^2 in 2020.

Vietnam has achieved rapid economic growth over the past decade, with the GDP per capita rising from 1,958 USD in 2011 to 3,552 USD in 2020. Rapid economic growth often has a negative impact on primary industry sectors, as Japan has experienced before. However, Vietnam has successfully maintained its primary industry sectors in Central Vietnam, as proven by the statistical data analysis presented in this section.

4.1.3 Natural Environment

Geography The north of Central Vietnam borders the Red river delta and the Northwest. The south borders the provinces of Binh Phuoc, Dong Nai, and Ba Ria–Vung Tau. Central Vietnam borders the Easter Sea in the east and

Table 4.4 Transition of fishery production in Thua Thien-Hue, Danang, and Quang Nam from 2011 to 2020 (ton)

	2011	2012	2013	2014	2015	2016	2017	2018	2019	2020
Thua Thien-Hue	43,183	45,724	47,593	50,807	54,300	45,406	52,079	55,632	57,820	56,750
Danang	34,628	33,555	33,203	33,590	34,980	34,569	36,210	37,241	38,682	38,162
Quang Nam	75,707	82,320	83,905	91,764	97,665	102,649	110,314	114,500	119,500	124,600
Total	153,518	161,599	164,701	176,161	186,945	182,624	198,603	207,373	216,002	219,512

Table 4.5 Transition of concentrated planted forest area in Thua Thien-Hue, Danang, and Quang Nam from 2011 to 2020 (km^2)

	2011	2012	2013	2014	2015	2016	2017	2018	2019	2020
Thua Thien-Hue	41.0	42.0	43.0	43.0	42.0	59.0	61.0	61.0	62.0	65.0
Danang	15.0	15.0	20.0	23.0	20.0	20.0	23.0	23.0	14.0	10.0
Quang Nam	112.0	138.0	132.0	159.0	146.0	154.0	177.0	189.0	195.0	200.0
Total	168.0	195.0	195.0	225.0	208.0	233.0	261.0	273.0	271.0	275.0

Laos and Cambodia in the west. The plains are surrounded by the Truong Son range along the west and a sandy coast along the east. The terrain of Central Vietnam has a lower elevation in the northwest to southeast directions. The western North Central coast region has intermediate and low mountains. The altitude of the mountainous area in Thanh Hoa is 1,000–1,500 m above sea level. The mountainous areas in Nghe An and Ha Tinh form the headwaters of the Truong Son range and have a very rugged terrain. Most of the high mountains are distributed in this region. The plains span a total area of nearly 0.62 million ha; of this, the Thanh Hoa delta, which is accreted by alluvial sources from the Ma and Chu rivers, accounts for nearly half and is the most expansive plain of the central region. The Central highlands are located to the west of the Truong Son range. Here, the terrain is diverse and complex, mainly plateaus with high mountains at an altitude of 250–2,500 m. The South Central coast comprises a coastal plain and low mountains. It harbors a system of short and steep rivers, a deep coastline with many bends, and a narrow continental shelf. The western mountains spread in the north–south direction, gradually approaching the sea and progressively narrowing the area. The plains are small and are accreted by rivers and seas, following the foothills of mountains (Le 1998, 2009).

Geology The territory of Vietnam is divided into five structural blocks: Northeast, Northwest, Truongson, Kontum, and Nambo. Central Vietnam constitutes the Truongson and Kontum blocks. The Truongson block is regarded as a northwest–southeast-trending Paleozoic-folded system with thick Paleozoic formations (>12,000 m). Precambrian strata are widespread in the Kontum block. Archean rocks are found only in the Kontum block, regarded as a stable massif without Paleozoic sedimentary rocks (Tran 1995).

Archean granulites in the Kontum massif are grouped into the Cannak complex, which is the oldest stratigraphic unit in Vietnam. They are closely connected to the autochthonous plutons of alkali charnokites (orthopyroxene-bearing granites). The thickness of the complex is approximately >4,000 m. Lower Proterozoic rocks found in the Kontum massif have been placed into different complexes with local names. The general thickness of the complexes is approximately 2,500–3,000 m. Upper Proterozoic metamorphic rocks with a stratal thickness of 1,000–1,200 m are widespread and form a 50–60 km-wide belt bordering the Kontum massif (Thi 1985, cited from Tran 1995).

Paleozoic strata in Central Vietnam are widespread in the Truongson block but are absent in the Kontum block. In the Truongson block, the 1,300–2,000 m-thick Middle Cambrian–Lower Ordovician strata primarily comprise limestone, mudstone, and sandstone and are called the Changpung type. The Changpung type has been regarded as a part of the Eastern Asia–Australia paleobiogeographical province. The 3,000–3,500 m-thick Upper Ordovician–Lower Silurian formations are largely distributed in the Truongson block. Rocks are terrigenous flysch and contain graptolite assemblages. The 2,000–3,500 m-thick Upper Silurian formations covering large areas in the Truongson block comprise many fossils, such as graptolites, brachiopods, trilobites, and acritarchs (Ngan et al. 1986, cited from Tran 1995). The Lower Devonian strata are characterized by a 2,000 m-thick formation of shale and sandstone, while the Middle Devonian strata are characterized by a 1,200 m-thick terrigenous formation in the lower part and a 500 m-thick carbonate formation in the upper part. The thin Upper Devonian strata are composed of two facies: the carbonate facies with Frasnian–Famennian stromatoporoids and conodonts distributed in the north part of the block; the clastic facies of the tidal–littoral zone are Frasnian alone in the south part (Thanh et al. 1986, cited from Tran 1995). The Carboniferous–Permian formations widely distributed in the Truongson block are composed of Lower Carboniferous shale, sandstone, conglomerate, limestone, and Carboniferous–Permian limestone with abundant foraminifer assemblages. The total thickness of the Carboniferous–Permian strata is approximately 1,800–2,200 m (Hung and Tien 1986, cited from Tran 1995).

Climate Central Vietnam is located in the tropical belt and is hot and humid throughout the year. It suffers from the Asian torrential monsoon regime, mainly as northeast and southeast monsoon. Along the North Central coast, the impact of the northeast monsoon is powerful, causing two distinct seasons in the region: the hot season lasts from May to October and the cold season lasts from November to April of the next year. Meanwhile, the South Central coast and Central highlands are mainly affected by the southeast monsoon with heat and wetness throughout the year. Due to the region's varied terrain, Central Vietnam has several sub-climate regions. For instance, the Lam Dong province in the Central highlands has a temperate climate, while the Ninh Thuan province has a tropical savanna climate. The mean annual temperature is 26–27 °C, mean rainfall is ~700–800 mm in Phan Rang, and ambient humidity is 75–77% (Le 1998; Socialist Republic of Viet Nam 2004; Babatunde et al. 2016). Central Vietnam witnesses 6–10 storms and tropical depressions, which cause heavy rains and floods. Storms and tropical depressions often occur from June to November, mainly in September and October (Socialist Republic of Viet Nam 2004).

Vegetation Vietnam is one of the 25 countries with rich biodiversity. It is home to many types of ecosystems and biological species and harbors diverse and endemic genetic resources. The total forest area as well as forest cover has increased from 2015 to 2020. In 2020, the total forest area was 14.7 million ha, of natural forests accounted for 10.3 million ha and plantation forests for 4.4 million ha; forest cover occupied 42% of Vietnam's total land area. Natural forests are mainly woody forests, bamboo forests, and mixed forests of wood and bamboo, accounting for approximately 8.9, 0.24, and 1.14 million ha of the total forest land, respectively. The *Areca*–coconut forests occupy a very small area (<5,000 ha) (Vietnam Ministry of Natural Resources and Environment 2021).

Central Vietnam is home to a large part of Vietnam's forested area and occupies three of the eight forest ecological zones on the mainland of Vietnam. The forest cover in the Central highlands, South Central coast, and North Central coast is 45.9, 50.4, and 57.4%, respectively.[1] In 2020, the natural forest area in Central

[1] Vietnam Ministry of Agriculture and Rural Development (2021). Decision 1558/QD-BNN-TCLN on announcing the national forest status in 2020.

Vietnam was 6.0 million ha (2.2, 1.6, and 2.2 million ha in the North Central coast, South Central coast, and Central highlands, respectively). The plantation forest area in Central Vietnam was 2.2 million ha, occupying 50% of that in the whole country. The plantation forest area distributed in North Central coast, South Central coast, and Central highlands is 0.92, 0.87, and 0.38 million ha, respectively. Acacia forests (1.35 million ha) are the largest plantation forests in the central region and mainland of Vietnam. This forest is mainly developed in the South Central coast (0.7 million ha) and North Central coast (0.55 million ha)[2] (To et al. 2021).

In Vietnam, biodiversity is concentrated mainly in special-use and protected forests, which play a vital role in watershed protection and biodiversity conservation, contributing to climate change response. According to the purpose of biodiversity protection and conservation, special-use forests are classified into (i) national parks; (ii) nature reserves; (iii) species and habitat conservation areas; (iv) landscape protection zones (forests preserving historical–cultural relics and scenic spots); (v) forest of beliefs; (vi) urban environmental protection forests, industrial zones, export processing zones, economic zones, and high-tech zones; (vii) scientific research and experimental forests; (viii) national botanical garden; and (ix) national seed forests (Vietnam Ministry of Natural Resources and Environment 2021). As of 2020, the central region had 14 national parks, 24 nature reserves, 7 species and habitat conservation areas, and 21 landscape protected areas, accounting for respectively 50, 42.8, 58.3, and 39.6% of the total areas in Vietnam. The special-use forest area of the three ecological regions is 0.61 million ha in the North Central Coast, 0.28 million ha in the South Central coast, and 0.48 million ha in the Central highlands (Vietnam Ministry of Agriculture and Rural Development 2021). Protected forest area distributed in the three ecological regions is 1 million ha in the North Central coast; 1.16 million ha in the South Central coast, and 0.64 million ha in the Central highlands (Vietnam Ministry of Natural Resources and Environment 2021).

[2]Vietnam Ministry of Agriculture and Rural Development (2021). Decision 1558/QD-BNN-TCLN on announcing the national forest status in 2020.

4.2 Transition of Fishery Activities in Coastal Areas in Central Vietnam: Within the Flow of Times and Economic Growth

4.2.1 Outline of Coastal Zone and Fisheries Industries in Vietnam

In Vietnam, fisheries constitute an important industry and approximately 4 million people depend on it for their livelihood (4% of total population in 2016, FAO 2017). Fisheries in Vietnam consists of 3 sectors: (1) marine capture, (2) inland capture, and (3) aquaculture. While inland capture is limited to small figures, marine capture (fish, crustaceans, and mollusks) reached 2,757,000 tons in 2015, which made Vietnam the eight biggest producer after Japan (FAO 2017). In addition, the aquaculture sector (fish, crustaceans, and mollusks) recorded 3,438,000 tons of production in the same year, which made Vietnam the fourth largest aquaculture producer in the world (FAO 2017). The main species included shrimp and prawn, such as *Penaeus monodon* and *Penaeus vannamei*, cultured in pond or shallow coastal zones along the shores in brackish river, lagoon, or bay. Thus, the fisheries industry in Vietnam is characterized by marine capture and aquaculture practiced along its long coastal line facing the ocean (South China sea). With shrimps and prawns being exported worldwide, Vietnam has become the third largest fisheries exporter in the world in recent times.

The ocean currents of South China sea change seasonally based on monsoon winds (dry and rainy seasons) (Hu et al. 2000). The regional differences are highlighted by seasons and terrestrial characteristics. In the northern part, Tonkin gulf blocks the current and winds from Hainan Island with inflows from rivers such as Red river (Hong river). The southern part is characterized by tropical ocean waters with large inflows from the Mekong river and its big delta. The central region sees complex current patterns owing to the seasons; the direction of the currents is from south to north in the dry season, originating from Karimata strait. In rainy season, the currents from East China sea and Luzon strait are distinguished, and the currents are from northeast to southwest direction (Tomczak and Godfrey 1994). As the continental shelf is well-developed along the coast, there are plenty of shallow beaches all along Vietnam. The width and area of continental shelf tend to be narrow in the central region and are distinguished in the Tonkin gulf and Mekong river mouth regions.

Such conditions produce rich marine ecosystems such as mangrove forests, coral reefs, tidal flats, seagrass beds, and sandy beaches along the coast, and thus various shore fishing activities can be found. Small-scale inshore fisheries are commonly observed all along the coast, where the community uses gillnet, longline, lift-net, push net, hook and line, surrounding net, cast net, and various other fishing methods. Some fishing activities do not require the use of a boat (walk and pick bivalves and crabs, push net to catch small fishes/shrimps, and so on). The type of fishing boats/vessels varies as well—from simple bamboo coracle boats without motors to wooden vessels with relatively big engines. Fixed traps are used/set in the lagoon/brackish zones where there are relatively weaker waves and winds, such as in bay, inlet, and lagoon areas.

In offshore fisheries, beam trawl, and bottom trawl are commonly practiced by single or paired boats to capture prawn and shrimp; surrounding nets such as purse seine are used to catch anchovies and some pelagic fishes; and longlines, certain other mobile traps, and squid fishing are also carried out. Details of fishing gears, the types of vessels, and the target species are included in SEAFDEC (2002). Offshore fisheries are affected by seasons, especially in Central Vietnam, and the main fishing season is in the dry season; thus, fishers engage in other jobs on land or switch to inshore fisheries in the rainy season. During the dry season at night, hook and line fishing along the beach for squids is also seen.

4.2.2 Livelihood in Local Small-Scale Fisheries Village: Based on Author's Experiences in Vietnam

The time schedule in fishing depends on how far the fishery area is, how to catch, and what the target species are. In addition, natural capture/fishing is closely related to the time of the tide/moon and feeding times around sunset and sunrise, and seasonality. Therefore, there are many cycles and customs that exist and affect the livelihood of fishers. In addition, the characteristic topography and location of fishery is also important (face to the open sea, nearby river, etc.). The place where the author stayed is a brackish lagoon closed by a sandbar, so the water surface is calm most of the season. Therefore, local residents fish using a bamboo boat with bamboo stick/wooden oar or a wooden boat with a small engine.

The time schedule of local residents starts from the evening. Around 5 PM, fishers have some early dinner, and move to the fishing area by boat, and set nets and traps one by one. Fishing is mostly carried out by males, but in some cases, both husband and wife fish together as well. Around 9 PM, the first catch is implemented, and a small catch is consumed by late dinner on the boat or available hut in the lagoon/bay. From 10 PM to 2 AM, fishers take a nap for a while on the boat/hut or talk with other fishers, and watch over the fishery area to monitor the species passing by them. From 2 AM onward, main capturing activities start and later the catch is sorted and landed to the shore by approximately 5 AM. After landing, the women (wife) take charge of selling to the local market (open only in early morning). In some cases, the products are sold to the neighbor/middle worker or relatives are asked to sell to the market in the city. After landing, the fishers take rest with breakfast and have a cup of coffee, before retiring for the day. Some simply go to sleep, some may repair the nets and boat, some fishers gather and drink, some may bring kids to school by bike, while some go to work for a side business in daytime. After 6–7 AM, the shape of fishers is diversified. Some families make processed foods such as dried fishes or fermented foods such as Mam Tom and Mam Ca Ro at a small scale. The aforementioned is the typical time schedule for fishers in a lagoon or near-shore coastal fishers. Of course, if the place changes, then the main target catch also changes (squid, crab, mackerel, and so on), and the product processing may be different, but the schedule remains the same in most instances nonetheless. The individual role of each family member is also different: usually, the marketing is managed by women/wife. This is because, as per Vietnamese culture, "A wife/woman has the wallet (means, husband/man may waste money)."

As per custom, there are certain rituals and practices for a good year/catch and safe fishing at least twice in a year (one at *Tết* holiday, and one at the best catching period of the year). On such occasions, it is common to decorate the boat and light incense sticks, and some residents even build an offering shelf at their fishery ground and prey for a good catch.

4.2.3 Importance of Family and Relatives

In Vietnam, the relationship within the family and relatives is at the core of activities, and it is difficult to ignore this aspect when we talk about human relations, culture, and work. This may represent by the biggest holiday in Vietnam called "*Tết* holiday," which is celebrated over 1–2 weeks around the Lunar new year, wherein most of the family gathers in one place/town (parents' or grandparents' house) and visit relatives, enjoy family moments, celebrate new year with elders, and prey for a good year to the ancestors. Due to the experiences of the war, some families and relatives live overseas (*Việt kiều*, "Overseas Vietnamese"). In the fisheries villages, it is common to encounter stories of family members going to the USA or Japan in the middle or after the war for a job. Such family members who live abroad send remittances to their parents to support their family and relatives and, if possible, they visit their family during *Tết* holidays. The family and relatives tightly knit and support each other and their relatives. In the family fisheries business, for example, one family may catch the fish, and a relative's family may process it, and then another relative may sell it in the city market (and not in the local market). In such a system, everything from production to consumption is within families. If they live close by, they often ask kids to take care, so if you are a stranger visiting their house, you may be confused as to which kid is whose child. Thus, the relationship among family and relatives plays a very important role in livelihood, work, and safety network, including financial support.

4.2.4 Fisheries at Brackish Lagoon in Central Vietnam and Its Environment

Central Vietnam has one of the biggest lagoons in Southeast Asia, named Tam Giang–Cau Hai lagoon, located in Thua Thien Hue province (Hue). It has approximately 22,000 ha of the water area (Fig. 4.4), with an average depth of 2 m brackish water (Brzeski and Newkirk 2002). During the nineteenth–twentieth century, Hue was the capital city of the Nguyen dynasty in Vietnam. During the period, set net called "*Nò Sáo*" (Bamboo staked trap) and "*Đáy*" (Fyke net/stow net) were used as fishing gears by each village/commune with certain management on

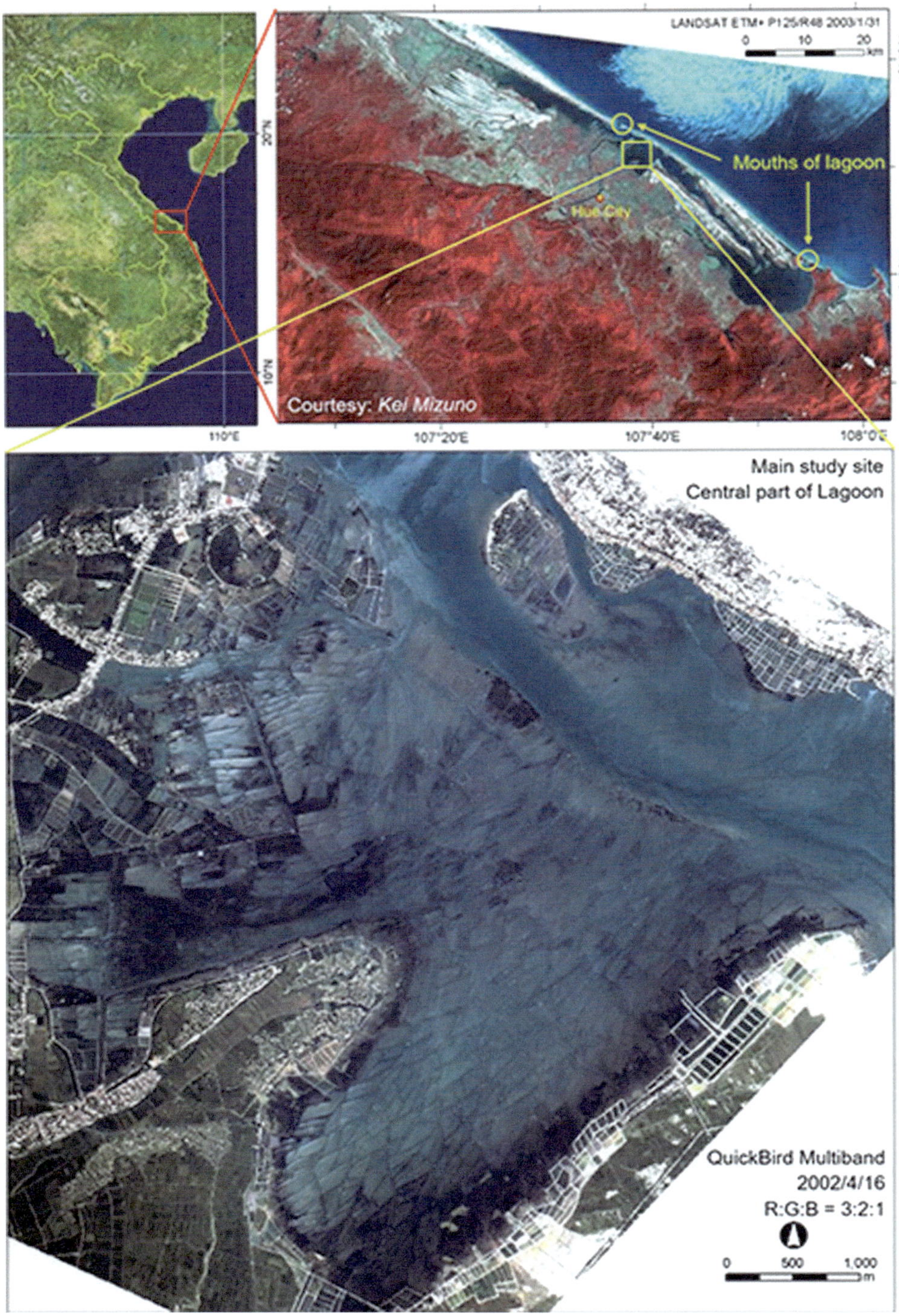

Fig. 4.4 Location of study site: central part of Tam Giang lagoon

usage of water surface via bid system and taxation, and certain products were offered to the king (Motohiro 2014). The same types of set nets are still in use at present (Fig. 4.5); however, the management system has changed after World War II and Vietnam War.

Village A was established in 1965, placed in a shallow tidal flat area in the central part of a lagoon, and started with 55 households at the beginning. Most local residents in the village are mostly engaged in fishery, and are called *Dân Thủy Diện*, or Sampan people. When the local residents settled into the current place, there was no agricultural land allocated by local government for them. Therefore, people kept relying on natural resources/fishery activities. Aquaculture in this region started as a research project on seaweed aquaculture in 1977 (Tra 1989). Ever since a private company was allowed to utilize aquaculture in 1984, it has been widely adopted. In addition, several policies on resettlement (boat to settle on the land), poverty reduction, and *Đổi Mới* (Renovation/Open Door Policy) encourage aquaculture activities. "Net-enclosure," which is specially used for aquaculture in this region, is built as a bamboo staked trap (*Nò Sáo*) and *Chuôm* (temporal bamboo fish reef/collecting device) surrounded/closed by net and bamboo sticks (Okamoto et al. 2008). The circumstances of the aquaculture introduction process, relevant regulations, and legal issues between district and local fishers have been covered by previous studies (Brzeski and Newkirk 2002). Fishers are required to form a unit called "*Tổ*" (a group formed by more than 2 fishers to conduct collaborative aquaculture) to obtain and start aquaculture activities with specific aquaculture methods by applying for a license from the district authority. However, actually only 4–5 such units have taken license and most work without them. In Village A, fishers pay a fee to use the water, called "*Tiền Thuế Mặt Nước*" (the

Fig. 4.5 "*Nò Sáo*" (Bamboo staked trap) and net-enclosure in lagoon

water surface tax/fee) to the commune/township every year to ensure their water surface. In addition, local fishers have formed several units called "*Nuôi Chung*" instead of "*Tổ*," which have the same function. As a result, there is a gap between the local government and actual response from local residents. Meanwhile, net-enclosure has increased and segmentation of each fishery area has been observed (Fig. 4.6). Figure 4.6 was created for current utilization of net-enclosure (aquaculture area) using GPS and verbal survey of each household to trace back the areas; therefore, the early stages of this figure depicting the fishery areas/*Nò Sáo* (bamboo staked trap) are not accurate because some places could not be traced by the survey as the former users had moved. The figure shows changes in water surface use from set net to net-enclosure or open area into net-enclosure. In addition, segmentation of the fishery area can be observed when inherited by sons in a family, with net-enclosures in the form of vertical long strips. In addition, trading of net-enclosure rights in the village has also contributed to their segmentation. Most contracts are individually carried out on paper among sellers and buyers, and fiat currency or gold can be used for payment. As to the record of present net-enclosures in June 2011, 39, 5, and 2% of the net-enclosures were obtained by local trade, trade twice, and inherited after trade respectively, and 40% of net-enclosures have been inherited by second, third, or fourth generations (Okamoto et al. 2012a). In this manner, rapid privatization and segmentation led to the current dense fishery area.

Aquaculture by net-enclosure is different from typical aquaculture using diked ponds, and the actual practice tends to be extensive due to the limitation/difficulties of management/control compared with intensive diked pond aquaculture. Since the 1990–2000s, aquaculture provided sufficient income for local households even for extensive aquaculture. The table shows examples of net-income from aquaculture activity in the year 2007 (Table 4.6) (Okamoto et al. 2008). According to the statistical data (Thua Thien Hue Statistical Office 2010), most of the net-income per household from aquaculture (2 times in a year) is higher than the average income in the agriculture sector of whole Vietnam, 20,200,000 VND (approximately 1,000 USD) in 2007. In addition, fishers also carry out natural catch/fishing by gill net/fish trap, and some other side businesses (as middleman, schoolteachers, bike repair shop owners, barbers, small shop owners, etc.) during daytime (Table 4.7); even though the aquaculture was conducted on a small scale in an extensive way, it contributed a lot for local residents. While aquaculture has been providing sufficient income, the risk of implementation of aquaculture has been rising due to the increasing of number of aquaculture enclosures year after year and due to the disease outbreaks (White spot syndrome virus) in 2002. The outbreak led to huge losses, and triggered a change from mono-culture to multi-specie aquaculture in the village. To maintain proper aquaculture, organic loads such as remains of feeds, residual dross should be maintained well; otherwise, the accumulation of organic matter will lead to decomposition and reduce the oxygen level in the water (Fig. 4.7). Acid-volatile sulfide (AVS) in the sediment is used as an index for evaluating the environment of aquaculture farm with respect to its organic load, and an increase and decrease in AVS affects the benthos including shrimps in the sediment (Uede 2008). Figure 4.8 shows the spatial increasing tendency of AVS values by the time of aquaculture season, while Fig. 4.9 shows the dissolved oxygen (DO) level graphs (Okamoto et al. 2012b). The minimum required DO is 3 mg/l at bottom water in summer season for aquatic species (Japan Fisheries Resource Conservation Association 2005), but one site recorded below this required value. These trends were seen in inner area, which tended to have slow water currents (Fig. 4.10), and they suggested the risk of massive death of aquatic life, stunted growth, or reduction in quantity. In addition, the reduction of AVS values in the surface sediment after the rainy season (following season) suggests the possible impact of flood and fallowing of

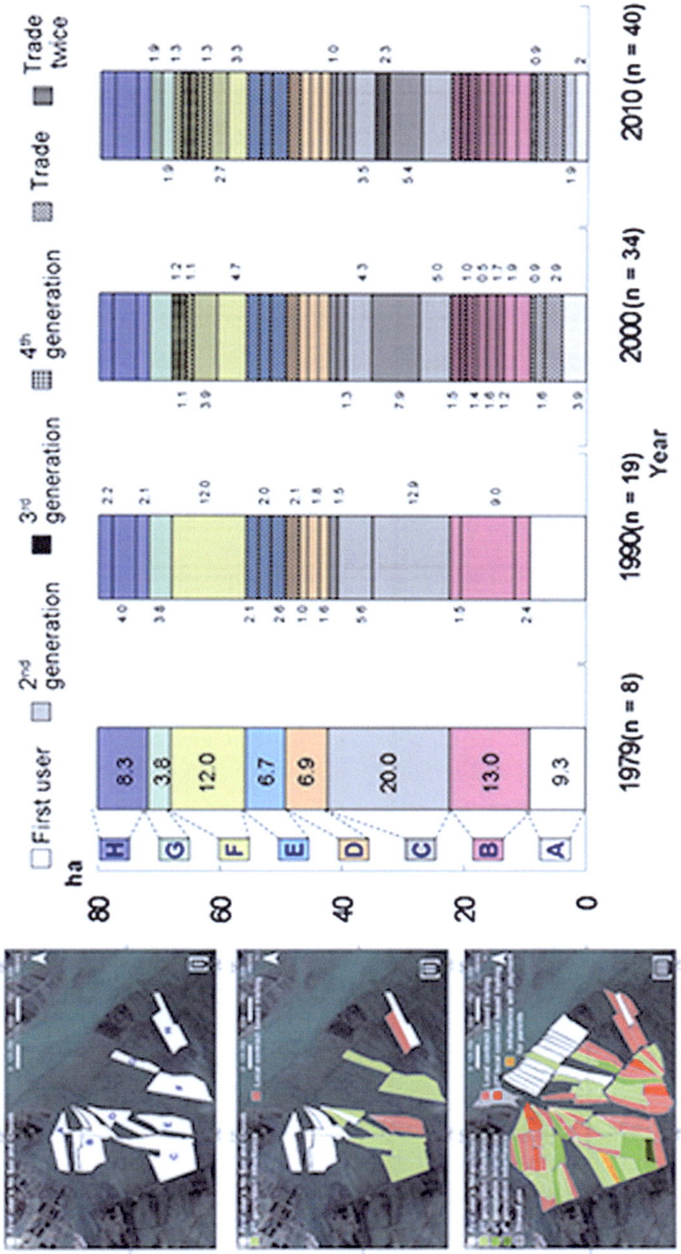

Fig. 4.6 Transition of water surface utilization and transfer of water surface to others

Table 4.6 Examples of economic balance of aquaculture activity in 2007

Type	Pond	Small pond and Net-enclosure		Net-enclosure		
Area (ha)	0.15	2.0	1.5	1.0	1.2	3.0
Cultured species	*P. monodon*	*P. monodon* *S. guttatus* *S. fluscescens* *Scylla spp.* *Gracilaria verrucosa*	*P. monodon* *S. guttatus* *S. fluscescens* *Scylla spp.* *Gracilaria verrucosa*	*P. monodon* *S. guttatus* *S. fluscescens* *Scylla spp.*	*P. monodon* *S. guttatus* *S. fluscescens* *Scylla spp.*	*P. monodon* *S. guttatus* *S. flusce-scens* *Scylla spp.*
Culture times/year	2	2	2	2	2	3
Production (US$)	2,500	2,250	2,000	1,600	1,050	2,700
Fee of feeds and maintainance	250	250	250	250	150	350
Fee for water surface usage	Pay to company	Pay to company	Pay to company	Pay to commune	Pay to commune	Pay to commune
Amount of payment	175	200	250	5	25	20
Fee for juvenile, fingerlings	400	250	250	250	250	500
Net profit (US$)	1,675	1,550	1,250	1,095	625	1,830

aquaculture (Fig. 4.11). The results outline the risks of high-density aquaculture in this region. Therefore, they were shared with each household in the village and some recommendations were made, such as reduction of feed, reduction of number of cultured shrimps, shift from monoculture to multi-specie aquaculture, and removal of some bamboo/nets during flood season. Some local residents showed interested in actual data for their own fields and sincere to apply some or own method to improve their areas adding his experiences (such as rotation of cultured area and change of feed). At the same time, the global price of shrimp dropped in 2009 and 2015 (IMF 2018), and some fishers stopped shrimp aquaculture (some residents considered that it is not worth paying for the larvae and feed at reduced price) and started seaweed, crab, and fish combination aquaculture without feeding (seaweed itself is a part of the feed and naturally available) or with reduced amounts of added feed. Thus, currently, the impact of aquaculture has reduced in this region.

With time, practices, management, interests, and living styles have shifted, but the brackish lagoon kept providing livelihood to the local residents. It is hoped the experience and knowledge of risks, management, historical and actual practices by local residents, and an understanding of their familial ties are passed over to the next generation, and it leads to sustainable harvesting of fishery products in this region and mitigation of the impact on the brackish ecosystem.

4.3 Spatial and Temporal Changes of Submerged Aquatic Vegetation Ecosystems in an Hoa Lagoon, Central Vietnam

4.3.1 Introduction

Submerged aquatic vegetation (SAV) plays an important role in aquatic fauna and human life (Tuya et al. 2014). This group of aquatic plants grows abundantly in marine environments and islands, and forms a typical ecosystem (Ca et al. 2011). SAV could treat several environmental

Table 4.7 Typical daily schedule of fisher (case of Mr. N)

Time	Activity
17	Early dinner and start for neew catch
18	Arrival of set net/cabin (collecting fishes from few set nets for night meal)
19	Set gillnet for main catch for early morning
20	
21	Night meal at cabin
22	Rest/chatting at cabin
23	Sleeping at cabin
0	
1	
2	
3	
4	Capturing fishes, removing fishes from nets, and sorting by species and size
5	
6	Arrival of house, toss product to family/middlemen, breakfast
7	Coffee/rest
8	Fixing net, removing garbage in the net/side job
9	
10	
11	Lunch
12	Taking nap
13	
14	
15	Chatting/free time/side job
16	
17	Early dinner and start for neew catch

pollutants through biological processes, support benthic organisms, reduce coastal erosion, and provide raw materials for several industries. SAV also absorbs carbon dioxide from the atmosphere, thereby contributing to reducing greenhouses gases in the atmosphere.

Quang Nam province belongs to the economic hotspot of Central Vietnam and is known to have diverse flora and fauna ecosystems. Therein, SAV constitutes one of the ecosystems that plays an essential role in the economy of people living in this area. However, SAV has been declining due to human activities such as destructive fishing and aquaculture development (Dai 2010). Besides, because of the seasonal variation character of SAV, the assessment of its distribution by traditional methods is time-consuming and costly, suggesting that modern methods should be applied to overcome this difficulty (Dai 2010; Nguyen et al. 2001).

In recent times, satellite remote sensing and geographic information systems (GIS) have been widely applied in different study fields. The integration of GIS technology and satellite remote sensing is one of the most time-saving and affordable geographic techniques. This method is not only convenient for information processing but also propriate for SAV change assessment (Nguyen et al. 2001).

There are several studies that have applied GIS to evaluate the distribution of SAV via water-column correction techniques. Hoang et al. (2015, 2016a) used a validated depth invariant index (DII) model for water-column correction and mapped *Sargassum* beds around Rottnest Island, Western Australia, with highly accurate classification results (Hoang et al. 2015, 2016a, b, c, d; Garcia et al. 2015). Koedsin et al. (2016) integrated field survey and remote sensing methods for mapping species composition, percentage cover, and aboveground biomass of seagrasses in monsoonal southern Thailand. Manuputty et al. (2015) evaluated and compared DII and principal component analysis to map the seagrass ecosystems of Kotok and Karang Bongkok islands, Indonesia.

In Vietnam, Thang and Son (2012) compared the bottom reflectance index (BRI) and DII methodology for mapping seagrass beds in Lap An lagoon, Thua Thien Hue province, Vietnam. The accuracy of the BRI method was found to be higher than that of the DII technique. This technique was found to be a robust approach for accurately estimating the distribution area of seagrass beds in Central Vietnam. Hoang et al. (2011) applied remote sensing technology and GIS to conduct a study on seagrass beds in the wetlands of Huong Phong commune, Huong Tra district, Thua Thien Hue province, to assess the distribution

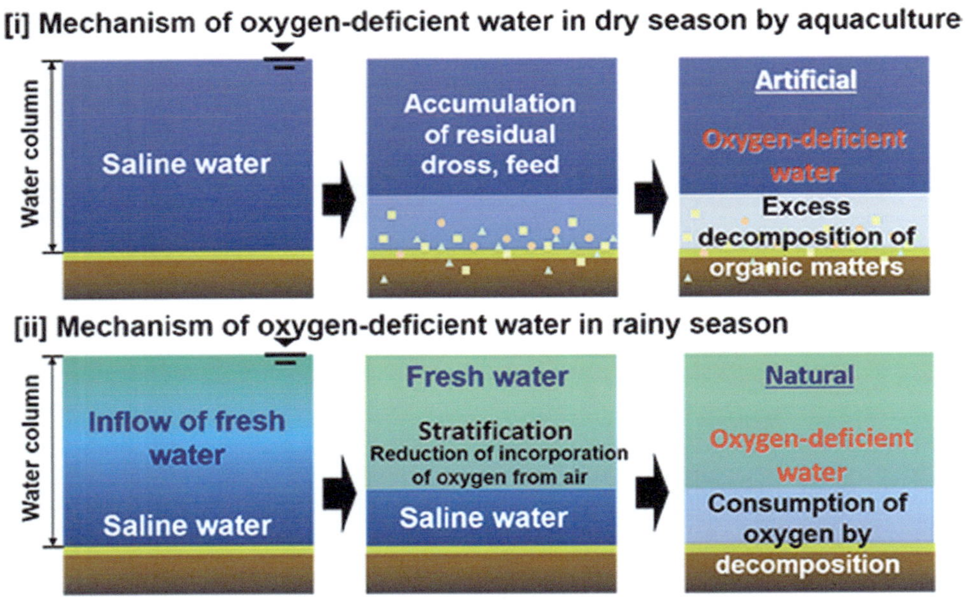

Fig. 4.7 Mechanism of oxygen-deficient water in fishery area

and spatial change in seagrass beds. Most studies have mapped SAV using only water-column correction technique, while a few have used two or more water-column correction techniques in data processing. Therefore, this work compared three water-column correction techniques to identify the appropriate method for the study area.

Thus, the objective of this research work was to evaluate the effectiveness of SAV management for sustainable development. GIS technologies and multi-temporal Landsat satellite image data were applied to assess SAV from 2000 to 2020 with the following specific objectives: (i) evaluating the classification accuracy of satellite remote sensing data; (ii) establishing the distribution maps of SAV in 2000, 2005, 2010, 2015, and 2020; and (iii) assessment of SAV changes over 2000–2005, 2005–2010, 2010–2015, and 2015–2020.

4.3.2 Materials and Methods

4.3.2.1 Study Materials

Multi-temporal remote sensing data were collected from Earth Explorer image databases (The United States Geological Survey (USGS), USA; https://earthexplorer.usgs.gov) to establish the distribution map of SAV in 2005, 2010, 2015, and 2020. A summary of remote sensing data sources is provided in Table 4.8.

4.3.2.2 Field Survey Methods

The field surveys in An Hoa, Quang Nam province were carried out in two phases. Phase 1 was from 25 to 28 February 2020 and Phase 2 was implemented between 2 and 4 May 2020. A motorboat was employed to carry out SAV detection in An Hoa lagoon. Diving along transects in the direction from the shore seaward, until SAV was no longer seen, we identified the SAV distribution and determined the substrate form for each survey site. Substrates were classified as SAV, sandy bottom, and mud bottom.

Finally, global positioning system (GPS) positions were recorded for each survey point. A total of 70 survey locations were established in An Hoa lagoon. These survey locations are shown in Fig. 4.12.

SAV Spatial Distribution and Area Change Mapping The process of mapping SAV distribution and temporal change in An Hoa lagoon is shown in Fig. 4.13.

Fig. 4.8 AVS value of sediment in aquaculture area by seasons

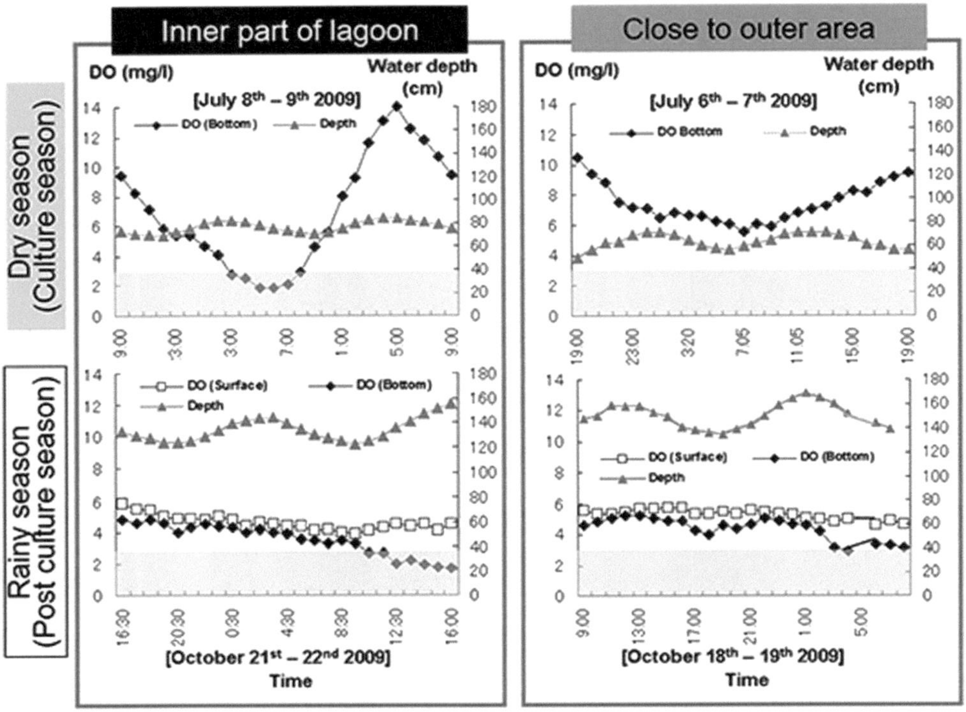

Fig. 4.9 Dissolved oxygen in the water in 2 sites of aquaculture area in dry and rainy season

Fig. 4.10 Relationship between water current and AVS value of surface sediment

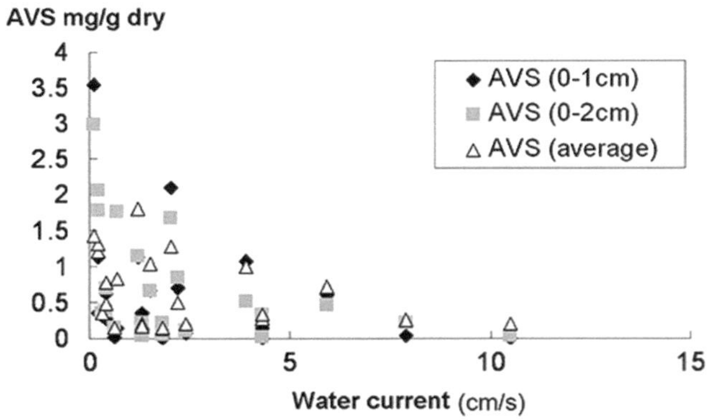

Radiometric and Atmospheric Correction The purpose of radiometric correction was to remove the diffraction effect and improve the quality of image pixel values (Moran et al. 1992). After radiometric correction, the possible effects of the atmosphere (aerosols, dust, gas, and air molecules) were removed (Schroeder et al. 2006). The Landsat data was corrected using the fast line-of-sight atmospheric analysis of hypercube method (ENVI 2009). Atmosphere-corrected Landsat data was used to produce more accurate surface reflectance.

Water-column Correction This is a technique to remove influencing factors stemming from particles in the water-column, which reduce light transmission from the surface to the euphotic depth, such as phytoplankton, total suspended solids (TSS), and colored dissolved

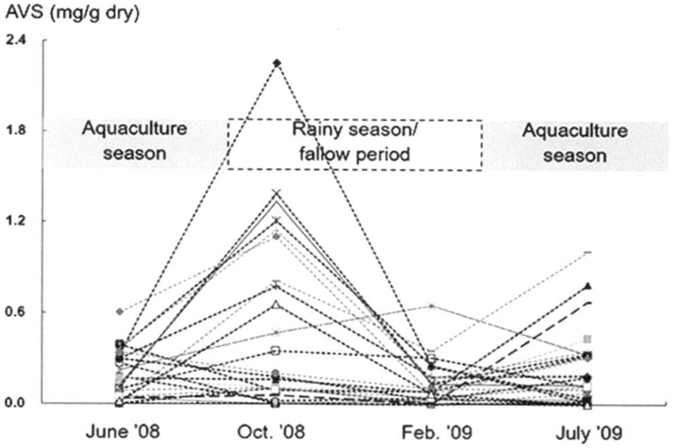

Fig. 4.11 Seasonal changes in AVS value at 24 sites

Table 4.8 Satellite image data used for SAV mapping and change detection

Year	Satellite sensor	Image ID	Acquired date	Time	Spatial resolution (m)	Path/Row	Cloud cover
2005	Landsat 7	LE07_L1TP_124049_20050521_20170115_01_T1	21/05/2005	02:55	30	124/049	3/5
2010	Landsat 7	LE07_L1TP_124049_20100706_20161213_01_T1	06/07/2010	02:58	30	124/049	0/0
2015	Landsat 8	LC08_L1TP_124049_20150610_20170408_01_T1	10/06/2015	03:05	30	124/049	1.73/0.8
2020	Landsat 8	LC08_L1TP_124049_20200506_20200509_01_T1	06/05/2020	03:05	30	124/049	0.77/0.67

organic matter (CDOM). The DII developed by Lyzenga (1981) modified DII by Son (2010) and the BRI were employed to correct the water-column.

The basis of the DII calculation is the linear relationship between the reflection spectra of Landsat band pairs. The linear relationship of band pairs uses randomly selected sand beds at different depths. The BRI technique, which is an advanced equation compared to DII, was applied to correct the water-column. BRI method is based on the radiance reflected of the bottom and an exponential function of the water depth.

Analyzing and Processing Images The Landsat data was processed by clipping, increasing quality, and masking. Clipping allowed us to remove non-study areas covered by the imagery while masking allowed us to hide land layers not involved in our work. These processing steps contributed to improving image quality for classification.

Supervised Classification The maximum likelihood method was used for interpreting Landsat images. This process directed training sites in the form to group all pixels on the Landsat image and classified bottom types, including SAV, sandy bottom, and mud bottom.

Assessing the Accuracy of the Classification For accuracy of classification classes, the overall accuracy (OA) and the Kappa coefficient (Ƙ) in the confusion matrix were chosen (Table 4.9). OA is the ratio between the total number of correct pixels and the total number of pixels in the confusion matrix. Kappa is used in classification

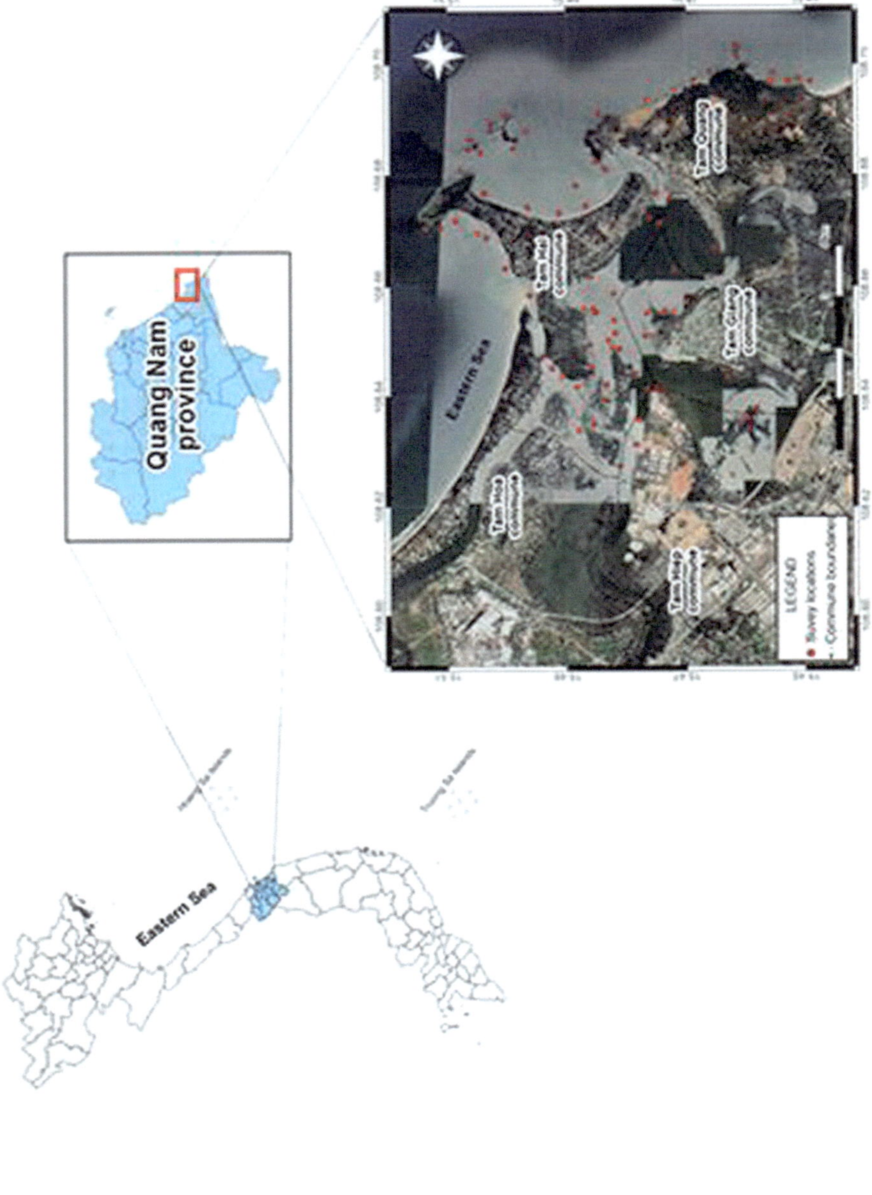

Fig. 4.12 Survey locations in An Hoa lagoon

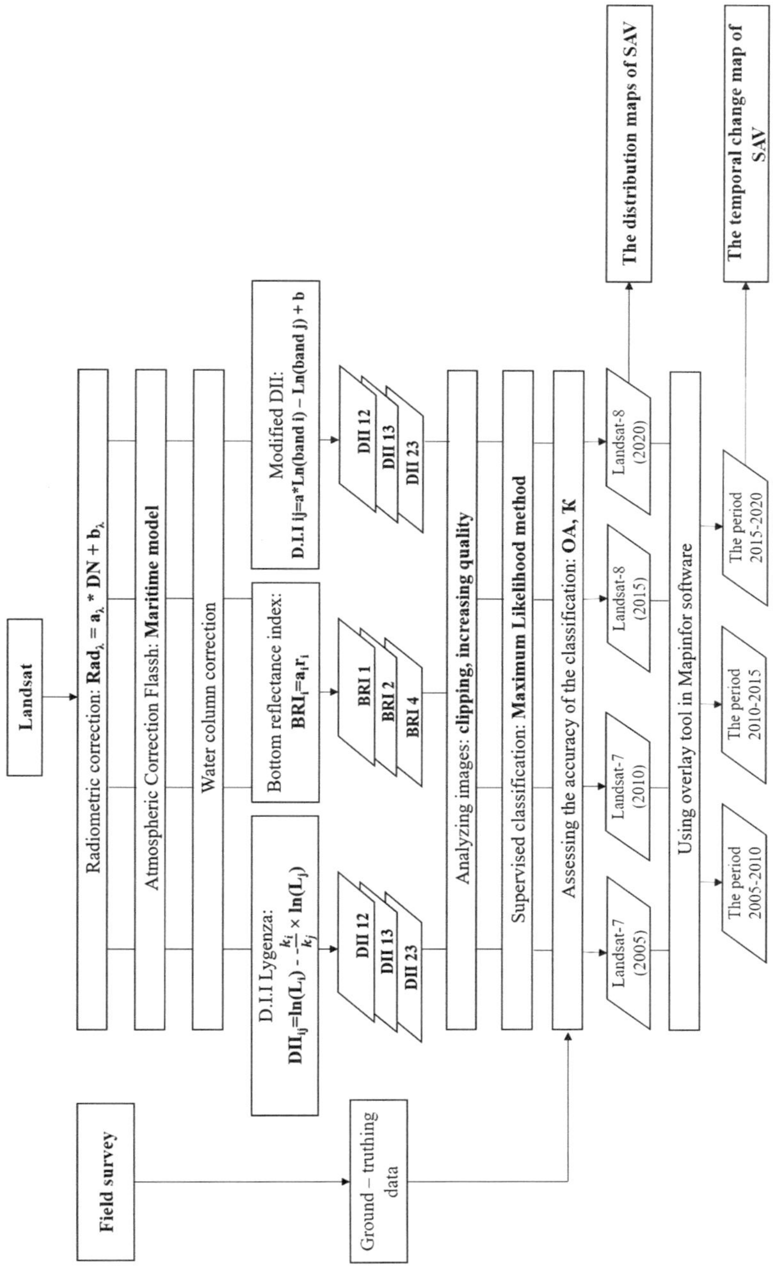

Fig. 4.13 Flowchart for establishing SAV distribution and temporal change mapping

Table 4.9 Classification results according to Kappa coefficient

Classification result	Kappa coefficient
Excellent	>0.81
Good	0.8–0.61
Average	0.60–0.41
Below average	0.4–0.21
Weak	0.20–0.00

as a measure of agreement between observed and predicted or inferred classes for cases in a testing dataset (Congalton and Green 1999).

The classification results according to Kappa coefficient are shown in Table 4.8

Assessment the Temporal Change of SAV Distribution After validation of the classification, QGIS version 3.20 was used to edit SAV distribution maps in An Hoa lagoon for 2005, 2010, 2015, and 2020 at a map scale of 1:25,000. After that, Landsat data were used to map the temporal changes in SAV over three periods: 2005–2010, 2010–2015, and 2015–2020.

4.3.3 Results

4.3.3.1 Assessing the Accuracy of Classification Results

Assessing the accuracy of Classification Results in 2020 The accuracy of benthic classification from three techniques for 2020 is shown in Table 4.10.

As shown in Table 4.11, three methods were used to provide accurate SAV distribution estimates. The Kappa coefficients and Overall Accuracy values exceeded 0.80 and 85%, respectively, with DII $_{\text{Lyzenga}}$ showing the best results (Ⱪ = 0.92; OA = 94.7%), followed by DII $_{\text{modified}}$ (Ⱪ = 0.85; OA = 89.5%), and then BRI (Ⱪ = 0.81; OA = 85.7%).

Table 4.10 Accuracy of benthic classification for the three methods

Method	Kappa coefficient	Overall accuracy (%)
DII $_{\text{Lyzenga}}$	0.93	94.7
DII $_{\text{modified}}$	0.86	89.5
BRI	0.81	85.7

Table 4.11 Accuracy of benthic classification from 2005 to 2020

Year	Kappa coefficient	Overall accuracy (%)
2005	0.83	87.2
2010	0.89	90.1
2015	0.90	92.9
2020	0.93	94.7

Assessing the Accuracy of Classification Results during 2005–2020 The accuracy of benthic classification using DII $_{\text{Lyzenga}}$ from 2005 to 2020 is shown in Table 4.10.

The classification results of all satellite images achieved good accuracy. Both Kappa coefficient and overall accuracy were higher than 0.80 and 85%, respectively. The classification results in 2020 had the highest accuracies compared to 15 other years (Ⱪ = 0.93; OA = 94.7%), followed by the results in the year of 2015 (Ⱪ = 0.90; OA = 92.7%) and 2010 (Ⱪ = 0.89; OA = 90.1%). Finally, the lowest classification result was in 2005 with Ⱪ = 0.83.

4.3.3.2 Establishing Spatial Distribution Map of SAV in an Hoa Lagoon

The result classification revealed that SAV is mainly distributed in Tam Hai, Tam Quang, Tam Giang communes, and scattered in Tam Hiep commune, with approximately 98.37 ha. The SAV distribution mapping results are depicted in Fig. 4.14, with the estimated SAV areas listed in Table 4.12.

Table 4.12 shows that the distribution area of SAV tended to decrease from 2005 to 2010. In 2005, the SAV area distributed was 230.6 ha, and the most distributed in Tam Hai commune with 93.8 ha, followed by Tam Hiep commune with 55.6 ha and Tam Giang commune with 48.1 ha. The lowest area was in Tam Quang commune (33.3 ha). By 2010, the SAV area reduced to 25.27% (decreasing to 172.3 ha).

In 2015, SAV areas continued to decrease in Tam Quang, Tam Hai, Tam Hiep, and Tam Giang communes; specifically, Tam Giang and

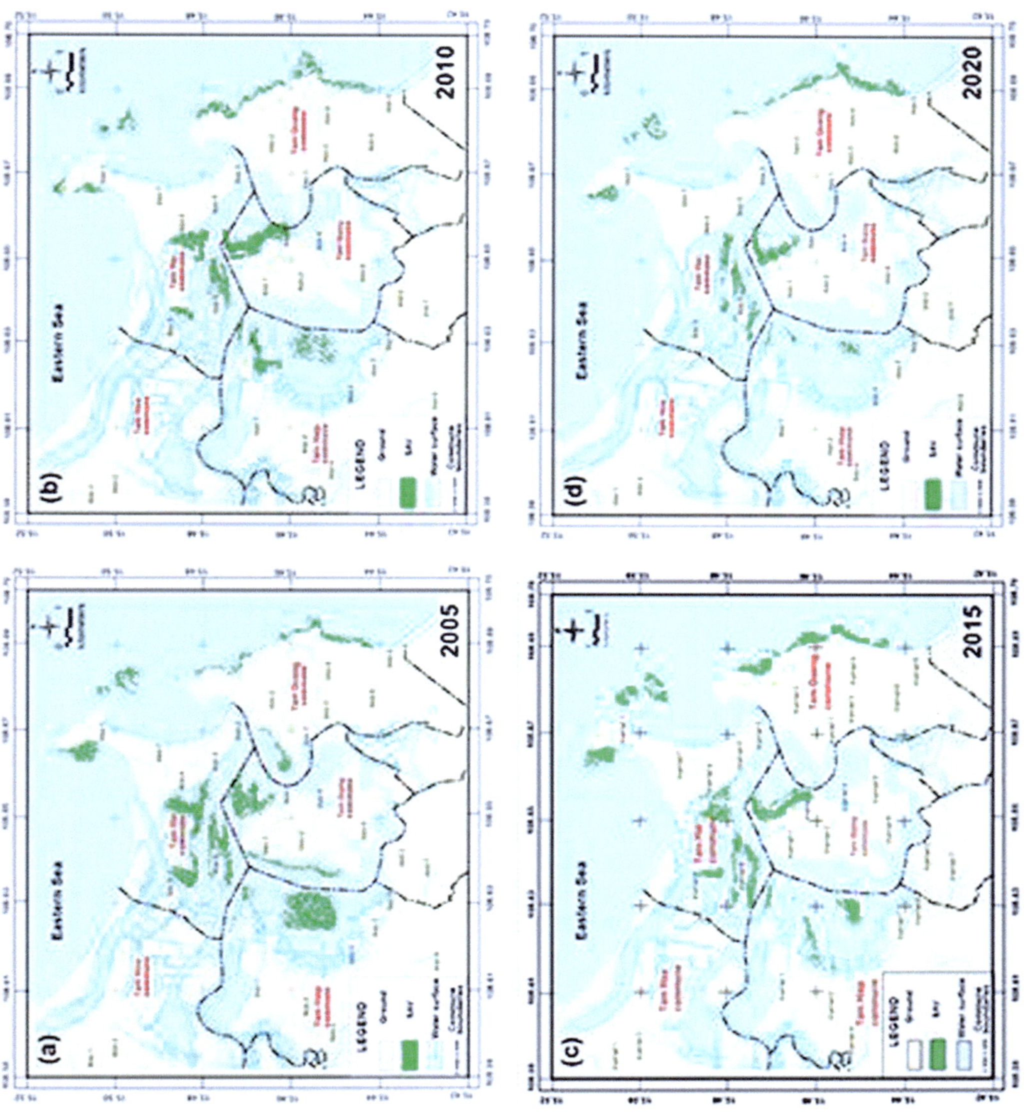

Fig. 4.14 SAV distribution maps in An Hoa lagoon: **a** 2005, **b** 2010, **c** 2015, and **d** 2020

Table 4.12 SAV areas in An Hoa lagoon over 2005–2020

Study location	Area of submerged aquatic vegetation (ha)				
	2005	2010	2015	2020	2005–2020
Tam Quang	33.3	31.6	33.4	30.2	−3.03
Tam Hai	93.8	61.9	54.3	38.5	−55.2
Tam Hoa	0.00	0.00	0.00	0.00	0.00
Tam Hiep	55.6	34.3	15.8	7.9	−47.7
Tam Giang	48.1	44.6	20.9	21.8	−26.3
Total	230.6	172.3	124.4	98.4	−132.3

Note "−": decrease; "+" increase

Table 4.13 SAV area change over 2005–2010

Study location	Area of submerged aquatic vegetation (ha)								
	2005–2010			2010–2015			2015–2020		
	Lost	Unchanged	Gained	Lost	Unchanged	Gained	Lost	Unchanged	Gained
Tam Quang	26.5	5.8	25.8	26.2	5.4	28.0	25.9	7.5	22.8
Tam Hai	70.2	23.5	38.3	51.7	10.1	44.1	43.3	11.0	27.6
Tam Hoa	0.00	0.00	0.00	0.00	0.00	0.00	0.00	0.00	0.00
Tam Hiep	47.9	7.7	26.6	32.1	2.2	13.6	12.5	3.3	4.6
Tam Giang	38.3	9.7	34.8	34.6	10.0	10.9	12.0	8.9	12.8
Total	182.9	46.7	125.6	144.6	27.7	96.6	93.7	30.7	67.7

Tam Hiep communes reduced by nearly half of the SAV areas compared to 2010. By contrast, SAV in the Tam Quang commune increased by 1.84 ha. By 2020, the SAV area in the Tam Quang commune decreased to 30.2 ha, and Tam Hai and Tam Hiep communes decreased to 38.5 and 7.9 ha. The area in Tam Giang commune did not change from 2015 to 2020 (Table 4.13).

In general, the SAV area decreased by 132.3 ha over 2010–2020, with average annual decrease of approximately 8.81 ha.

4.3.3.3 Assessment of Temporal Changes to SAV in an Hoa Lagoon from 2005 to 2020

The results of the SAV change maps are shown in Figs. 4.15, 4.16, 4.17, and Table 4.13.

Table 4.13 shows that SAV area lost is more than gained SAV. For 2005–2010: the total SAV area lost is approximately 182.9 ha, and the total SAV area gained in An Hoa is 125.6 ha. The SAV area continued to decrease over 2010–2015, with the total SAV area of 144.6 ha lost and approximately 96.6 ha gained. From 2015 to 2020, the loss of SAV was approximately 93.7 ha, while SAV gained was 67.7 ha.

The classification results revealed that land cover in An Hoa lagoon, Quang Nam province, witnessed significant fluctuations during the observed period (Table 4.14).

The area of mangrove forest and vegetation decreased sharply as well. In 2005, the mangrove area was approximately 92.0 ha and decreased to 38.5 ha by 2020. Vegetation also decreased from 6008.9 ha (2005) to 2531.0 ha (2020). In 2005, aquaculture increased rapidly to 2463.1 ha in 15 years—an average increase of approximately 51.58 ha annually. Mangrove

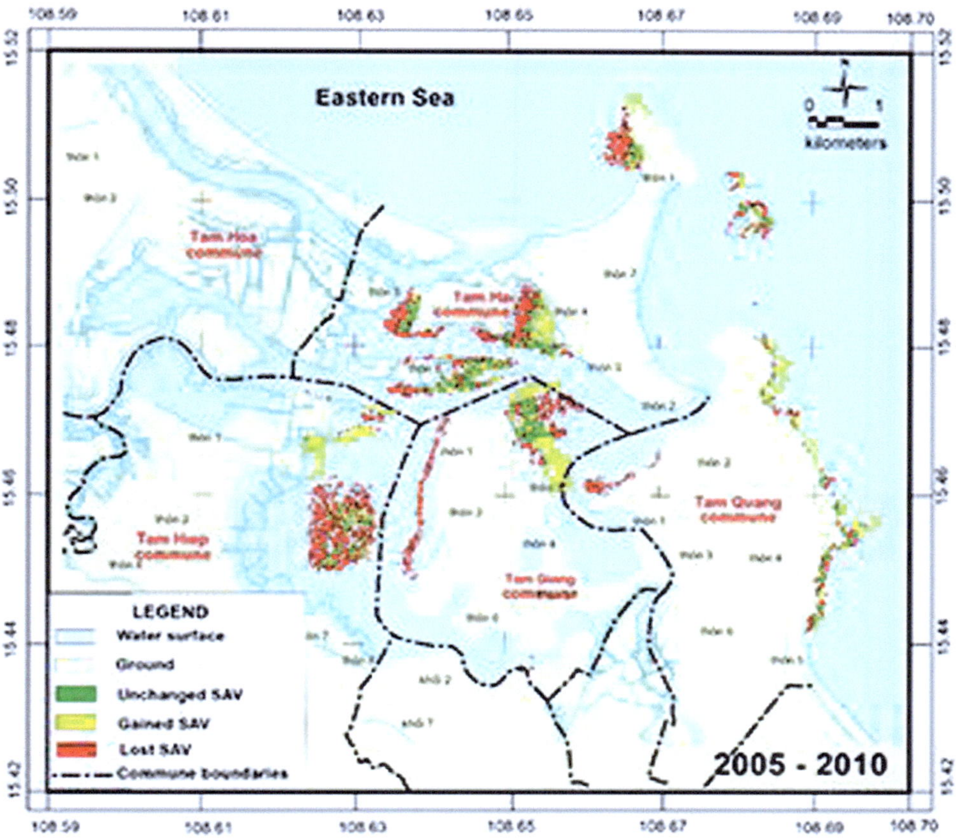

Fig. 4.15 SAV change map in An Hoa lagoon from 2005 to 2010

areas, vegetation, or other types of vegetation may be suitable topography for the development of SAV, which typically does not appear near bare ground, residential areas, and aquaculture. In Tam Giang and Tam Hiep communes, SAV disappeared due to the change of land-use purpose from planting trees to aquaculture or residential areas (Fig. 4.18).

Several areas in Tam Giang and Tam Hiep communes were lost due to aquaculture development or residential construction (Fig. 4.19), which is consistent with the data collected from interviews with local villagers. There can be a number of other reasons that could have caused fluctuation in the SAV area over 2005–2020.

Specifically, An Hoa lagoon is facing sever human impact due to economic development activities. The main reason for the SAV reduction was building construction that could have polluted the area. A field survey found, at several areas near the industrial park, waste deposits contributing to sediments and smelly and unpleasant odor. This impact caused the SAV area to decline in the past 15 years. Aquaculture is another reason. Since 2005, after 15 years of aquaculture development, the local people have focused on aquaculture development, leading to significant decrease in area of SAV (Fig. 4.19).

4.3.4 Discussion

Based on the above, it can be seen that DII show higher score than BRI due to the effect of depth. Several studies also show that BRI is suitable for areas with 10–20 m deep. However, this work mainly focuses on water depths ranging 0.5–10 m. This is shown through a linear

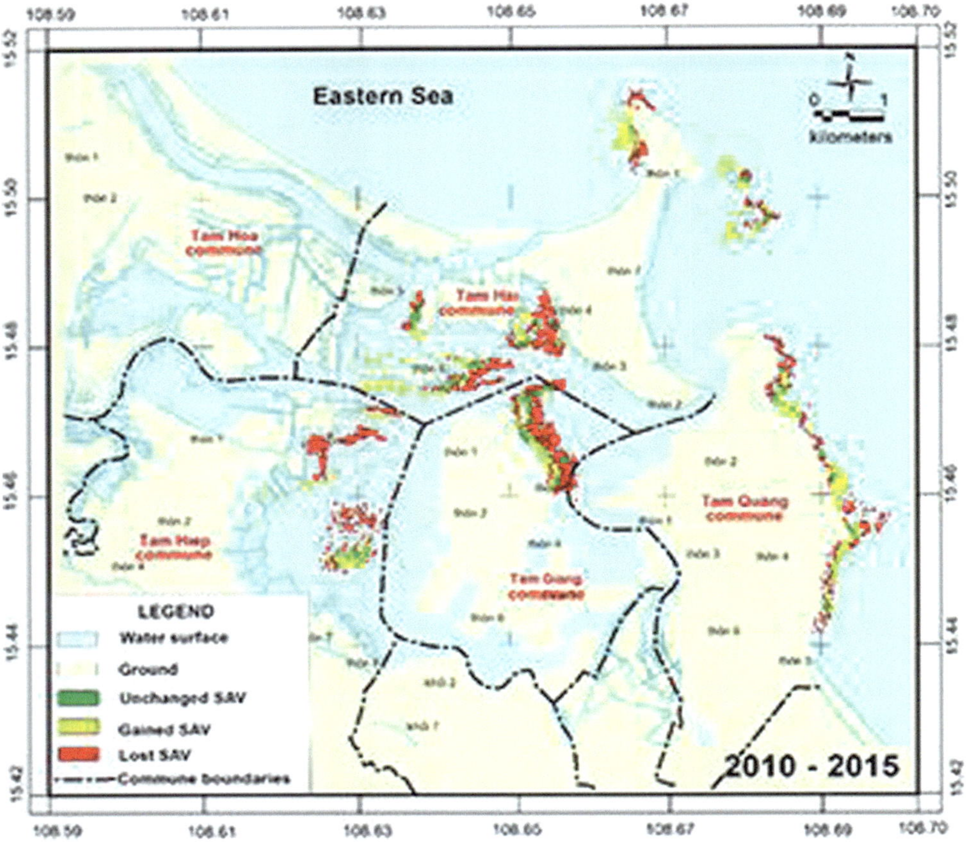

Fig. 4.16 SAV change map in An Hoa lagoon from 2010 to 2015

relationship between the bands and the depth: R^2 coefficient between bands and depth reaches 0.71–0.84, while each pair of bands was greater than 0.94. Therefore, this could be the reason why DII achieved higher accuracy than BRI. Therefore, DII Lyzenga was chosen to adjust the water-column for the remaining years. Similar to this study, other works established SAV overlay with high accuracies—such as Thang and Trang (2015) classified ALOS images with an accuracy of 81.8% in Lap An lagoon (Thang and Trang 2015). Comparisons show that three higher accuracy methods were used because water-column adjustment methods are more suitable for the coastal water of An Hoa lagoon. In general, the application of DII Lyzenga gave the best results, and the modified DII and BRI had quite high bottom classification results. However, there were some deviations observed in the classification process. Several reasons have been identified, one of which is the average spatial resolution of the remote sensing data. The Landsat image with a 30-meter resolution is not sufficiently high to allow for the detection of SAV areas smaller than 0.09 hectares (less than 1 pixel in Landsatimage). In addition, the influence of environmental factors such as turbidity and depth also affected the image interpretation results.

The results for the DII Lyzenga method showed Kappa coefficients of 0.93 in 2020, 0.9 in 2015, and 0.89 in 2010, with the lowest value in 2005 of 0.83. Because the distribution map of 2020 was built based on the results of GPS location collection and satellite

Fig. 4.17 SAV change map in An Hoa lagoon from 2015 to 2020

Table 4.14 Land cover area in An Hoa lagoon from 2005 to 2020

Land cover type	Land cover area (ha)			
	2005	2010	2015	2020
Bared ground	1183.1	3304.4	2119.0	2132.6
Vegetation	6008.9	3059.9	1465.7	2531.0
Mangroves	92.0	69.3	23.0	38.5
Other vegetation	3057.5	3976.9	3736.4	4366.2
Residential areas	725.1	922.5	2086.0	2063.1
Aquaculture	1689.4	1821.8	2295.8	2463.1

image interpretation with high image quality, it had the highest Kappa coefficient and overall accuracy. As it is impossible to collect GPS locations for 2005, 2010, and 2015, the Kappa coefficients and overall accuracies of these years were lower than 2020, although the Kappa coefficients were still high. For 2015 and 2020, both used Landsat 8 image source with marine. The Aerosol band (band 1) from Landsat 8 is commonly used in marine resource assessment, while the remaining two years were taken from Landsat 7. Therefore, the accuracy values for 2020 and 2015 were higher than that of 2005 and 2010. Other studies have identified SAV distribution, and comparisons show that our study area had more SAV than both the Lap An lagoon (47.3 ha) and Huong Phong coastal areas (76.8 ha) (Hoang et al. 2011).

Over a period of 15 years (2005–2020), SAV area tended to decrease sharply; the area lost was 421.2 ha and the area gained was 289.9 ha. The cause of this rapid decline was exploitation, aquaculture, and infrastructure development. In

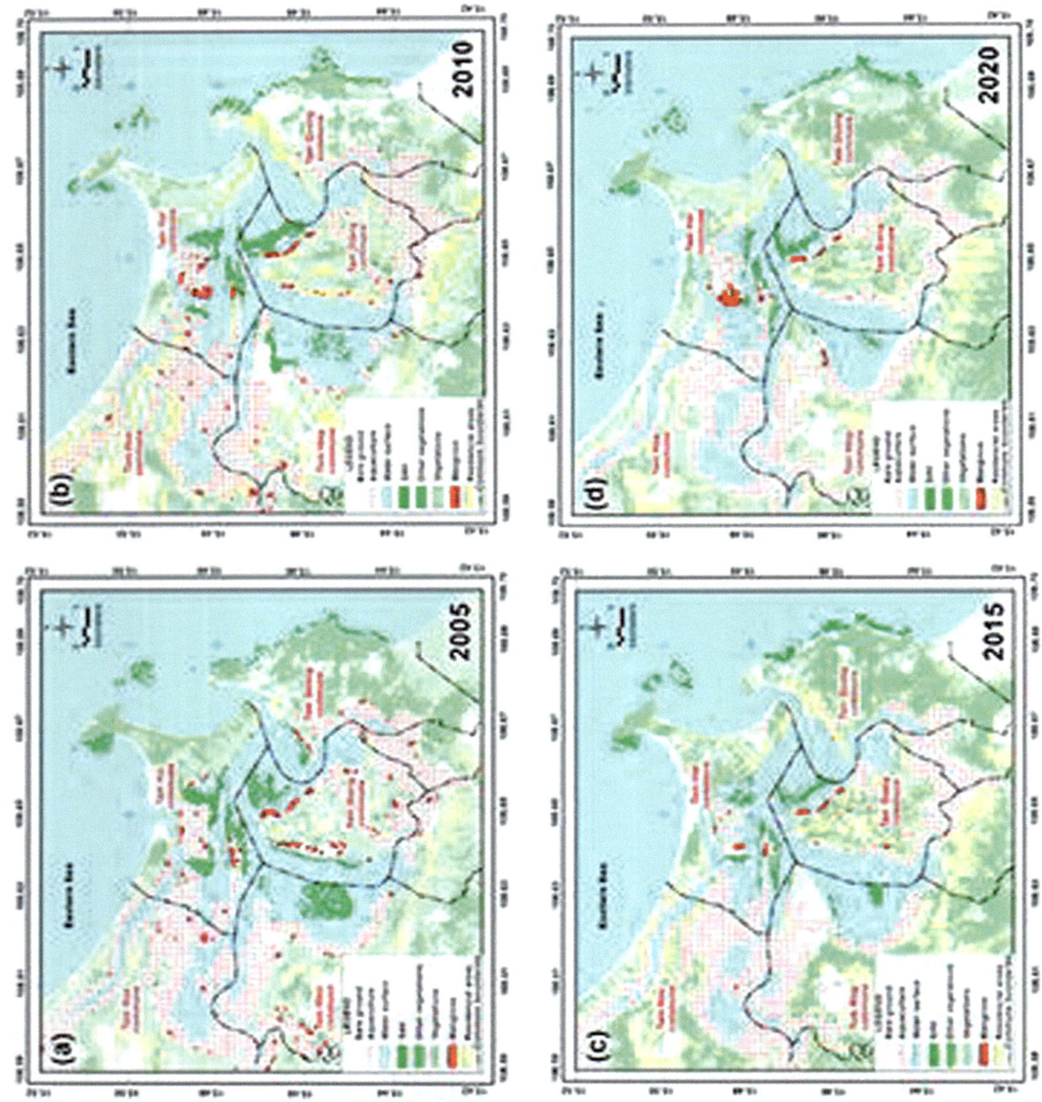

Fig. 4.18 Map of land cover types in An Hoa lagoon over 2005–2020

Fig. 4.19 Causes of rapid decline of SAV in An Hoa lagoon were over-exploitation (**a**), local people harvesting *Gracilaria* sp. seaweed (**b**), harvesting *Sargassum* sp. (**c**) dried commercial seaweeds (**d**), facing human impact due to aquaculture development activities (**e**), and several areas near the industrial park (**f**). Photos were captured in February 2020 (**a**, **b**, **f**) and May 2020 (**c**, **d**, **e**)

addition, destructive fishing activities, such as the use of electrical impulses, are one of the reasons for the loss of SAV. Researchers discovered that the submerged aquatic vegetation (SAV) areas exhibited signs of damage, including scratched roots and leaves pulled up in clusters along the shore. These issues contributed to a significant decline in the health of the SAV areas (Fig. 4.19). According to Mr. Pham, a fisherman, local people of Tam Giang commune said that when the season of fish such as grouper and snapper arrives, fishermen would use all available tools to scrape the surface in the SAV areas. If these activities continue without admonishment and good management, SAV will surely be lost and the recovery will not be easy.

Harvesting seaweed can allow coastal communities to increase income. Coastal communities do not need to invest capital, and their income is approximately (8.75 USD) per day when they harvest seaweed. However, seaweed is being over-exploited, leading to a decrease in income of coastal communities. Similar to trees in the forest, if we cut down all the roots in an area, it would not recover. Besides, the degradation of seagrass could affect the economy of coastal communities. Through nursery habitat provision for fish and prawns/shrimps, seagrass meadows support the productivity of 20% of fisheries. Any loss in seagrass would lead to decreased fish/shrimp population, leading to a direct impact on commercial fishers.

4.3.5 Conclusions

Multi-temporal Landsat satellite data was interpreted to establish the temporal change map in An Hoa lagoon over the period of 2005–2020. Field surveys were conducted to collect ground-truthing points and representatives of households were interviewed to identify the major reasons causing changes in SAV areas. The results showed that the three techniques could provide high accuracy, with Kappa coefficient (Ꝁ) values above 0.81. Herein, the DII Lyzenga method achieved the highest (Ꝁ = 0.93). Therefore, DII Lyzenga was applied to reduce the water-column effect in 2005, 2010, 2015, and 2020. The total area of SAV in 2020 was estimated to be approximately 98.37 ha.

The degradation of seagrass beds could affect marine ecosystems and aquatic resources in Central Vietnam. Therefore, seagrass beds are necessary to conserve, restore, and manage this ecosystem for sustainable socio-economic management and development. Seagrass ecosystem restoration is critical, i.e., pressure on this ecosystem needs to be reduced so that seagrass beds have enough time to recover naturally. In the current economic conditions of Central Vietnam, the priority task with regard to seagrass beds is conservation to protect seagrass from the damaging effects of human activities.

In addition to seagrass conservation, remote sensing is one of the useful techniques to map the distribution of SAV in the coastal areas. Mapping the status and ecological characteristics of seagrass could provide a comprehensive overview of the spatial and temporal distribution, thereby providing a baseline for improved monitoring, management, and conservation of seagrass ecosystems in Central Vietnam.

4.4 Sustainable Cropping Systems in Lowland Sandy Soil in Thua Thien Hue Province, Central Vietnam

4.4.1 Sandy Soils in Vietnam and in Thua Thien Hue Province

In Vietnam, the soil survey method was introduced by French soil scientists (P. Morange, 1898–1902, E. M. Castagol, 1934, 1935, 1936, 1937, cited from Chieu 1994). For a long period, the soils of Vietnam were classified based on soil genesis theory using three groups of criteria: factor of genesis, soil formation process, and soil properties in different classification levels.

The first soil classification in North Vietnam (1959) included 5 groups and 18 units; later on, in 1964, 5 groups and 28 units were added to the system. Finally, in 1976, the revised general soil classification in Vietnam comprised 13 groups and 30 units. Since around 1990, numerous studies on soil classification have been undertaken following FAO-UNESCO, Russian classification, and soil taxonomy. The FAO-UNESCO classification was applied to more than 500 soil profiles from all over Vietnam (Chieu 1994), taking into account the existing experience in soil classification in Vietnam, which provided a soil map of Vietnam at 1/1,000,000 scale (Fig. 4.20).

BẢN ĐỒ ĐẤT VIỆT NAM
SOIL MAP OF VIET NAM

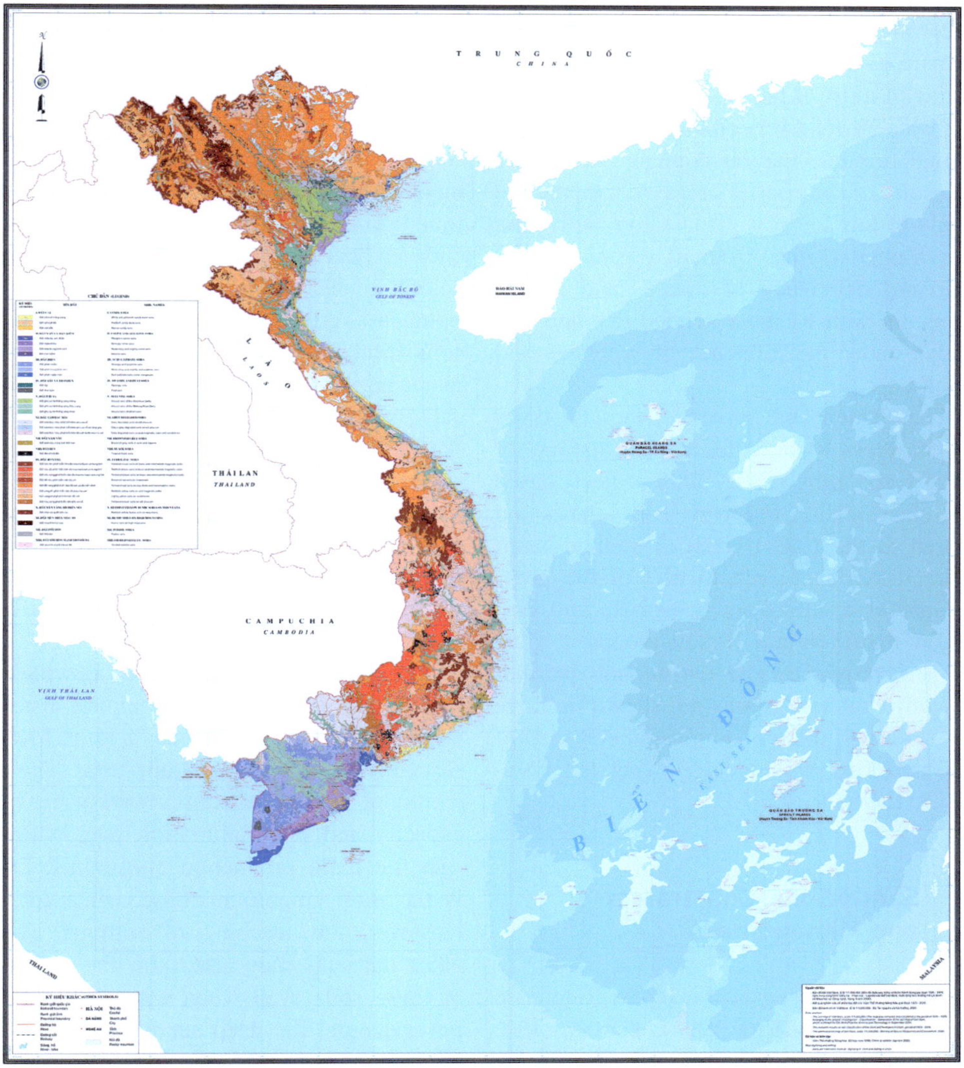

Tỷ lệ 1 : 1.000.000 (Scale 1 : 1,000,000)

Fig. 4.20 Map of soils in Vietnam (NIFS 2002)

According to Vietnam Society of Soil Science (1996), the Vietnamese group of sandy soils may be classified mainly into 3 units: white and yellow sand dune soils, red sand dune soils, and sandy marine soils. More than 36% of agricultural soils in Vietnam are classified as light textured. Communities that are economically marginalized and have few viable options available to address issues of food insecurity, poverty, and unsustainable resource management often dominate these soils. This has a direct negative impact on the economic and social fabric of communities that are dependent on natural resources for goods and services in order to sustain already tenuous livelihoods.

According to Toan (2004), marine sandy soils in Vietnam have the largest sand content (>80%) and, as with most of the degraded soils, are distinguishable by relatively small clay content and also small organic carbon content. The sandy marine soils of the coastal areas of Vietnam are generally of low fertility status. Crop cultivation here is limited owing to small organic matter content and deficiency of macro- and micronutrients, as well as the extended dry and wet situations under the influence of typhoons and hot winds.

In Thua Thien Hue province, the soil types have been recently classified by the Institute of National Geography (2002–2008 project) based on soil classification following soil genesis and compared to the FAO-UNESCO method-based soil classification at each district of the province. General results from 8 districts and Hue city show that there are 10 soil groups in Thua Thien Hue province. The extent and distribution of different soil groups are shown in Table 4.15. At present, there is no general soil map of Thua Thien Hue province following the FAO-UNESCO method.

4.4.2 Sandy Soils: Constraints and Management[3]

Sandy soils, as defined in the World Reference Base (FAO-ISRAC-ISSS 1998), contain<18% clay and>65% sand in the first meter of the solum. Generally, the main Reference Group for sandy soils is Arenosol (FAO-ISRAC-ISSS 1998). The nearest Soil Taxonomy equivalent is the Psamments sub-order (Van Wambeke 1992). In the World Reference Base, sandy soils may also occur in Regosols, Leptosols, and Fluvisols.

Tropical sandy soils are prevalent in landscapes where felsic volcanic or siliceous sedimentary rocks and their erosional products are found. They are also prevalent in desert regions and as beach deposits and dune features in coastal zones. The lower Mekong river basin in Southeast Asia has an extensive area of sandy soils (Mekong River Commission 2002). Other major regions covered with sandy soils in tropical environments include the Sahel zone of West Africa, the Kalahari basin (covering two-thirds of Botswana and Angola), and Northern Australia. However, smaller but still significant areas of sandy soils occur in most tropical regions, including parts of Central Vietnam, Pakistan, Saudi Arabia, Iran, and Brazil (FAO 2005).

Tropical sandy soils have a wide range of limiting factors for agricultural use: these include nutrient deficiencies, acidity, low water storage, and poor physical attributes. The coarse texture of sandy soils causes a low water holding capacity and high infiltration rate, which present the main production constraints. Drought is the most important consequence for crops. However, the leaching of nitrogen (N) and other nutrients may also limit the productivity of these soils even when water is not limited. For example, the deep sandy Prey Khmer soil in Cambodia has lower potential productivity for rice even with higher fertilizer application than the other major sandy lowland soil (Prateah Lang), which has a significant nutrient content in the subsoil (White et al. 1997).

Low nutrient levels are common in sandy soils, and crops grown on these soils commonly express multiple nutrient disorders, which limits crop productivity (Bell et al. 1990). Acidity is also common in sandy soils. Where lowland rice is the main crop, flooding alleviates acidity (Kirk 2004). However, for rainfed crops, acidity may give rise to a range of fertility constraints, including aluminum (Al) and manganese (Mn)

[3] This subsection is based on the article by Hoang Thi Thai Hoa, one of the authors of this section and has been deposited in the repository of Murdoch University at https://researchportal.murdoch.edu.au/esploro/outputs/conferencePaper/Sandy-soils-in-south-central-coastal/991005546037707891#file-0. It does not violate any policy of the repository.

Table 4.15 Extent and distribution of 10 soil groups in Thua Thien Hue province (Thanh et al. 2005)

No	Soil group	Area (ha)	% of the total area	Location (district)	Soil properties
1	Sand dune and coastal sandy soils (Arenosols)	43,962	8.70		
1.1	White and yellow sand dune soils (Luvic Arenosols)	24,358	4.82	Phong Dien, Quang Dien, Huong Tra, Phu Vang, and Phu Loc	Low organic carbon (OC) content, low nutrients, low CEC, and water holding capacity (WHC)
1.2	Coastal sandy soils (Haplic Arenosols)	19,604	3.90	Quang Dien, Huong Thuy, Huong Tra, Phu Vang, Phong Dien, and Phu Loc	Low OC content
2	Saline soils (Salic Fluvisols)	6,290	1.24		
2.1	Strongly saline soils (Hapli-salic Fluvisols)	145	0.03	Phu Vang	pH from less acid to neutral, low OC, low total N and P, high exchangeable cations
2.2	Medium and slightly saline soils (Moll-isalic Fluvisols)	6,145	1.22	Phu Loc, Phu Vang, Huong Tra, and Quang Dien	neutral pH, medium OC
3	Acid sulfate soils (Thionic Fluvisols)	6,888	1.36	Phong Dien, Quang Dien, Huong Tra, Huong Thuy, Phu Vang, and Phu Loc	High OC and Al^{3+}
4	Alluvial soils (Fluvisols)	41,002	8.11		
4.1	Umbric Fluvisols	2,661	0.53	Phu Loc, Phong Dien, and Hue city	High OC with light soil texture and good water drainage
4.2	Dystric Fluvisols	20,635	4.11	Phu Loc, Huong Thuy, Huong Tra, Phu Vang, Quang Dien, Phong Dien, and Hue city	Low pH and poor OC
4.3	Gleyic Fluvisols	5,955	1.18	Phu Loc, Huong Tra, Phu Vang, Quang Dien, and Phong Dien	Low pH, medium total N, high total P
4.4	Arenic Dystric Fluvisols	4,115	0.81	Phu Vang, Huong Tra, Quang Dien, Phong Dien, and Phu Loc	Medium/light soil texture, low pH
4.5	Stagnic Dystric Fluvisols	2,200	0.44	Low-elevation sites	Very poor water drainage
4.6	Dystric Plinthosols	4,846	0.96	Hue city, Huong Thuy, Huong Tra, Phu Vang, Quang Dien, Phong Dien, and Phu loc	Medium OC, high total N, low available P
4.7	Alluvial soils locate along the streams	590	0.12	Phong Dien, Phu Loc, Nam Dong, and A Luoi	Low pH

(continued)

Table 4.15 (continued)

No	Soil group	Area (ha)	% of the total area	Location (district)	Soil properties
5	Swampy and peat soils (Gleysols and Histosols)	100	0.2	Phong Dien and Phu Loc	High OC, low pH, high content of Al^{3+}
6	Grey degraded soils (Acrisols)	800	0.16	A Luoi and Phong Dien	Light soil texture, low pH and OC
7	Yellowish red soils (Ferralsols)	347,431	68.7		
7.1	Yellowish-brown soils on gabbro/diorite (Xanthic Ferralsols)	4,934	0.98	Nam Dong	Low pH, high nutrients content
7.2	Yellowish red soils on clay shales and metamorphic, igneous rocks	159,114	31.48	A Luoi, Phong Dien, Huong Thuy, Nam Dong, Huong Tra, and Phu Loc	Low pH, low base saturation (BS)
7.3	Redish yellow soils on acid magmatic rocks	135,450	26.80	Nam Dong, A Luoi, Phu Loc, Huong Thuy, Huong Tra, and Phong Dien	Low pH, low BS, low nutrient content
7.4	Light yellow soils on sandstones	37,523	7.42	Phong Dien, A Luoi, Huong Tra, and Huong Thuy	Low pH, OC, and nutrient content
7.5	Yellowish-brown soils on old alluviums	9,880	1.95	A Luoi, Phong Dien, Nam Dong, Phu Loc, Huong Tra, and Huong Thuy	Low pH and nutrient content
7.6	Yellowish red soils changed due to rice cultivation	530	0.10	Phong Dien, Phu Loc, and Huong Tra	Low nutrient content
8	The soil located in the valleys (Dystric Gleysols)	640	0.13	Phong Dien, Phu Loc, and Huong Thuy	Low OC and pH, high Al^{3+} content
9	Redish yellow humic soils in mountains (Humic Ferralsols)	15,942	3.15	A Luoi, Nam Dong, and Phu Loc	Soil depth is 0.6–1.2 m; light soil texture
10	Eroded skeletal soils (Leptosols)	5,220	1.03	Phong Dien, Huong Tra, Huong Thuy, Phu Loc, Nam Dong, and A Luoi	Very low nutrient content

toxicities and deficiencies of nutrients (Dierolf et al. 2001). Poor N fixation in legumes is often a consequence of soil acidity due to the low tolerance of Rhizobia to acid pH and low levels of plant-available molybdenum (Mo). Salinity is also a common constraint on sandy soils wherever irrigation is used in semi-arid environments. In sandy coastal zones, salinity is associated with seawater intrusion (White et al. 1997). Less well-recognized is dryland salinity, which arises from perturbation of the hydrological balance in rainfed environments.

Sandy soils represent an ongoing challenge for water and nutrient management at landscape and field scales. Productivity of these soils tends to be low; however, there are a number of promising avenues for sustainable development in sandy soils.

Sandy soils generally have lower organic matter levels than heavier textured soils given similar rainfall, temperature, land use, and tillage. Clay levels on deep sandy soils may be too low to protect organic matter from oxidation (Baldock and Nelson 2000). Reducing tillage may help maintain organic matter levels. Perennial crops also tend to maintain higher organic matter levels. However, there are practical limits to the levels of organic matter that are achievable on sandy soils (Shirato et al. 2005). Research to boost organic matter levels in sandy soils often fails to recognize this limit and, as a result, there have been many futile investments in organic matter management. Organic matter levels can be enhanced in sandy soils through minimum tillage, zero burning and retaining crop residues, and use of farmyard manure. Returning crop residues to soil improves soil physicochemical properties significantly. Slowly decomposing litter was found to build organic matter levels over several years on sandy soil in Ubon, Northeast Thailand (Naklang et al. 1999). Regarding the context of this work, improving agricultural practices in sandy soils by optimizing the organic material cycle in farms is expected to lead to sustainable soil fertility, as shown in Fig. 4.20.

Choice of suitable crops and cropping sequences is often very delicate. Casuarinas (*Casuarinas equisetifolia*), eucalyptus (*Eucalyptus* sp.), photina (*Phitinia prunifolia*), kapok tree (*Alba pentandra*), guava (*Psidium guajava*/*Myrtacea*), jackfruit (*Artocarpus heterophyllus*); and vestiver (*Vestiveria* sp.) are frequently cited as plant species that can firstly grow on sandy soils of Vietnam and can be used as fixing trees for wooden, fuel, fruit, or medicinal purposes. Cashew (*Anacardium occidentale* L./*Anacardiacea*), mango (*Mangifera indica* L.), coconut (*Cocos nucifera* L.), dragon fruit (*Hylocereus undatus*), and citrus/lemon/orange (*Citrus reticulata Blanto*) have also been adapted and grown in some coastal areas. These trees develop very well on sandy soils with good cultural practices such as fertilization for cashew and lighting regulation for dragon fruits. Permanent dry sandy soils may be used for cash crops such as peanut, maize, and sesame, while seasonal or permanently flooded areas are very well adapted for rice crops (Ha et al. 2005).

Where a soil constraint is widespread and difficult to overcome through conventional agronomy, there is a strong case for breeding for tolerance to the stress. In sandy soils, this would commonly mean breeding for drought tolerance first, followed by tolerance of mineral disorders, including soil acidity.

In general, management of these soils requires integrated practices that can increase their fertility as well as nutrient and water holding capacities. To improve their quality, soil management that facilitates an increase in soil biological activity, i.e., systematic use of farmyard manure, crop residues, green manures, and alley cropping, can prove to be effective. In addition, the effective management of these soils needs careful consideration of appropriate techniques that not only address the issue of low productivity but also that of environmental protection. These soils are prone to significant nutrient losses through leaching; thus, any intensification of production needs to recognize this potential adverse effect and develop management strategies that minimize off-site pollution. These technologies need to be assessed in pilot demonstration plots under local conditions prior to their adoption by the wider agricultural community in coastal areas.

4.4.3 Cropping Systems in Lowland Sandy Soil

The coastal zone of Thua Thien Hue encompasses 8.2% of the total provincial area and 45% of its population. It is largely dominated by the Tam Giang–Cau Hai lagoon, the largest lagoon in Vietnam. With its area of 21,600 ha and 5 districts related to the lagoon system, and 300,000 people depending directly on the lagoon's resources, the Tam Giang–Cau Hai lagoon plays a very important role in the province's economic activities. It might be said that the lagoon determines the rate and pattern of socio-economic development in Thua Thien Hue. The growing urban population of Hue is exerting increasing pressure on the lagoon system through increased agriculture, aquaculture, fishing, transportation, and cargo handling, in a limited and confined area. Figure 4.21 shows the climate features and the seasonal calendar of the main crops in the coastal sandy zone (Hoang et al. 2010).[4]

Like the lowland zone, there are many diversified cropping systems in the coastal sandy zone. However, due to various elevations from one commune to another, the cropping systems are not the same even in the same sub-ecological area. Rice–rice, rice–cash crop, groundnut + corn, and spring vegetables are always implemented in average elevation area, and (rice + fish)–fallow and (rice + fish)–(rice + fish) are used in low-elevation area. In high-elevation areas, due to lack of water in the summer season, popular cropping patterns are spring rice–fallow, spring groundnut–fallow, and cassava. Apart from food self-sufficiency, cash crops such as groundnut, vegetables, and chili are important in terms of income generation for households in the lowland and coastal sandy zones. The crop yield is not as high as in lowland areas. Based on rainfall amount and temperature in a year, farmers arrange different crop calendars to manage crops with the best growth and yield, i.e., rice–rice crop is often followed from January to May and June to September.

4.4.4 Characteristics of Sandy Soils in Thua Thien Hue Province

The properties of the 300 soil samples that were collected from 7 communes (representative of 7 coastal districts) in Thua Thien Hue province with an area of 4,310 ha are presented in Table 4.16. The type of distribution varies greatly according to properties. For pH_{H2O}, pH_{KCl}, total N, and total phosphorus (P), the mean values are nearly equal to the median values. Generally, good agreement between the mean and median values suggests that the data are normally distributed about the mean (Spiegel 1992), which can be visually assessed from the histograms in Fig. 4.22. On the contrary, EC, organic C, soil texture (sand, silt, clay), and related properties (cation exchange capacity (CEC), exchangeable cations) have discrete frequency distribution.

Based on this general examination of data, it appears that though collected in the coastal area of the province, some samples cannot be considered as sandy soils. Indeed, if we plot all data in a textural triangle (Fig. 4.23), we notice that most samples are clustering in the classes of sand, loamy sand, and light-textured sandy loam.

In fact, 33 samples are scattered through other textural classes with higher silt and clay content. As the scope of this study is to concentrate on sandy soils of the coastal zone of Thua Thien Hue, we limited further statistical study to the remaining 267 samples with sand fraction greater than 70% and clay fraction lower than 15%. It is interesting to notice that all 33 eliminated samples are coming from rice fields in the communes of Phong Hoa (Phong Dien district) and Phu Luong (Phu Vang district). It is very likely that in the sandy coastal zone, priority is given to rice cultivation on the scarce surfaces covered by less sandy soils because of their lower permeability and higher nutrient storage capacity.

Soil textural classes: Percentages of clay and sand in the main textural classes of soil; the remainder of each class is silt.

[4] This paragraph in the subsection is based on the article by Hoang Thi Thai Hoa, one of the authors of this section and has been deposited in the repository of Murdoch University at https://researchportal.murdoch.edu.au/esploro/outputs/conferencePaper/Sandy-soils-in-south-central-coastal/991005546037707891#file-0. It does not violate any policy of the repository.

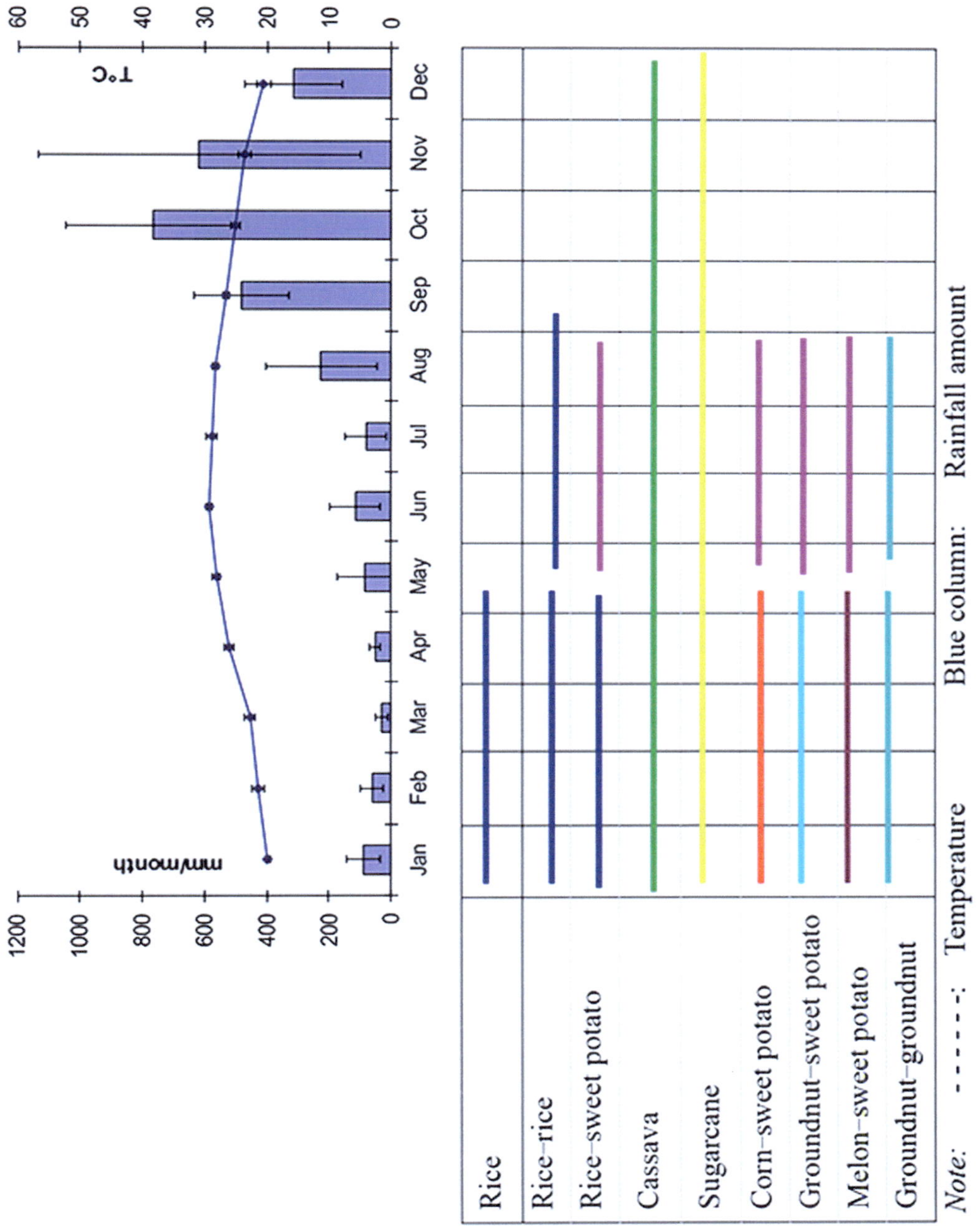

Fig. 4.21 Climate and crop calendar in the coastal sandy zone

Table 4.16 Descriptive statistics of physical and chemical properties of the 300 soil samples

Parameter	Mean	Median	Min.	Max.	Std. dev.
Sand (%)	82.57	87.73	10.12	96.72	14.79
Silt (%)	11.06	7.97	1.91	35.49	8.24
Clay (%)	6.37	3.52	0.12	56.53	8.74
pH_{KCl}	4.26	4.20	3.24	5.53	0.31
pH_{H2O}	5.03	5.04	3.66	6.15	0.45
EC (μs/cm)	35.34	23.00	1.00	205.00	32.29
C (%)	1.09	0.89	0.12	4.17	0.68
N (%)	0.06	0.05	0.02	0.33	0.03
P_{tot} (% P_2O_5)	0.04	0.04	0.01	0.11	0.02
P_{av} (mg P_2O_5/100 g)	6.08	5.76	1.25	13.51	2.18
Ca_{ex} ($cmol_c$/kg)	0.69	0.45	0.05	4.59	0.81
Mg_{ex} ($cmol_c$/kg)	0.09	0.05	0.02	0.65	0.11
Na_{ex} ($cmol_c$/kg)	0.05	0.04	0.01	0.25	0.04
K_{ex} ($cmol_c$/kg)	0.03	0.03	0.01	0.12	0.01
H_{ex} ($cmol_c$/kg)	0.69	0.53	0.04	3.56	0.65
Al_{ex} ($cmol_c$/kg)	0.06	0.03	0.000	0.64	0.09
CEC ($cmol_c$/kg)	3.11	1.78	0.50	37.36	4.66

4.4.5 Characteristics of Sandy Soil Following Cropping Patterns

4.4.5.1 Properties of Soils from Paddy Fields and Other Fields

Paddy fields include those growing 1 rice crop, 2 rice crops, and 1 rice crop followed by another crop (sweet potato or peanut). Other fields include those growing non-rice crops such as peanut, sweet potato, and cassava. In Fig. 4.24, we notice obvious discrimination between rice and other soils. Rice soils are scattered on the left-hand side of the PC1 axis and the other soils are on the right, which confirms that rice fields are preferably established on less sandy soils with higher organic matter content. A possible difference in the distribution of the two populations along the PC2 axis is less obvious from that figure.

In fact, all 17 properties are significantly different for the two types of soil use. This results not only from differences in intrinsic soil properties, of which texture and organic matter content are the most important to govern other characteristics, but likely also from cropping practices. Indeed, waterlogged conditions in paddy fields cause soil reduction and set in motion a series of physicochemical and biological processes. After submersion, the pH of acid soils increases but soluble ferrous iron released from Fe (III)-oxyhydroxides competes with exchangeable bases for adsorbing sites. When paddy soils are re-aerated, Fe (II) re-precipitates as Fe (III), and the liberated sites are re-protonated or loaded with aluminum. This can explain why the pH of paddy soils, as measured after drying, is significantly smaller than that in the other soils. Nevertheless, the base saturation percentage of both types of soils is similar (31% for rice soils and 33% for the others), but as the CEC of rice soils is higher, they can store more exchangeable bases. In any case, the content in exchangeable bases is very low; the storage of major nutrients such as potassium is crucial, approximately 1 mg/100 g.

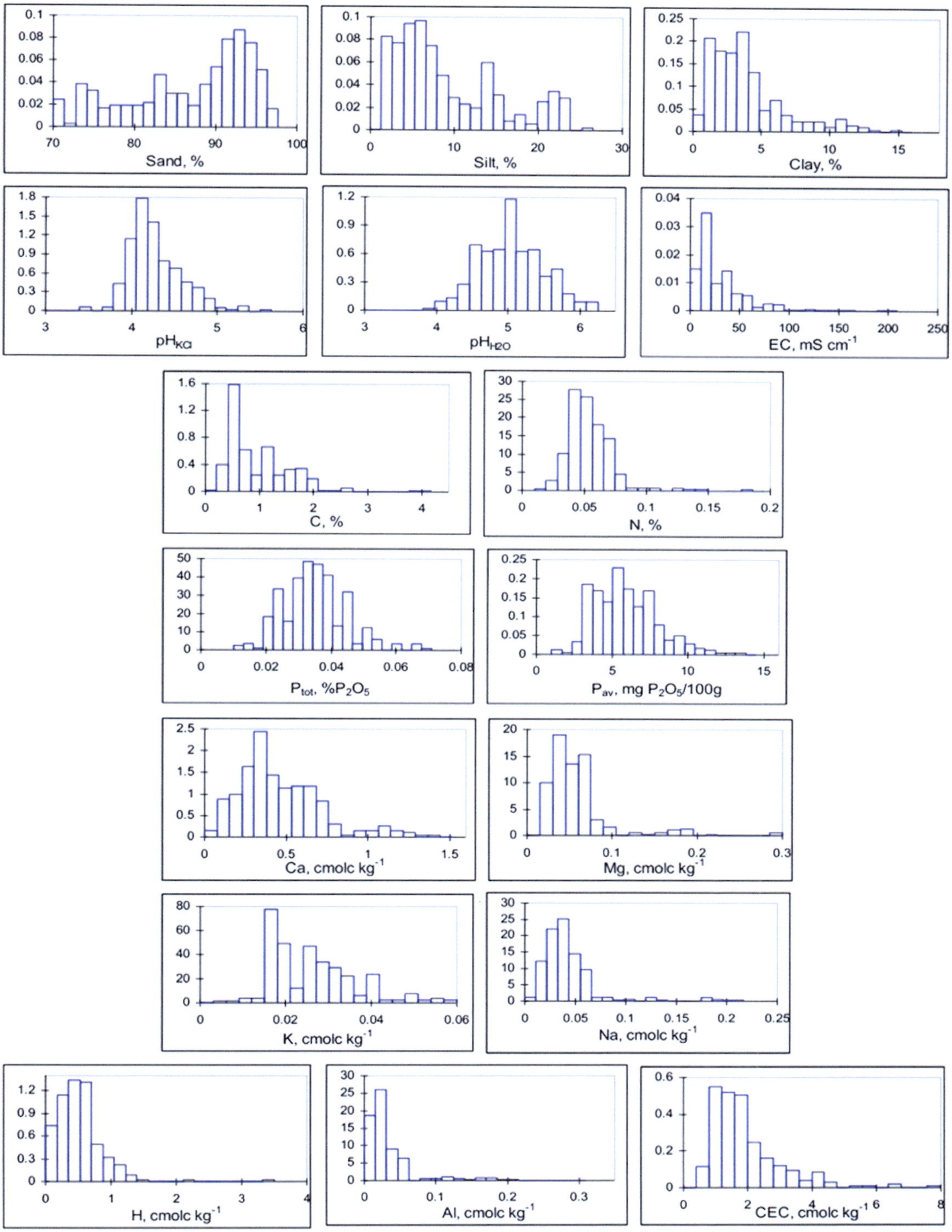

Fig. 4.22 Frequency distribution of physicochemical properties of the 300 soils

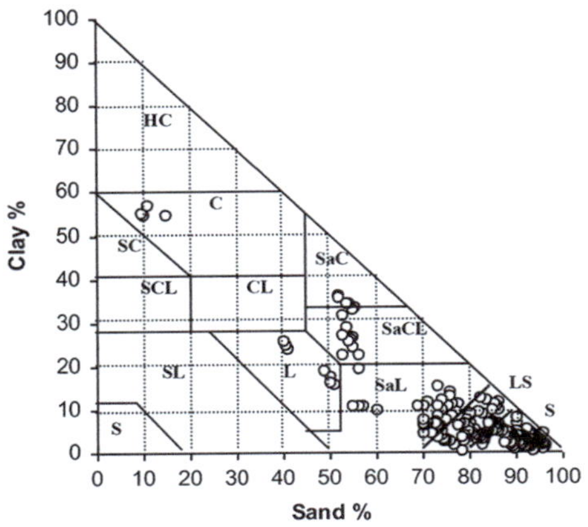

Fig. 4.23 Range of soil texture of the 300 soil samples. *Note* HC: heavy clay; C: clay; SC: silt clay; CL: clay loam; SCL: silt clay loam; SL: silt loam; S: silt; L: loam; SaC: sandy clay, SaCL: sandy clay loam; SaL: sandy loam, LSa: loamy sand; Sa: sand

The decomposition of soil or added organic matter is relatively fast under aerobic conditions where oxygen is the electron acceptor. However, under submerged conditions, the supply of free oxygen is low or absent, and the decomposition of organic matter depends on the availability of electron-acceptors such as ferric iron or sulfate. Moreover, the alternate electron-acceptors (ferric oxyhydroxides or sulfate) are less efficient in destroying organic matter than oxygen. Consequently, the decomposition of organic matter is comparatively slow and incomplete under flooded or anaerobic soil conditions (Sahrawat 2005). This can be one reason for the higher content in organic C observed in paddy soils, other reasons being the higher clay content, which stabilizes humic compounds.

4.4.5.2 Properties of Paddy Soils with One or Two Rice Crops per Year

There are two rice seasons in the study zone. The first season extends from December to May (called Dong Xuan season), and the second continues from May to September (called He Thu season). The single rice crop traditionally occurs in Dong Xuan season and is the dominant cropping system in the coastal sandy area. In the summer season, most of these lands are fallowed, or a part of them is utilized to grow sweet potatoes. Double rice cropping is mainly practiced in fields with lower topography, allowing irrigation during the dry summer season.

In Fig. 4.25, we selected only paddy soils with single rice crop and double rice crops per year, and we noticed small discrimination between single and double rice crop soils, with a tendency of most single rice soils to concentrate on the right-hand part of the cluster of points along the PC1 axis, i.e., soils with a coarse texture and less organic carbon content; however, there were no significant differences in texture between the single and double rice crops.

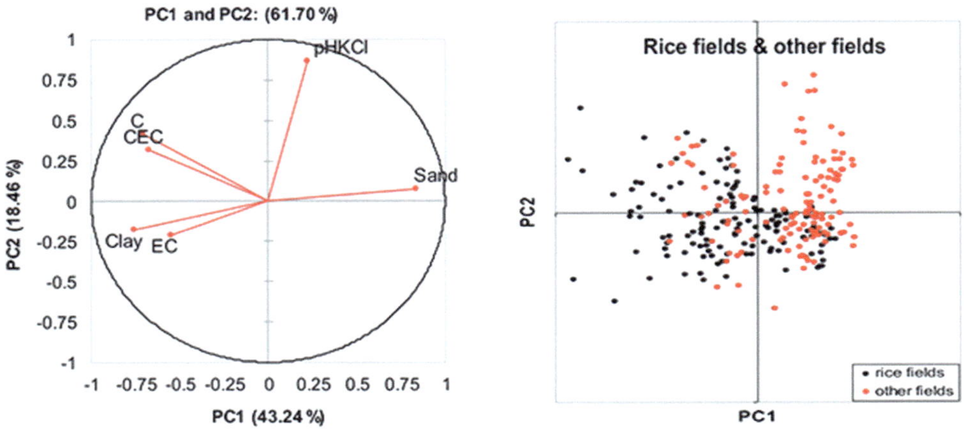

Fig. 4.24 Scatter of soil samples in the PC1–PC2 plane following the cropping pattern: rice fields and others

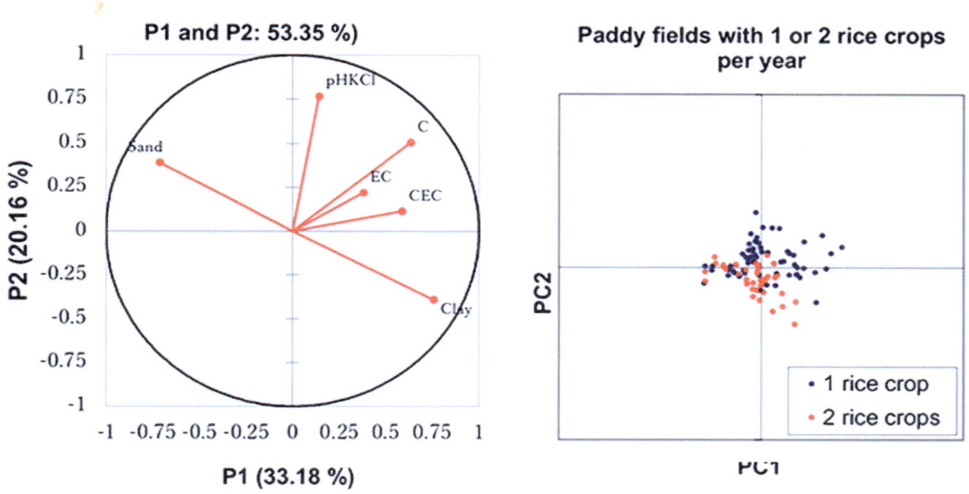

Fig. 4.25 Scatter of paddy soil samples in the PC1–PC2 plane following the number of crops per year

There are only small relative differences between most of the soil properties in the two cropping patterns. While no significant difference is observed in terms of texture, the soils with double rice crops are, on average, less sandy and more silty than the soils with a single rice crop. This can result from the lateral transport of fine particles along the topography, but a more targeted and specialized study is necessary to support this hypothesis. The content in organic C is higher in fields with double rice crops (1.6%) than in fields with single rice crop (1.0%). This might be due to the higher annual return of crop residues (rice straw, rice roots) in soils that are cultivated over two cropping seasons as compared to soils that are left fallow during all dry seasons. Indeed, Verma and Bhagat (1992) reported that adding organic matter to the soil in the form of straw can increase soil organic matter content in rice systems. We also noticed that the mean N content is not different in the two cropping patterns, so that mean C/N value for double rice crops (C/N = 24) is higher than for single rice crop (C/N = 16). This indicates that the humification process is more complete in soils with a single rice crop, a process that is favored by fallow periods (aeration of soils) and a lower annual return of rice straw with a very high C/N value. The soluble salt concentration, as inferred from EC, is greater in soils with double rice crops than in soils with only one rice crop. This may be both a consequence of the lower topographic situation of soils with double rice crops and a consequence of higher organic matter content in these soils, as explained above.

4.4.5.3 Properties of Soils Among Main Cropping Systems

Major crops in the study region include rice, peanut, sweet potato, cassava, banana, and vegetables (home garden). Traditional cropping systems vary since they have evolved in response to prevailing soil and climatic conditions as well as social and ethnological preferences (Tu et al. 2005). Traditional farmers often plant more than one crop species in a small field. Intercropping, the practice of growing two or more crops simultaneously in the same field, is common throughout the tropics and often practiced in the study zone. Multi-story homestead gardens, where more than three annual crop and vegetable species are mixed with tree crops, are common in humid regions.

In the present analysis, we compared four types of land use: rice (130 samples), peanut (39 samples), home garden crops (42 samples), and other crops (54 samples), i.e., mainly sweet potatoes and cassava that are often cultivated in the same field during winter season and summer season, respectively, on raised beds. Peanut is cultivated in

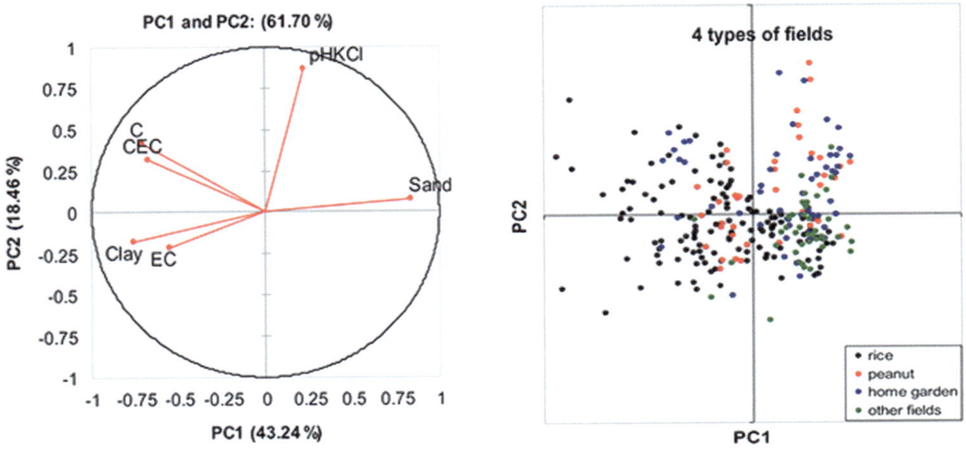

Fig. 4.26 Scatter of soil samples in the PC1–PC2 plane following four types of cropping patterns

leveled fields, except in some plots of the Quang Loi region. Rice soils in this section include single rice crop, double rice crop, and single rice crop with other crops (peanut, sweet potato).

Figure 4.26 illustrates that the rice soils are scattered on the left-hand side of PC1, the peanut and home garden soils are scattered in the center of this axis, while the other fields are on the right-hand side of the graph, i.e., more sandy soils with lower organic matter content.

Soils under rice and peanut cultivation are less sandy and more silty than the others, as was discussed in the section comparing soil preparation techniques. Now we can discriminate soils under rice from soils under peanut. Rice soils are still finer textured than peanut soils, and it is interesting to notice that the mean clay content of peanut soils is as small as that of soils under cropping systems with raised beds.

In other words, as compared to the most sandy soils of the study zone cultivated with root crops and vegetables on raised beds, the peanut soils are less sandy and more silty, and the rice soils are less sandy, more silty, and more clayed. Mean organic C content is slightly higher in rice soils (C = 1.2%) than in peanut soils (C = 1.0%) and in other soils (C = 0.5%) except in home garden soils (C = 0.8%). Despite their very coarse texture, home garden soils show higher humus content than that in other soils with similar textures, likely because of the higher return of organic residues due to the soil near family houses and due to continuous cropping over the year. The mean pH values are slightly smaller in rice soils than in peanut soils, which can be explained by previous comments on the effects of waterlogging and organic matter content on soil acidity. As far as CEC is concerned, the only significant feature is the higher mean value in rice soils than in the other soils, which is coherent with the higher clay and C contents observed in rice soils.

4.4.6 Conclusion

The major conclusion is that different cropping patterns are usually linked to significant differences in soil properties. This interpretation is not always straightforward, but we believe that the cropping systems were historically adapted to the soil constraints. It is not surprising that paddy fields are established on less sandy soils in this very sandy land because of their lower permeability and greater natural fertility. Indeed, these paddy soils also have higher organic matter content than the others. The practice of double or single rice crop in a year is related to the topographic situation and access to irrigation; nevertheless, the soils under double rice crops are less sandy and more silty and have higher organic C content than the soils under single rice crop. Peanut is also cultivated on less sandy soils other than root crops such as cassava and sweet potatoes, but the mean clay content of peanut soils is similar to that of the sandy soils with root crops cultivated on raised beds.

These results can be used as a sound base for further studies on soil management, including organic and inorganic fertilization for sustained productivity. In this context, optimization of the organic matter cycle within farming systems should aim at maximizing organic return to soils in order to ensure the progressive release of nutrients by mineralization and higher storage capacity of exchangeable bases as well.

4.5 Improvement of Rice Cropping in Saline Soils in North-Central Coastal Region of Vietnam[5]

4.5.1 General

Vietnam is a humid tropical country with a long coastline and large coastal area. The mainland stretches from N08°10′ to N23°24′ and widens from E102°09′ to E109°30′. Vietnam has two fertile river deltas, Red and Mekong. The total saline soil area of the country is approximately 971,356 hectares, covering approximately 14% of the entire agricultural land (Vietnam Society of Soil Science 1996).

At present, salinization is the second most widespread soil problem in rice (*Oryza sativa* L.)-growing countries (after droughts). It is considered a severe constraint on rice production worldwide (Ghafoor et al. 2004: 110–111). In Vietnam, particularly the north-central coastal region of the country, soil salinization is considered one of the most serious problems affecting rice and other crops. Salinization is mainly due to the inundation of seawater during high tidal action and ingress through creeks (Lang et al. 2010). As a result of the effect of salinization, the average yield of rice and maize in the north-central coastal region is low, approximately 0.8–1.5 and 1.0–2.0 tons ha^{-1}, respectively (Lang et al. 2010). The national average yield of rice and maize in 2010 was 5.3 and 4.1 tons ha^{-1}, respectively (General Statistics Office of Vietnam 2014a).

In the past, problems associated with saline soils were not given much attention by the Vietnamese government because of the country's small population. However, in recent years, the rapid increase of population has created a growing demand for food and put high pressure on food security. From 2000 to 2012, the population of Vietnam increased from 77.0 to 88.8 million people, while rice cultivation areas increased just slightly from 7.7 to 7.8 million ha (General Statistics Office of Vietnam 2014b). Hence, reclamation of saline soils and its potential to aid socio-economic development in north-central coastal zones are gaining attention.

Among the provinces, north-central coastal region of Vietnam, agricultural land in the province of Thua Thien Hue, is seriously affected by soil salinization. Approximately 2,500 ha of agricultural soils adjacent to the Tam Giang lagoon (the biggest lagoon in Southeast Asia; approximately 22,000 ha) suffer from high salinity (Dan et al. 2006). Therefore, there is an urgent need to develop practical solutions to alleviate the adverse effects of soil salinization on agricultural production in Thua Thien Hue.

Since the livelihood of the farmers in the coastal region of Thua Thien Hue is supported mainly by agricultural production on saline soils, where the rice yield is very low, as stated above, the farmers' income is not always high enough to introduce expensive countermeasures to mitigate salinity hazard. We therefore must seek feasible technologies that the farmers in the area could choose. For these reasons, we try two approaches in the present study; one is to establish site-specific management based on the knowledge of actual spatiotemporal variations of soil salinity in the region, and the other is to seek the possibility of introducing high-yielding,

[5] This section is based on the article by Nguyen Ho Lam, one of the authors of this section and has been deposited in the repository of Kyoto University at https://repository.kulib.kyoto-u.ac.jp/dspace/handle/2433/195980. It does not violate any policy of the repository.

salt-tolerant rice varieties that have been bred and cultivated in different areas of Vietnam.

In the first approach, spatial and temporal variability in soil salinity and its effects on rice production needs to be investigated because information on spatial and temporal variability in soil salinity can be used to select the appropriate site-specific methods for the amelioration of saline soils (Qadir et al. 2000; Zheng et al. 2009). Li et al. (2007) suggested that both spatial and temporal variations in soil salinity need to be considered to achieve the ultimate goal of sustainable cropping systems. The critical role of elevation in salt accumulation in soils has been reported by many authors, including Funakawa et al. (2000), Haruyama et al. (2006), Funakawa and Kosaki (2007), Li et al. (2007), Sugimori et al. (2008), Zheng et al. (2009), and Moral et al. (2010). However, there is limited information on the influence of minor differences in elevation (on the scale of centimeters to decimeters) on soil salinization, especially for saline soils adjacent to lagoons, such as those in the Thua Thien Hue region.

To develop solutions to limit the unfavorable effects of soil salinization and promote rice production, we examined the spatial distribution and temporal variability of soil salinity in paddy fields, in which there are slight differences in elevation. We evaluated the influence of relative elevation, electrical conductivity (EC) of soil saturation paste (ECe), the electrical conductivity of soil solution on the fields (ECw), and soil properties on rice yield (production).

4.5.2 Basic Information of Study Site

The research site was located in the Quang Phuoc commune (N16°35′, E107°32′), located in the Quang Dien district, Thua Thien Hue province, in the north-central region of Vietnam (Fig. 4.27). The commune is 14 km northeast of the city of Hue, stretching approximately 5 km along the Tam Giang lagoon, and has a flat landscape. People in this commune started to cultivate rice and other crops for a living in the sixteenth century (Dan et al. 2006). The total area is 1,048 ha, of which 509.3 ha is agricultural land, 161.4 ha is shrimp farming, 367.3 ha is non-agricultural land, and 10 ha is unused land (Dan et al. 2006). Two main irrigation rivers (Dien Hong and Bao Kho) provide fresh water to cultivate rice and other crops. The dike stretching along the lagoon was constructed in

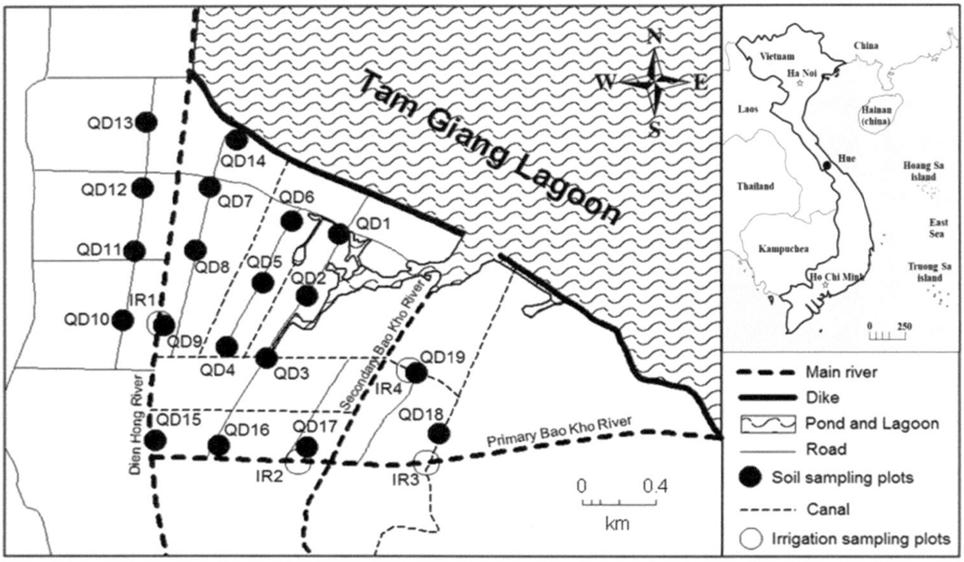

Fig. 4.27 Location of 19 research plots in the Quang Phuoc commune

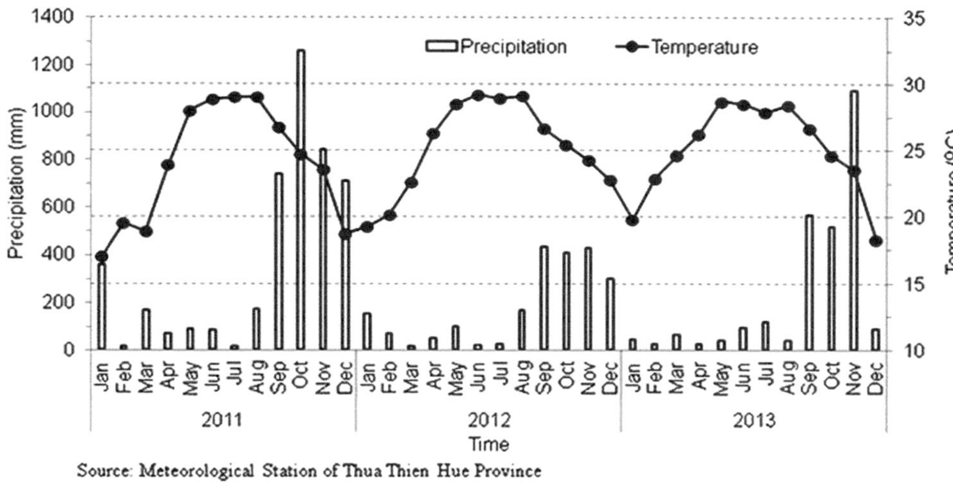

Fig. 4.28 Monthly precipitation and air temperature of research site from 2011 to 2013 (Mereological Statics of Tua Thien Province)

1994–1995 to prevent seawater intrusion from the lagoon into the land. However, certain sections of the dike have not yet been constructed because of financial constraints (Fig. 4.27).

The average air temperature and monthly precipitation values of the studied site over three years (2011–2013) are shown in Fig. 4.28. The climate of this region can be categorized as tropical monsoon climate with two distinct seasons: dry (from April to August) and rainy (September–December). Annually, this region receives a large amount of precipitation; the total precipitation amounts in 2011, 2012, and 2013 were 4,540, 2,190, and 2,730 mm, respectively. Because precipitation is unevenly distributed over the months and is mainly focused on September to December, this area is flooded annually during this period. The temperature increases gradually from January to May, then stays stable around 28 to 29 °C during May to August (the hottest period), and then decreases until the end of the year.

Rice fields in the central part of the research site (QD2, 3, 4, and 5), along the lagoon (QD1, 6, and 14), and the secondary Bao Kho river (QD17 and 19) (Fig. 4.27) always have water on the surface. Water table levels are approximately 10 to 60 cm from January to August and approximately 60–200 cm from September to December (flooding time). In the areas along the two main irrigation rivers of IR1 and IR2 (QD7, 8, 9, 10, 11, 12, 13, 15, 16, and 18), there is no water on the surface of the fields from late April to the middle of May and from the middle of July to late August. Water table levels are approximately 40–60 cm from September to December and approximately 20–30 cm during the rest of the time.

Rice Cropping Calendar There are two rice cropping seasons: winter—spring cropping lasts from January to May and summer–autumn cropping from May to September. Because Quang Phuoc commune is usually flooded in the rainy season (from the end of September to the end of December), there is no rice cultivation during this period because of water submergence (Dan et al. 2006).

General Soil Properties The physicochemical properties of soils at 19 representative plots (Table 4.17) showed that the fertility of these soils in this region was very low, indicating that these soils were not suitable for rice growth. Most of the soils were high in sand and low in silt and clay, and most of the soils were classified as sandy loam. Most soils had pH (H_2O) from 5.3 to 6.6 and pH (KCl) from 4.3 to 5.8, and cation exchangeable capacity (CEC), total carbon (C), and total nitrogen (N) were very low,

Table 4.17 Cation concentrations, sodium adsorption ratios, pH, and EC of water from four different irrigation rivers in the study area

Collection date	Location	Na^+	K^+	Mg^{2+}	Ca^{2+}	NH_4^+	Cl^-	NO_3^-	SO_4^{2-}	SAR†	pH	EC	Irrigation water class[b]	
		($mmol_c$ L^{-1})								($mmol\ L^{-1})^{0.5}$		($dS\ m^{-1}$)		
21 Jan 13	IR1	0.19	0.09	0.09	0.14	0.02	0.22	0.08	0.04	0.55	7.22	0.07	C1-S1	Low salinity
	IR2	0.21	0.08	0.11	0.18	0.02	0.16	0.07	0.08	0.55	7.09	0.06	C1-S1	Low salinity
	IR3	1.40	0.09	0.41	0.38	0.05	1.61	0.02	0.16	2.22	7.35	0.27	C2-S1	Medium salinity
	IR4	2.25	0.09	0.54	0.25	0.06	2.71	0.02	0.28	3.58	7.19	0.38	C2-S1	Medium salinity
2 Jun 13	IR1	0.13	0.04	0.10	0.15	0.00	0.24	0.02	0.04	0.37	7.31	0.09	C1-S1	Low salinity
	IR2	0.19	0.04	0.11	0.15	0.00	0.13	0.02	0.05	0.54	7.01	0.06	C1-S1	Low salinity
	IR3	0.20	0.05	0.12	0.16	0.00	0.28	0.02	0.07	0.53	7.32	0.09	C1-S1	Low salinity
	IR4	0.17	0.04	0.10	0.15	0.00	0.17	0.02	0.05	0.47	7.16	0.07	C1-S1	Low salinity
11 Sep 13	IR1	0.41	0.15	0.16	0.25	0.03	0.52	0.00	0.10	0.89	7.24	0.11	C1-S1	Low salinity
	IR2	0.48	0.12	0.16	0.19	0.01	0.55	0.00	0.10	1.16	7.20	0.11	C1-S1	Low salinity
	IR3	0.32	0.21	0.10	0.17	0.04	0.37	0.00	0.13	0.86	7.14	0.10	C1-S1	Low salinity
	IR4	2.54	0.19	0.60	0.32	0.08	3.17	0.00	0.32	3.74	7.15	0.45	C2-S1	Medium salinity

IR1, IR2, IR3, and IR4 located at Dien Hong river, primary Bao Kho river section adjacent plot of QD17, primary Bao Kho river section near plot of QD18, and canal adjacent plot of QD19, respectively

[a]SAR: Sodium adsorption ratio

[b] United States Salinity Laboratory Staff (1954)

from 2.0 to 6.4 $cmol_c\ kg^{-1}$, 7.8 to 15.2 $g\ kg^{-1}$, and 0.8 to 1.6 $g\ kg^{-1}$, respectively. In soils at the plots in the central part (QD2, 3, 4, and 5), those along the lagoon (QD1, 6, and 14), and those along the secondary Bao Kho river (QD17 and 19) (Fig. 4.27), the dominant exchangeable cations were Na^+, Mg^{2+}, and Ca^{2+}. In soils at the plots along two main irrigation rivers of IR1 and IR2 (QD7, 8, 9, 10, 11, 12, 13, 15, 16, and 18), the dominant exchangeable cation was Ca^{2+}, followed by Mg^{2+} and Na^+. The exchangeable cation K^+ of all soils was small.

4.5.3 Spatiotemporal Variability in Soil Salinity and Its Effects on Rice Production

Experimental Design Experiments were conducted on 19 plots (QD1–QD19) distributed widely in the Quang Phuoc commune (Fig. 4.27). The distance between plots was approximately 400 m. To evaluate the influence of relative elevation on salinity distribution, the elevation of each point was measured via the static surveying technique with a handheld Global Navigation Satellite System (GNSS) receiver and field controller (GRS-1 GG, TOPCON, Tokyo, Japan) and computer software (GNSS-Pro, TOPCON, Tokyo, Japan) (one of the sampling plots was designated as an elevation of 0 m and the elevations of the 18 remaining plots were measured relative to this). To investigate the spatiotemporal variability of soil salinity, soil samples at the 19 plots were collected three times a year in both 2012 and 2013, from a depth of approximately 0–20 cm, using an auger. Soil sampling was conducted in January (after flooding and at the beginning of winter–spring cropping), in the middle of May (the transitional period between the winter–spring cropping harvest and the beginning of the sowing/transplanting time of the summer–autumn cropping), and in September (at the end of the summer–autumn cropping and before flooding). To measure the EC of the soil solution in the paddy fields (ECw), 19 porous cups (DIK-8390–11, Daiki, Saitama, Japan) were installed in the 19 plots to collect soil solution samples at an interval of approximately 10 d (from February 2012 to September 2013) at a depth of approximately 20 cm from the soil surface. We did not measure ECw from the end of September to the end of December because all research plots were flooded during this period.

Irrigation water sampling was conducted three times in 2013 at almost the same time as soil sampling. The sample collection points were the same as the pump setting stations of the farmers. There were four sample collection points (sites IR1 through IR4) located at the Dien Hong river, the primary Bao Kho river section adjacent to QD17, and the primary Bao Kho river section near QD18, and the canal adjacent to QD19, respectively (Fig. 4.27).

The rice yield of five representative square meters near each plot in the winter–spring and winter–spring cropping seasons of 2013 was measured and then averaged to examine the influences of elevation, salinity, and major soil properties on rice production, including pH, CEC, clay content, total C, and total N contents.

Analytical Methods Before analysis, soil samples were air-dried, then crushed, and passed through a 2-mm mesh sieve. Then the following items were analyzed: particle size distribution (Gee and Bauder 1986), CEC and exchangeable cations (under 1 mol L^{-1} and pH 7.0) (Miller and Curtin 2006), exchangeable cations such as Na^+, K^+, Mg^{2+}, and Ca^{2+}, and total C and total N contents (with a CN analyzer [VarioMax CHN, Elementar, Hanau, Germany]). Soil solution was obtained as follows: 100 g of air-dried soil was mixed with 43 mL of deionized water (ratio 1:0.43) to prepare the saturation paste, which was used to measure pH (pHe), ECe, and concentrations of water-soluble cations and anions. The soil: water ratio of 1:0.43 was predetermined to satisfy the saturation paste criteria; these include the requirements that the soil paste should glisten and flow slightly when the container is tipped, and there is no free water accumulation on the surface after standing for at least 4 h (Miller and Curtin 2006). To compare the soil parameter values over time, the ratio of soil and water was fixed for all samples and for all soil collection times. pHe was determined using a glass electrode meter. ECe was measured using a conductivity meter (CM-30S, TOA Electronics Ltd, Tokyo, Japan). Water-soluble cations Na^+, K^+, Ca^{2+}, Mg^{2+}, and NH_4^+ and water-soluble anions Cl^-, SO_4^{2-}, and NO_3^- in the extracts were measured via high-performance liquid chromatography (LC-20AT, Shimadzu, Kyoto, Japan) equipped with a shim-pack IC-C3 for cations, a shim-pack IC-A1 for anions, and a CDD-10A conductivity detector.

Because soluble cations may account for a substantial proportion of the extractable cations, exchangeable cations were calculated as the total extractable cations minus the concentration of soil solution (soluble) cations (United States Salinity Laboratory Staff 1954). To evaluate the hazards of sodium in solution, sodium adsorption ratio (SAR) was calculated using the formula:

$$\text{SAR} = [\text{Na}^+]/\left([\text{Ca}^{2+}] + [\text{Mg}^{2+}]\right)^{1/2}$$

and the concentrations of Na^+, Ca^{2+}, and Mg^{2+} were presented in $mmol_c\ L^{-1}$ (Miller and Curtin 2006). ECw was measured by the conductivity meter.

The correlations between rice yield and elevation, ECe, pHe, CEC, clay content, total C, and total N concentrations were analyzed using Sigma Plots 11 statistics software (SSI, San Jose, CA, USA). Because the data for ECe was not normally distributed, the Spearman rank-order correlation was applied for analyzing correlations between the yield and this parameter. Data for pHe, CEC, clay content, total C, and total N were normally distributed; thus, Pearson product-moment correlation was used for analyzing correlations between the yield and these parameters.

Chemical Compositions of Irrigation Water The chemical composition of irrigation water at the study sites is given in Table 4.17. The value of pH was between 7.01 and 7.35. Although the irrigation water samples were collected at different times, the fluctuation in pH was slight. The EC values of IR4 in January and September and IR3 in January were higher than those of the other locations, 0.38, 0.45, and 0.27 dS m^{-1}, respectively. Meanwhile, the EC of IR1 and IR2 were low in all seasons, ranging from 0.06 to 0.11 dS m^{-1}. Concentrations of major cations and anions occurred in the following order: Na^+, Mg^{2+}, Ca^{2+}, and Cl^-. Concentrations of NH_4^+ and NO_3^- were negligible.

According to the criteria used to evaluate the quality of irrigation water based on EC and SAR (United States Salinity Laboratory Staff 1954), irrigation water at IR4 in January and September, and IR3 in January, was classified as C2–S1 (EC from 0.25 to 0.75 dS m^{-1} at 25 °C and SAR from 0 to 10). The classification for the two main irrigation rivers of IR1 and IR2 was C1–S1 (EC from 0.10 to 0.25 dS m^{-1} at 25 °C and SAR from 0 to 10) in all seasons.

Topography Based on the values of relative elevations (Table 4.18) and total ion concentrations and ECe of the soil solution collected from 19 representative plots (Tables 4.19 and 4.20), the research site can be divided into low- and high-elevation plots. Soils with an ECe higher than 1.9 dS m^{-1} in at least one of the samples were in the low-elevation plots. In contrast, soils with an ECe smaller than 1.9 dS m^{-1} in all the samples were in the high-elevation plots. An ECe of 1.9 dS m^{-1} was used for the division because rice growth starts to get influenced when ECe is higher than 1.9 dS m^{-1} (Grattan et al. 2002). The low-elevation plots were in the central part of the research site (QD2, 3, 4, and 5), along the lagoon (QD1, 6, and 14), and secondary Bao Kho river (QD17 and 19), while the others surround the two main irrigation rivers of IR1 and IR2 (QD7, 8, 9, 10, 11, 12, 13, 15, 16, and 18) (Fig. 4.27). However, the difference between the high- and low-elevation plots was not significant; relative elevation in the low-elevation plots was between −0.52 and 0.07 m, and that in the high-elevation plots was between 0.26 and 0.86 m.

Physicochemical Properties of Soils The physicochemical properties of soils (Tables 4.18, 4.19, 4.20, Fig. 4.29) showed that the fertility of these soils was very low, indicating that they were not suitable for rice growth. All the soils were high in sand and low in silt and

Table 4.18 Cation concentrations, sodium adsorption ratios, pH, and EC of water from four different irrigation rivers in the study

Plots		Elevation relative to QD1	Sand	Silt	Clay	CEC	Exchangeable cations[a]				Total C	Total N
							Na^+	K^+	Mg^{2+}	Ca^{2+}		
		(m)	(%)			($cmol_c$ kg^{-1})	($cmol_c$ kg^{-1})				($g\ kg^{-1}$)	
Low-elevation plots	QD3	−0.52	54	28	18	4.5	0.71	0.12	1.07	1.08	12.7	1.2
	QD5	−0.44	77	11	13	3.7	0.21	0.08	0.79	0.73	9.9	0.8
	QD2	−0.41	67	19	14	4.7	0.54	0.16	1.06	0.82	12.5	1.1
	QD4	−0.41	74	15	10	4.0	0.44	0.15	0.98	0.37	9.1	0.8
	QD6	−0.35	82	7	11	3.6	0.35	0.09	0.70	0.73	10.5	0.9
	QD14	−0.14	67	18	15	3.9	0.37	0.11	1.07	0.97	9.7	1.0
	QD19	−0.02	80	13	8	2.5	0.11	0.07	0.30	0.31	8.7	0.8
	QD1	0.00	68	17	15	3.9	0.58	0.19	1.13	0.79	10.1	1.0
	QD17	0.07	61	28	12	3.7	0.31	0.07	0.72	0.59	10.8	1.0
High-elevation plots	QD16	0.26	44	42	14	5.5	0.15	0.05	0.44	0.77	10.9	1.0
	QD18	0.26	44	40	16	4.0	0.14	0.07	0.38	0.65	15.2	1.6
	QD15	0.28	45	39	15	6.4	0.10	0.05	0.09	0.35	13.5	1.4
	QD7	0.37	34	40	26	4.9	0.11	0.10	0.22	0.73	12.7	1.4
	QD8	0.42	58	26	17	3.0	0.04	0.06	0.04	0.32	11.1	1.2
	QD9	0.73	27	57	16	6.4	0.14	0.10	0.21	0.54	10.3	1.0
	QD10	0.75	69	19	12	2.9	0.06	0.06	0.04	0.71	12.0	1.3
	QD11	0.81	69	20	12	3.6	0.09	0.10	0.07	0.93	12.1	1.3
	QD13	0.84	70	21	9	2.0	0.04	0.07	0.03	0.37	8.1	0.8
	QD12	0.86	73	18	9	2.2	0.07	0.05	0.01	0.37	7.8	0.8

[a] This value have been subtracted water-soluble cations concentration

Table 4.19 Total cation and anion concentrations in soil solution prepared from saturation paste of surface soils (0–20 cm) from the 19 plots in the study area at different sampling times

Plots		Total cations ($cmol_c$ kg^{-1})						Total anions ($cmol_c$ kg^{-1})					
		2012			2013			2012			2013		
		1 Jan	27 May	9 Sep	21 Jan	14 May	4 Sep	1 Jan	27 May	9 Sep	21 Jan	14 May	4 Sep
Low-elevation plots	QD3	1.73	2.12	2.43	2.00	3.86	3.18	2.00	2.22	2.71	2.00	4.01	3.00
	QD5	0.95	2.57	1.69	1.16	1.76	1.54	1.09	2.59	1.92	1.05	1.89	1.55
	QD2	1.74	3.53	2.69	1.60	2.99	2.2	2.01	3.92	3.05	1.61	3.49	2.08
	QD4	2.34	4.31	4.21	1.91	2.49	2.31	2.73	4.63	5.04	1.90	2.85	2.16
	QD6	1.48	3.14	2.86	1.36	1.95	2.28	1.72	3.40	3.30	1.34	2.19	2.27
	QD14	1.42	3.88	3.82	1.58	2.21	1.68	1.46	4.14	3.87	1.34	2.54	1.76
	QD19	0.29	0.75	1.41	0.69	1.40	0.96	0.32	0.80	1.67	0.79	1.54	1.10
	QD1	2.66	4.50	2.91	2.22	3.57	2.94	2.99	5.06	3.38	2.27	3.92	2.07
	QD17	0.28	0.77	0.54	0.41	0.99	1.05	0.23	0.75	0.53	0.50	0.99	1.15
High-elevation plots	QD16	0.15	0.21	0.29	0.15	0.39	0.33	0.18	0.22	0.31	0.14	0.44	0.35
	QD18	0.15	0.31	0.18	0.19	0.29	0.23	0.12	0.18	0.18	0.12	0.30	0.23
	QD15	0.14	0.15	0.07	0.12	0.18	0.16	0.11	0.15	0.06	0.10	0.20	0.16
	QD7	0.10	0.21	0.10	0.08	0.17	0.14	0.08	0.20	0.09	0.04	0.21	0.14
	QD8	0.09	0.16	0.12	0.09	0.13	0.08	0.08	0.14	0.12	0.06	0.16	0.07
	QD9	0.11	0.08	0.11	0.09	0.12	0.10	0.10	0.06	0.09	0.07	0.13	0.08
	QD10	0.13	0.37	0.24	0.20	0.26	0.26	0.09	0.28	0.19	0.07	0.27	0.26
	QD11	0.30	0.30	0.27	0.18	0.23	0.20	0.14	0.16	0.19	0.09	0.22	0.20
	QD13	0.11	0.15	0.15	0.11	0.14	0.17	0.09	0.11	0.15	0.08	0.16	0.20
	QD12	0.14	0.16	0.16	0.10	0.10	0.11	0.11	0.09	0.15	0.05	0.12	0.11

Table 4.20 ECe and sodium adsorption ratio (SAR) of soil solution prepared from saturation paste of surface soils (0–20 cm) from the 19 plots in the study area at different sampling times

Plots		ECe (dS m^{-1})						SAR[a] ($mmol\ L^{-1}$)$^{0.5}$					
		2012			2013			2012			2013		
		1 Jan	27 May	9 Sep	21 Jan	14 May	4 Sep	1 Jan	27 May	9 Sep	21 Jan	14 May	4 Sep
Low-elevation plots	QD3	4.26	4.14	4.69	4.48	7.08	6.60	9.41	6.03	5.50	9.30	12.11	12.53
	QD5	2.22	4.49	3.51	2.62	3.81	3.38	5.78	3.48	5.52	7.39	6.02	6.45
	QD2	4.16	6.68	5.13	3.84	6.06	4.45	9.58	6.03	6.63	8.92	8.19	9.68
	QD4	5.16	8.35	9.11	4.45	5.31	4.86	14.31	8.74	11.81	10.23	9.09	13.57
	QD6	3.53	5.57	5.68	3.50	4.32	3.39	6.62	4.58	7.89	6.84	6.37	4.66
	QD14	3.10	7.34	5.57	3.43	4.56	3.31	7.55	6.60	5.72	9.54	6.73	4.66
	QD19	1.01	1.75	2.39	2.00	3.39	2.31	2.57	2.86	3.52	6.47	8.00	7.98
	QD1	6.20	8.16	6.23	4.89	6.74	5.99	9.59	6.10	7.31	8.77	6.74	8.77
	QD17	0.87	1.94	0.98	1.56	2.32	2.99	7.32	6.51	4.79	6.00	8.23	12.79
High-elevation plots	QD16	0.46	0.54	0.58	0.42	1.00	0.82	2.20	1.93	1.53	3.18	2.09	2.85
	QD18	0.45	0.72	0.37	0.46	0.76	0.56	2.02	1.02	1.05	1.88	1.64	1.75
	QD15	0.26	0.40	0.53	0.32	0.52	0.43	2.63	1.34	0.76	1.89	1.28	1.57
	QD7	0.27	0.57	0.24	0.18	0.56	0.40	1.43	1.04	0.85	1.05	1.13	1.30
	QD8	0.25	0.49	0.30	0.20	0.42	0.24	0.75	0.53	0.80	1.29	0.99	1.04
	QD9	0.31	0.18	0.26	0.25	0.37	0.26	1.57	0.96	1.27	2.20	1.42	1.64
	QD10	0.34	0.89	0.47	0.45	0.69	0.62	0.90	0.57	0.52	0.75	0.67	0.90
	QD11	0.64	0.68	0.56	0.44	0.63	0.55	1.57	0.60	0.55	0.52	0.62	0.40
	QD13	0.36	0.39	0.37	0.28	0.44	0.42	0.77	0.74	0.71	1.90	1.34	1.83
	QD12	0.35	0.41	0.38	0.25	0.37	0.33	1.64	0.68	0.36	0.49	0.39	0.71

[a] SAR: Sodium adsorption ratio

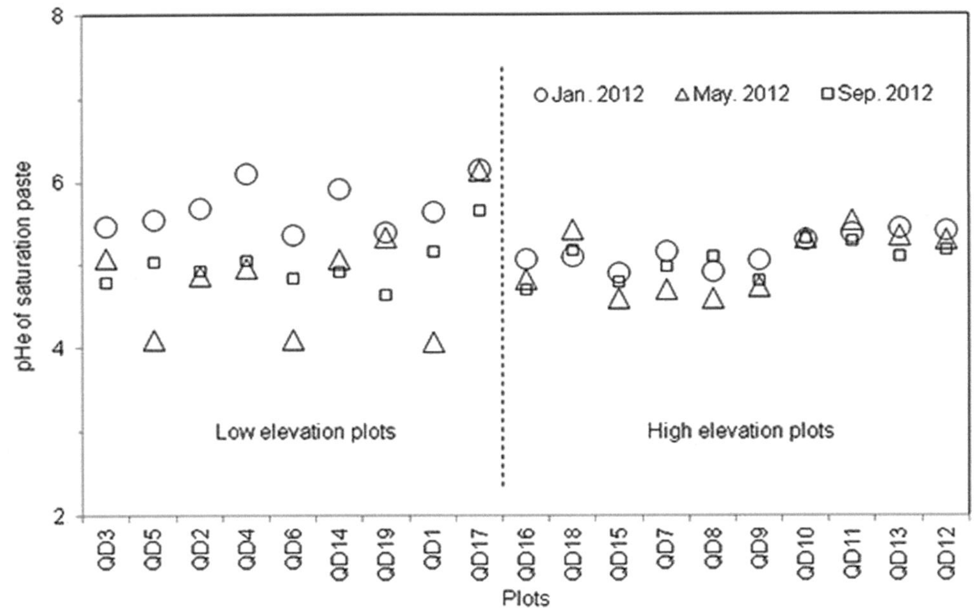

Fig. 4.29 PHe of saturation paste of 19 research plots in 2012

clay; most soils were classified as sandy loam. CEC, total C, and total N were low, from 2.0 to 6.4 $cmol_c$ kg^{-1}, 7.8 to 15.2 g kg^{-1}, and 0.8 to 1.6 g kg^{-1}, respectively. Most of the soils had a pHe from 4.0 to 6.0 (Fig. 4.29). In the low-elevation plots, pHe values in May and September were much lower than in January; while in the high-elevation plots, pHe was stable throughout the seasons. In soils at the low-elevation plots, the dominant exchangeable cations were Mg^{2+} and Ca^{2+}, followed by Na^+ (Table 4.18). The exchangeable cation K^+ was low. In soils at the high-elevation plots, the dominant exchangeable cation was Ca^{2+}, followed by Mg^{2+} and Na^+. The exchangeable cation K^+ was present in negligible concentrations. Because an increase in ECe and total soluble cations coincided with an increase in SAR, Na^+ was mainly responsible for salt accumulation in the surface paddy soils (Funakawa and Kosaki 2007; Funakawa et al. 2000). Moreover, a highly significant positive correlation between the water-soluble cation Na^+ and water-soluble anion Cl^- indicated an influence of NaCl from seawater on the ECe, ECw, and total salt content (Haruyama et al. 2006). There were positive correlations between ECe and SAR, total cations, and total anions. There were no correlations between ECe and pHe, clay content, CEC, total C, and total N.

Spatiotemporal Variability of Salinity in the Surface Soils The spatial variability in ECe in the surface soils is shown in Tables 4.19, 4.20 and Figs. 4.30, 4.31, and temporal variability is shown in Tables 4.19, 4.20 and Figs. 4.28, 4.32. ECe and total soluble cation concentrations were high at low-elevation plots (Tables 4.19, 4.20 and Fig. 4.30). Compared with the high-elevation plots, ECe and ECw in the low-elevation plots were much higher. At the high-elevation plots, ECe was less than 1.0 dS m^{-1}. At the low-elevation plots (except for soils at QD17 in January 2012 and 2013 and September 2012, and QD19 in January and May 2012), ECe was higher than the threshold for rice growth of 1.9 dS m^{-1} (Grattan et al. 2002). ECw in the high- and low-elevation plots increased gradually from January to September, though the increase was slight in the high-elevation plots (Fig. 4.32).

Influence of Salinity on Rice Production The relationship between rice yield and ECe is shown in Fig. 4.33. The results demonstrate that rice yield in summer–autumn cropping was

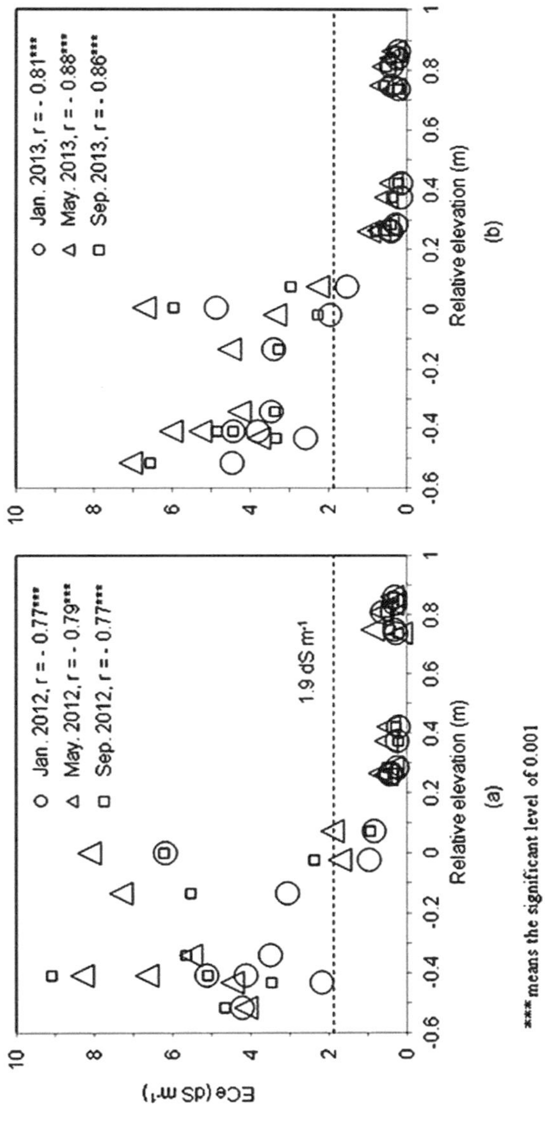

Fig. 4.30 Relationship between ECe of saturation paste and relative elevation **a** in 2012 and **b** in 2013

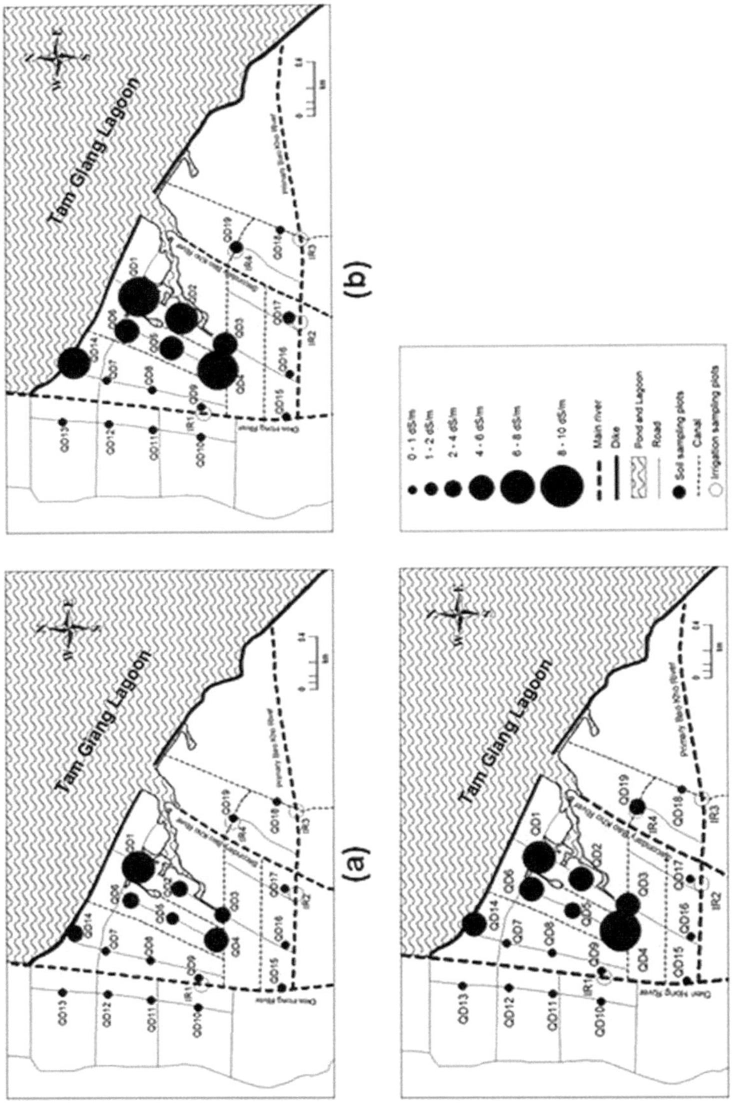

Fig. 4.31 Distribution of salinity of saturation paste (ECe) in (**a**) January, **b** May, and **c** September 2012

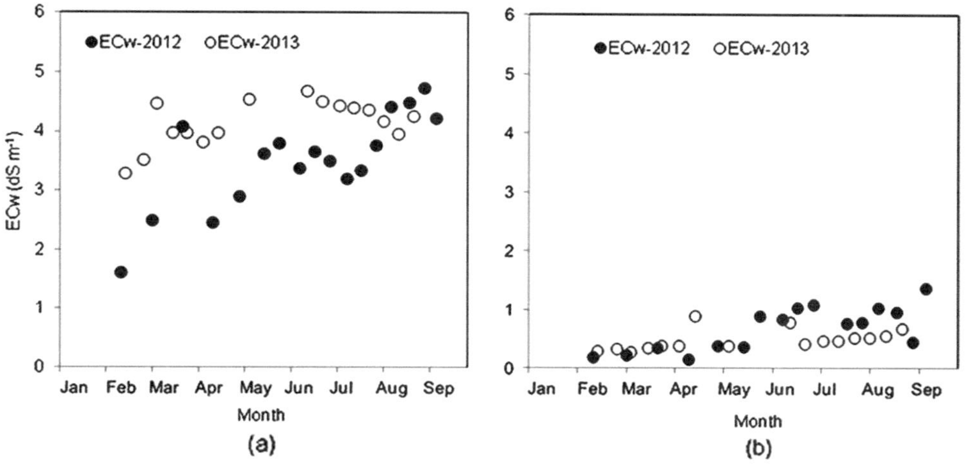

Fig. 4.32 Temporal variability of ECw of soil solution throughout the year at **a** the low-elevation plot QD3 and **b** the high-elevation plot QD10

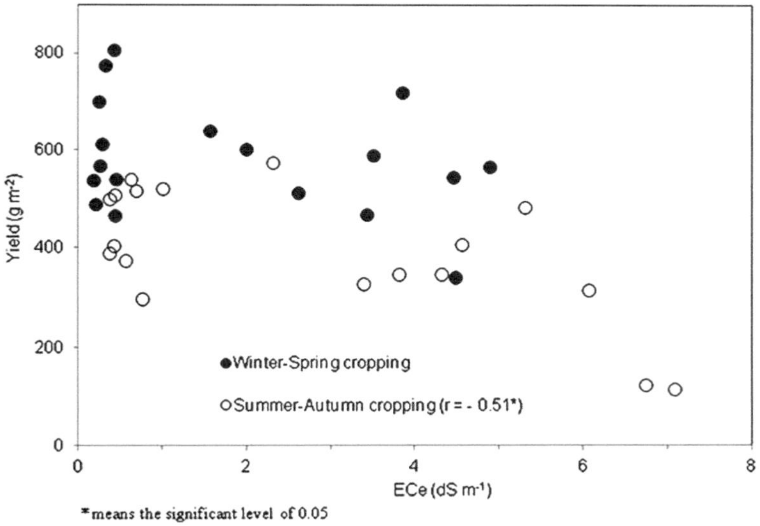

Fig. 4.33 Relationship between rice yield and ECe of saturation paste * Indicates a significant level at 0.05

significantly negatively correlated with ECe ($r = -0.51$, $P < 0.05$), while a nonsignificant correlation was observed in winter–spring cropping. There was a nonsignificant correlation between yield and elevation, pHe, clay content, CEC, and total C and N contents. Figure 4.33 also shows that rice yield in winter–spring cropping was much higher than in summer–autumn cropping.

Discussion The EC of the two main irrigation water sources (IR1 and IR2) was low and stable throughout the year, being less than 0.1 dS m^{-1} and classified as C1–S1 (United States Salinity Laboratory Staff 1954) (Table 4.17). This implies that the intrusion of seawater does not influence IR1 and IR2. Therefore, IR1 and IR2 water can be used for the irrigation of most crops and soils with little risk of causing harmful levels of salinity and exchangeable sodium (United States Salinity Laboratory Staff 1954). Because IR3 and IR4 are located adjacent to the secondary Bao Kho river that connects directly to the lagoon (Fig. 4.27), seawater from the lagoon easily seeps into these two irrigation sources. As a result, the EC of IR3 (in January) and IR4 (in January and September) was medium level (0.27, 0.38, and 0.45 dS m^{-1}, respectively) and was classified as C2–S1 (United States Salinity Laboratory Staff, 1954). However, the EC of IR3 and IR4 was low (less than 0.1 dS m^{-1}) at the beginning of June; because the beginning of June is the beginning of the second rice cropping season, farmers take extra care to ensure the sluice gates are closed to prevent seawater intrusion during this period. Irrigation water sources IR3 (in January) and IR4 (in January and September) are classified as C2–S1; thus, they can also be used for the irrigation of crops and soils with a moderate amount of leaching occurring.

ECe and total soluble cation concentrations were high at the low-elevation plots (Fig. 4.30). This indicates insufficient leaching of salts or salt re-accumulation in the low-elevation plots. ECe and ECw in the high-elevation plots were much higher than ECe and ECw in the low-elevation plots. In the low-elevation plots, ECe and ECw were higher than 1.9 dS m^{-1} in all seasons (except for ECe at QD17 in January 2012 and 2013 and September 2012, and at QD19 in January and May 2012) (Tables 4.19, 4.20 and Figs. 4.30 and 4.32a), indicating the intrusion of seawater from the lagoon into these plots. Certain dense canals and ponds are connected directly to the lagoon (Fig. 4.27) and are flooded annually from September to December in the depressed region; seawater, therefore, easily inundates the inland area via the canals and ponds at the time of rising tide or during flooding. In the high-elevation plots, ECe values were low and less than 1.0 dS m^{-1} (Tables 4.19, 4.20 and Fig. 4.30), indicating that soils in these plots were not invaded by seawater. The high-elevation plots are located along the two main irrigation rivers of IR1 and IR2; therefore, these plots are supplied frequently with fresh water with low salinity (Table 4.17). Though the relative elevation measurements (Table 4.18) indicate that the differences in elevation between the 19 plots were minimal (tens of centimeters), they still led to significant differences in soil salinity levels (Figs. 4.30, 4.31 and 4.32). Thus, salt accumulation was strongly influenced by elevation. The influences of elevation on salinity accumulation have also been reported by Funakawa and Kosaki (2007), Sugimori et al. (2008), and Moral et al. (2010), Grattan et al. (2002) suggested that if the ECe of soil is greater than 1.9 dS m^{-1}, rice growth (production) starts to be adversely affected. Thus, rice growth (production) in the low-elevation plots (except for soils at QD17 in January 2012 and 2013 and September 2012, and at QD19 in January and May 2012) was adversely affected by salinity because ECe and ECw at these plots were much higher than 1.9 dS m^{-1} (Tables 4.19, 4.20 and Figs. 4.30 and 4.32a). In contrast, rice growth (production) in the high-elevation plots was not affected by salinity because ECe and ECw were much lower than the threshold of 1.9 dS m^{-1} (Tables 4.19, 4.20 and Figs. 4.30 and 4.32b).

Besides being influenced by elevation, soil salinity also varied seasonally. ECe and ECw from May to September were much higher than in January, and the differences in ECe and ECw between May and September were small (Tables 4.19, 4.20 and Figs. 4.31, 4.32). Because the flood or the rainy season ends at the beginning of January, and the hot season with higher temperature and surface water evaporation occurs from May to September, salt concentrations in soils were diluted by high amounts of fresh water supplied by flooding in January and became more condensed from May to September.

Rice yield in the summer–autumn cropping season tended to be lower than in the winter–spring cropping season (Fig. 4.33). High ECe and ECw may cause these results in the summer–autumn cropping season (May 2013) (Figs. 4.30 and 4.32). Rowell (1994) and Ghafoor et al. (2004) suggested that soil salinity directly affects rice growth through osmotic stress and anionic toxicity caused by Na^+, Cl^-, and SO_4^{2-}, which may also promote imbalances in rice nutrient metabolism. The negative correlations between yield and ECe were also reported by Sugimori et al. (2008).

In conclusion, though the differences in relative elevations between the 19 plots were minimal (tens of centimeters), they can lead to significant differences in soil salinity. Elevation and ECe measurements show that the research site could be divided into low- and high-elevation plots. The low-elevation plots were located in the central part of the research site, along the lagoon and the secondary Bao Kho river, and the high-elevation plots surrounded the two main irrigation rivers. ECe and ECw in the low-elevation plots were much higher than in the high-elevation plots. In the high-elevation plots, soils were irrigated by fresh water with low EC and not subject to seawater intrusion; thus, ECe and ECw were low throughout the year. Because ECe levels were much lower than the threshold for rice growth of 1.9 dS m^{-1}, rice growth (production) was not influenced by salinity. However, soils had poor fertility, as indicated by the low pHe, CEC, total C, and total N; soil fertility is therefore also considered one of the main factors limiting rice production in the high-elevation plots. In the low-elevation plots, soils were subject to seawater intrusion; thus, their ECe was much higher than the threshold for rice growth of 1.9 dS m^{-1}. Moreover, they also had poor fertility; hence, rice yield (production) at the low-elevation plots was influenced not only by high salinity but also by low soil fertility. ECe and ECw from May to September were the highest in the year. Based on these results, some solutions are recommended to limit the unfavorable effects of soil salinization and promote rice yield. First, a dike with sluice gates stretching along the lagoon was constructed from 1994 to 1995 to prevent saltwater intrusion from the lagoon onto land; however, certain sections of the dike have not yet been constructed because of financial constraints. Therefore, a complete dike system has to be constructed along the lagoon to prevent seawater intrusion into the land completely. Second, because there are difficulties associated with drainage in the depressed plots, it is necessary to construct adequate facilities to promote the rapid drainage of water into the canals during and after flooding and irrigation in these plots. Third, because ECe and ECw increase from May to September, adequate fresh water for irrigation should be supplied frequently for the rice fields to lower ECe and ECw during this period. In addition, the irrigation freshwater quantity in the rivers must be provided appropriately to push back saltwater intrusion from the lagoon; thus, farmers can use fresh water for their fields. Fourth, since soil fertility was poor with low organic matter content and low pHe, the application of organic fertilizers and lime should be repeated annually over a long period (Corwin and Lesch 2003). Finally, regarding improving rice breeding: farmers in the Quang Phuoc commune are cultivating the local rice varieties 4B, KD, and TH5, which have a low salt-tolerant capacity and low yield. Hence, to increase rice yield and income, these local varieties need to be replaced with better-quality varieties (high salt-tolerant varieties).

4.5.4 Performance of Newly Introduced Salt-Tolerant Rice Cultivars in Saline Soils

Recently, farmers in the Mekong River Delta and the south-central coastal regions of Vietnam started cultivating certain new high-yielding, salt-tolerant rice varieties, including AS996, OM4900, OM5900, MNR3, and CM2, on the saline soils in these regions. These varieties were bred by the Cuu Long Delta Rice Institute (Can Tho province) and the Southern Central Coastal Agricultural Science Institute (Binh Dinh province). However, because they are new varieties, there is no information on their growth and yield when cultivated on the low fertility, low pH, saline soils in the north-central coastal region of Vietnam.

Analyzing the relationships between yield and various yield components can reveal the best indicators for yield and the significance of their contributions to yield. However, because the correlations between agronomic traits of rice and yield are strongly affected by the environmental conditions and rice genotypes (Senanayake and Wijerathen 1988), correlation analyses should be conducted for each specific variety in field experiments in the region, in which it will be cultivated. Although the correlations between rice agronomic traits and yield have been intensively studied, there have been no reports on these relationships for the new rice varieties cultivated in the north-central coastal region of Vietnam. Therefore, it is essential to analyze the relationships between various rice traits and yield to identify the best yield indicators and their contributions to yield. This information can be used to select the best breeding lines to cultivate in the north-central coastal region of Vietnam.

This study had two objectives: to evaluate the performance of 13 newly introduced salt-tolerant rice varieties cultivated in the north-central coastal region of Vietnam and to identify the best yield indicators and quantify their contributions to yield. To achieve these objectives, the 13 highest yielding new salt-tolerant rice varieties that are widely cultivated on saline soils in the Mekong river delta and south-central coastal regions of Vietnam (Table 4.21) were cultivated at three sites in the north-central coastal region of Vietnam in the winter–spring and summer–autumn cropping seasons of 2012 and 2013.

Rice Cultivars Fourteen rice varieties (including one local variety, KD) were used in this research. Twelve varieties were bred by the Cuu Long Delta Rice Institute, located in Can Tho province, South Vietnam. One variety was bred by the South Central Coastal Agricultural Science Institute, located in Binh Dinh province, south-central Vietnam). The local variety KD was used as the control. The names and major characteristics of the 14 rice varieties are shown in Table 4.21.

Experimental Design The field experiments were conducted in three locations with varying degrees of soil salinity: low (Plot A), medium (Plot B), and high (Plot C) (Fig. 4.34). In total, seven field experiments were conducted in 2012 and 2013: one in Plot B, in which the EC of the soil solution (ECw) was 6.6 before transplanting and 7.3 dS m^{-1} at harvest, in the 2012 winter–spring cropping season; three in Plots A, B, and C with ECw ranges of 3.9–4.3, 6.3–6.5, and 8.3–9.6 dS m^{-1}, respectively, in the 2012 summer–autumn cropping season, and three in Plots A, B, and C with ECw value ranges of 3.9–4.5, 6.2–6.5, and 9.1–10.7 dS m^{-1}, respectively, in the 2013 winter–spring cropping season (Table 4.22).

The field experiments had a randomized complete block design with three replicates. In the 2012 winter–spring cropping season, rice was planted in rows with 20×20 cm spacing and two plants per hill. In the other six experiments, rice was planted in rows with 10×20 cm spacing and one plant per hill. The area of each plot was 5.0 m^2 (2.5 m $\times$ 2.0 m). The fertilizer application rate for all experiments was the same as that used by local farmers (200 kg ha^{-1} lime, 100 kg ha^{-1} N as urea, 60 kg ha^{-1} P_2O_5 as superphosphate, and 60 kg ha^{-1} K_2O as potassium chloride). For basic application, 100% of P_2O_5 and 30% of N were applied. The remaining fertilizer was applied twice as a dressing. The first application was at 2 weeks after sowing (50% of K_2O and 40% of N), and the second application was at 20 days before flowering (50% of K_2O and 30% of N). The rice plots were weeded twice during the cultivation period, simultaneously as to p-dressing. The climatic conditions during the experimental period are shown in Table 4.23.

Agronomic Traits Soil samples were collected and analyzed in the same manner as stated in the previous section. Then five plants in each plot were randomly selected to record agronomic traits. At 1 week before harvest, plant height was determined by measuring the distance from the soil surface to the end of the longest panicle. At harvest, five plants were pulled out, their roots were removed, and then the number of panicles per plant was counted. Five panicles from the five measured plants were randomly selected and then bagged individually before measuring yield components. Panicle length, number of grains per panicle, number of filled grains per panicle, 1,000-grain weight, and panicle weight were measured. After removing unfilled seeds and drying the

Table 4.21 Details of 14 rice varieties used in this study

Designation	Growth duration (days)	Plant height (cm)	1,000-grain weight (g)	Parentage	Origin
MNR3	100–105	100–105	27–28	OM6073/DS20/DS20	Cuu Long Delta Rice Institute
OM5629	95–100	95–100	28–29	C27/IR64/C27	Cuu Long Delta Rice Institute
OM2395	95–100	90–100	27–28	IR63356-6B/TN1	Cuu Long Delta Rice Institute
MNR4	100–105	100–105	25–26	AS996/JAS-MINE85	Cuu Long Delta Rice Institute
AS996	90–95	95–100	26–27	IR64/ORYZA RUFIPOGON	Cuu Long Delta Rice Institute
CM2	100–105	90–100	27–28	–[b]	Southern Central Coastal Agri-cultural Science Institute
OM6L	90–95	95–100	27–28	OM1490/HOA-LAI/HOALAI	Cuu Long Delta Rice Institute
OM5900	90–95	95–100	27–28	AS996/IR50404	Cuu Long Delta Rice Institute
OM4900	95–105	100–110	28–29	C53/JASMINE85/JAPONICA	Cuu Long Delta Rice Institute
OM2718	90–100	100–105	22–25	OM1738/MCRDB	Cuu Long Delta Rice Institute
OM8104	92–95	100–110	27–28	OM2718/OM2717	Cuu Long Delta Rice Institute
OM9922	90–95	95–100	27–28	IR68523-61-1-1–2-2/OM1570	Cuu Long Delta Rice Institute
OMCS2000	90–95	95–110	27–28	OM1723/MRC19399	Cuu Long Delta Rice Institute
KD[†]	100–105	90–195	22–25	KD18 mutation	Thua Thien Hue Seed Company

[a] KD: Khang Dan, the local variety

[b] Information about the parentage of CM2 was not available from the Southern Central Coastal Agricultural Science Institute

filled seeds, yield per hill was calculated from the weight of dried filled seeds. To determine the aboveground biomass (without roots) of each plant, the shoots were oven-dried at 70 °C for 1 week and then weighed. Grain fertility (%) was calculated by dividing the number of filled grains per panicle by the number of grains per panicle (Zeng and Shannon 2000). The harvest index was calculated as the total dry grain weight divided by total aboveground biomass (Zeng and Shannon 2000).

Statistical Analysis Data were analyzed using Statistix 9.0 (Analytical Software, Tallahassee, FL, USA). To compare the differences in agronomic traits and yield per plant among rice varieties at each EC level and at combinative three EC levels altogether, respectively, we conducted one-way ANOVA and Tukey's test. To evaluate the effects of different EC levels on agronomic traits and yield per plant in each cropping season, we used two-way ANOVA and Tukey's test. To test the effects of the two cropping seasons on

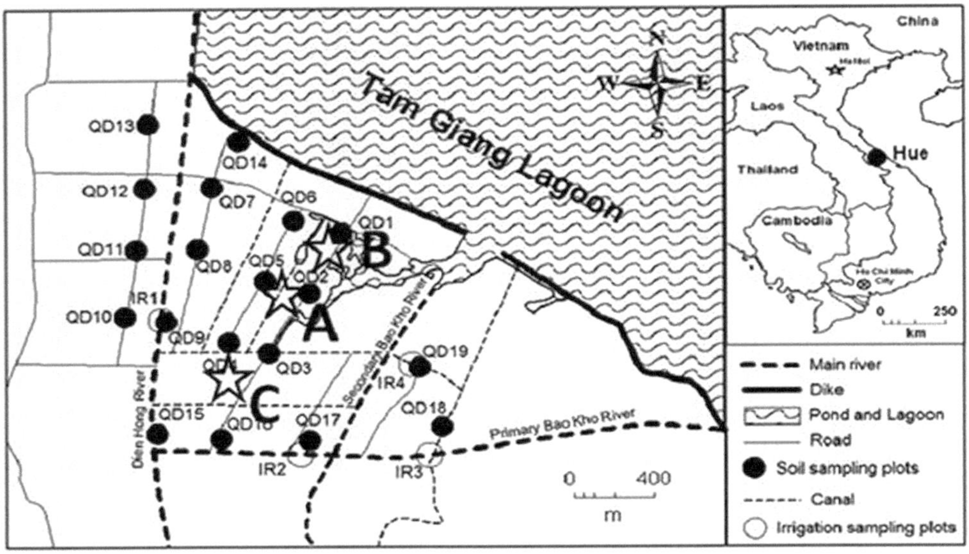

Fig. 4.34 Location of experimental plots A, B, and C

agronomic traits and yield per plant at each EC level, we conducted ANOVA and Fisher's Least Significant Difference test. In all cases, $P<0.1$ was considered to be significant. The correlations between agronomic traits and yield were analyzed after evaluating the normality of the data by Pearson's test.

Electrical Conductivity of Soil Solution at the Experimental Plots To measure the electrical conductivity in the field (ECw) during the rice cultivation period, five porous cups (DIK–8390–11, Daiki, Saitama, Japan) were installed at the center and the four corners of the experimental fields to collect soil water samples. The water samples were collected twice per month from approximately 20 cm depth during the rice cultivation period. ECw values were measured with a conductivity meter.

The ECw values recorded during the growing period in the seven experiments are shown in Fig. 4.35. The ECw was quite stable during the rice-growing period in most field experiments. However, in experimental Plot C in the winter–spring cropping season in 2013, ECw increased gradually from January until May.

Physicochemical Properties of Experimental Soils The physicochemical properties of soils indicated that they were not suitable for rice growth (Table 4.22). The soils in experimental Plots A, B, and C were classified as sandy loam with a sand fraction of 45, 68, and 61%, respectively. The pH (H_2O) of soil in experimental Plots A, B, and C was lower in the 2012 summer–autumn cropping season than in the 2013 winter–spring cropping season; that may be because of higher SO_4^{2-} concentration in the 2012 summer–autumn cropping season and lower concentration in 2013 winter–spring cropping season. All the soils showed very low values for CEC (3.5–5.0 $cmol_c$ kg^{-1}), total C (10–13 g kg^{-1}), and total N (0.9–1.3 g kg^{-1}). As shown in Table 4.19, the major water-soluble cations were Na^+, Ca^{2+}, and Mg^{2+}, while the amounts of the water-soluble cations K^+ and NH_4^+ were negligible. The primary water-soluble anions were SO_4^{2-} and Cl^-, while the amount of NO_3^- was insignificant.

Performance of Studied Rice Varieties in Terms of Agronomic Traits and Yield per Plant Table 4.24 shows the performance of the 14 rice varieties, in terms of ten main agronomic traits and yield per plant, in experimental Plot A and three experimental Plots A, B, and C altogether in the 2013 winter–spring cropping season. Among the 14 studied rice varieties in the seven field experiments, only MNR3

Table 4.22 Physicochemical properties of soils in experimental plots

	Winter–spring cropping season 2012	Summer–autumn cropping season 2012			Winter–spring cropping season 2013		
Experimental plot	B	A	B	C	A	B	C
ECw[a] (dS m^{-1})	6.6–7.3	3.9–4.3	6.3–6.5	8.3–9.6	3.9–4.5	6.2–6.5	9.1–10.7
Sand (%)	68	45		61			
Silt (%)	17	31		26			
Clay (%)	15	24		14			
pH (H_2O)	6.0	5.3	4.7	5.2	6.3	6.6	6.2
pH (KCl)	5.4	4.6	4.3	4.7	5.3	5.6	5.8
ECe[b] (dS m^{-1})	6.2	4.0	6.7	9.6	3.9	5.9	7.9
CEC ($cmol_c$ kg^{-1})	5.0	3.7		3.5			
Total C (g kg^{-1})	10	13		10			
Total N (g kg^{-1})	1.0	1.3		0.9			
Exchangeable cations ($cmol_c$ kg^{-1})							
Na^+	2.03	0.74		2.55			
K^+	0.24	0.10		0.38			
Ca^{2+}	1.20	0.88		1.63			
Mg^{2+}	1.72	1.01		2.37			
Water-soluble cations ($cmol_c$ kg^{-1})							
Na^+	1.40	0.82	1.45	2.76	0.70	1.21	2.00
K^+	0.05	0.05	0.07	0.07	0.02	0.06	0.04
Ca^{2+}	0.41	0.40	1.05	0.85	0.22	0.36	0.22
Mg^{2+}	0.58	0.46	0.97	1.13	0.19	0.53	0.47
NH_4^+	0.04	0.04	0.04	0.05	0.04	0.06	0.06
Sum of cations	2.50	1.76	3.57	4.86	1.16	2.22	2.78
Water-soluble anions ($cmol_c$ kg^{-1})							
Cl^-	1.82	0.94	1.65	3.63	0.86	1.35	2.56
SO_4^{2-}	0.97	0.94	2.80	2.41	0.19	0.91	0.37
NO_3^-	0.02	0.01	0.01	0.01	0.01	0.01	0.00
Sum of cations	2.81	1.89	3.92	6.04	1.05	2.27	2.93

[a] ECw: Electrical conductivity of soil water-solution in the field
[b] ECe: Electrical conductivity of saturation paste

and OM5629 in experimental Plot A (2013 winter–spring cropping season; ECw from 3.9 to 4.5 dS m^{-1}) showed significantly higher yields per plant than that of the local variety KD (16.5 g and 15.3 g compared with 8.3 g, respectively). The other varieties showed low yields per plant (9.0–13.5 g). In the statistical analysis performed using all the values of three

Table 4.23 Climatic conditions from January 2012 to May 2013

Month	Year	Temperature			Humidity	Precipitation		Sunlight
		T_{ave}	T_{max}	T_{min}	(%)	(days)	(mm)	(h)
		(℃)						
January	2012	19.3	29.8	15.7	95	21	155.9	30
February		20.1	35.2	15.0	92	13	71.6	62
March		22.6	35.7	16.2	90	10	17.3	111
April		26.3	37.5	18.5	85	8	51.1	180
May		28.7	39.5	21.0	79	9	43.4	261
June		28.5	37.8	20.5	79	10	96.0	226
July		27.9	36.7	23.0	83	15	118.3	206
August		28.4	37.3	23.4	80	7	39.3	182
September		26.6	37.2	22.5	89	19	569	129
January	2013	19.8	28.0	14.5	91	15	47.3	90
February		22.9	32.0	16.5	90	6	27.0	144
March		24.6	35.0	16.8	88	10	64.0	159
April		26.2	39.7	20.4	86	13	25.5	159
May		28.0	37.0	22.0	80	13	87.5	205

T: Temperature; ave: average; max: maximum; min: minimum; h: hours

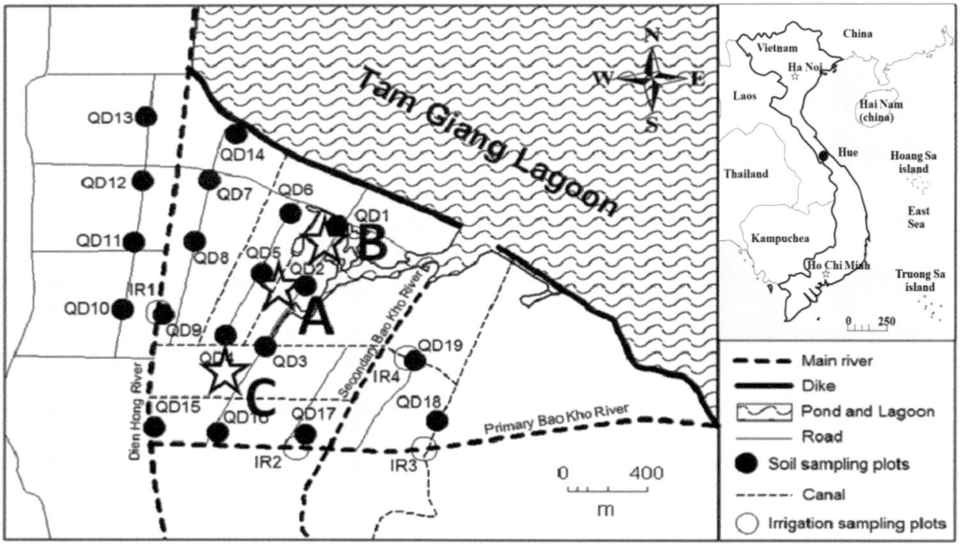

Fig. 4.35 Seasonal fluctuation of electrical conductivity of soil solution in experimental plots

Table 4.24 Performance of rice varieties, in terms of agronomic traits and yield, in experimental Plot A in winter–spring 2013

Traits	Plant height		Number of productive tillers		Panicle length		Panicle weight		Number of grains		Number of filled grains		Grain fertility		1,000-grain weight		Total biomass		Harvest index		Yield	
Varieties	(cm)		(plant^{-1})		(cm)		(g)		(panicle^{-1})		(panicle^{-1})		(%)		(g)		(g)		(%)		(g plant^{-1})	
MNR3	93.9	abc	7.3	ab	24.7	a	2.8	a	120.7	a	101.9	a	84.8	ab	27.9	ab	30.4	ab	54.4	a	16.5	a
OM5629	100.8	a	8.0	ab	22.2	ab	2.3	ab	109.1	ab	86.2	a	78.3	abcd	26.3	ab	30.5	a	49.9	abc	15.3	ab
OM2395	90.9	abc	7.6	ab	23.3	ab	2.3	ab	94.8	ab	82.4	a	86.0	a	27.7	ab	24.8	ab	54.5	a	13.5	abc
MNR4	92.0	abc	6.5	ab	22.6	ab	2.3	ab	114.3	ab	92.1	a	78.4	abcd	27.7	ab	24.3	ab	54.2	a	13.3	abc
AS996	87.9	bc	8.1	ab	22.6	ab	2.2	ab	94.5	ab	77.7	a	82.1	abcd	28.0	ab	24.9	ab	53.6	a	13.3	abc
CM2	90.7	abc	8.5	ab	20.9	bc	2.0	ab	91.0	b	65.2	a	73.1	bcd	31.1	a	25.4	ab	51.4	abc	13.0	abc
OM6L	84.6	c	8.7	ab	23.1	ab	2.2	ab	99.9	ab	87.9	a	87.4	a	25.8	b	24.4	ab	53.2	ab	12.9	abc
OM5900	87.5	bc	8.9	a	19.2	c	1.7	b	68.5	b	60.9	a	88.9	a	28.7	ab	24.1	ab	51.0	abc	12.4	abc
OM4900	94.7	abc	6.6	ab	22.6	ab	2.5	ab	119.1	a	100.8	a	83.6	abc	25.7	b	24.3	ab	50.4	abc	12.2	abc
OM2718	93.8	abc	7.4	ab	22.7	ab	2.3	ab	114.8	ab	83.9	a	71.7	d	27.1	ab	23.9	ab	49.4	abc	11.9	abc
OM8104	96.8	ab	6.9	ab	21.2	bc	2.2	ab	101.3	ab	80.7	a	79.1	abcd	27.7	ab	25.2	ab	43.9	abc	11.1	abc
OM9922	96.7	ab	6.1	ab	21.8	bc	2.0	ab	88.7	ab	71.7	a	80.4	abcd	28.1	ab	23.9	ab	42.2	c	10.1	bc
OMCS2000	92.1	abc	6.8	ab	22.0	abc	1.6	b	93.9	ab	65.8	a	71.8	cd	24.0	b	20.9	ab	42.4	bc	9.0	c
KD	93.1	abc	5.9	b	21.3	bc	1.9	ab	94.2	ab	76.5	a	80.6	abcd	25.4	b	20.1	b	41.2	c	8.3	c
Mean	92.5		7.4		22.1		2.2		100.3		81.0		80.4		27.2		24.8		49.4		12.3	
SD	4.2		1.0		1.3		0.3		14.2		12.5		5.5		1.7		2.8		4.9		2.2	
CV (%)	4.5		12.9		5.9		14.1		14.1		15.5		6.9		6.3		11.4		9.9		18.0	

Notes Values followed by different letters are significantly different (Tukey's test, $P < 0.1$)
CV: Coefficient of variation

experimental plots in the winter–spring cropping season, the five newly introduced rice varieties of MNR3, MNR4, OM4900, AS996, and OM2395 recorded significantly higher yields per plant, that is, 16.9, 16.2, 15.6, 15.5, and 15.4 g, respectively, than that of the local variety KD, 10.9 g (data not shown). In contrast, the analysis including all the experimental plots did not give any statistically significant difference for yields per plant between the 13 newly introduced rice varieties and local variety KD in the summer–autumn cropping season (data not shown). The yield per plant of the 14 studied rice varieties in the other six experimental plots (experimental Plot B in the winter–spring cropping 2012, three experimental Plots A, B, and C in the summer–autumn cropping 2012, and two experimental Plots B and C in the winter–spring cropping 2013) was not significantly different from that of the local variety KD (data not shown).

Comparisons of Agronomic Traits and Yield per Plant among Three Experimental Plots (A, B, and C) in Each Cropping Season Table 4.25 shows the mean performance of the 14 rice varieties, in terms of ten main agronomic traits and yield per plant, compared among the three Plots A, B, and C in each cropping season. In the 2012 summer–autumn cropping season, the mean yield per plant was similar among the three plots (ranged 12.4–13.0 g). In the 2013 winter–spring cropping season, the highest mean yield per plant was in Plot B (15.5 g) and the lowest was in Plot A (12.3 g). The mean plant height of the 14 studied rice varieties was less than 110 cm, ranging from 89.8 to 95.9 cm in the seven experiments. The mean grain fertility of the 14 varieties in Plots A, B, and C ranged from 77.7 to 82.6% in the 2013 winter–spring cropping season and from 59.3 to 72.3% in the summer–autumn cropping season. The mean values of the other agronomic traits of rice in the seven experiments were low.

Comparisons of Agronomic Traits and Yield per Plant for Each Experimental Plot between Two Seasons (Summer–Autumn 2012 and Winter–Spring 2013) The mean values for the ten main agronomic traits and yield per plant for the 14 rice varieties in each plot between two seasons are shown in Table 4.25. The mean yield per plant across the 14 studied rice varieties in Plots A and C in the 2012 summer–autumn cropping season (13.0 and 12.3 g, respectively) were similar to those in the 2013 winter–spring cropping season (13.0 and 14.1 g, respectively). The mean yield per plant in Plot B in the 2012 summer–autumn cropping season (12.4 g) was lower than that in the 2013 winter–spring cropping season (15.5 g). Mean grain fertility in Plots A, B, and C was lower in the 2012 summer–autumn cropping season (66.0, 72.3, and 59.3%, respectively) than in the 2013 winter–spring cropping season (80.4, 82.3, and 77.7%, respectively). In each experiment, the four agronomic traits of panicle weight, grain fertility, 1,000-grain weight, and harvest index were higher in the 2013 winter–spring cropping season than in the 2012 summer–autumn cropping season. In contrast, the four traits, i.e., number of panicles per plant, panicle length, number of grains per panicle, and total biomass, were lower in the winter–spring cropping season than in the summer–autumn cropping season. The number of filled grains per panicle did not differ between the two cropping seasons.

Correlations between Agronomic Traits and Yield The correlations between each of the ten main agronomic traits and yield per plant, based on data from the seven field experiments, are shown in Table 4.26. Yield per plant or hill was positively correlated with five traits: the number of productive tillers, panicle weight, number of filled grains, total biomass, and harvest index. The number of grains per panicle and grain fertility were also positively correlated with yield per plant or hill, but the correlations were weaker ($r = 0.2$ to 0.4), except for grain fertility in experimental Plot C ($r = 0.5^{***}$ in the 2012 summer–autumn cropping season). Plant height, panicle length, and 1,000-grain weight values were not significantly correlated with yield per plant.

Table 4.25 Comparisons of agronomic traits and yield of rice varieties among three experimental plots (A, B, and C) and between two cropping seasons (summer–autumn 2012 and winter–spring 2013)

Plots	Plant height			Number of productive tillers			Panicle length			Panicle weight			Number of grains			Number of filled grains			Grain fertility			1,000-grain weight			Total biomass			Harvest index			Yield		
	(cm)			(plant^{-1})			(cm)			(g)			(panicle^{-1})			(panicle^{-1})			(%)			(g)			(g)			(%)			(g plant^{-1})		
Summer–autumn 2012																																	
A	95.9	a	A	8.3	b	A	23.5	a	A	1.9	b	B	127.4	a	A	82.2	b	A	66.0	b	B	23.8	a	B	30.4	b	A	42.9	a	B	13.0	a	A
B	91.6	b	A	10.6	a	A	23.3	a	A	2.2	a	B	131.5	a	A	93.7	a	A	72.3	a	B	23.5	a	B	35.3	a	A	35.1	b	B	12.4	a	B
C	89.8	b	B	11.2	a	A	22.7	b	A	1.5	c	B	123.5	a	A	72.9	c	A	59.3	c	B	21.1	b	B	34.4	a	A	37.0	b	B	13.0	a	A
Winter–spring 2013																																	
A	92.5	a	B	7.4	c	B	22.1	a	B	2.2	b	A	100.3	b	B	81.0	b	A	80.4	a	A	27.2	a	A	24.8	c	B	49.4	a	A	12.3	c	A
B	91.0	b	A	8.2	b	B	21.9	a b	B	2.3	a	A	108.2	a	B	89.0	a	A	82.3	a	A	26.4	b	A	32.6	a	B	47.6	a	A	15.5	a	A
C	92.5	a	A	8.7	a	B	21.5	b	B	2.0	b	A	100.6	b	B	78.5	b	A	77.7	b	A	24.2	c	A	28.5	b	B	49.2	a	A	14.1	b	A

Notes
Different lowercase letters after values indicate significant difference among experimental plots (A, B, and C) in each cropping season (Tukey's test, $P<0.1$)
Different uppercase letters after values indicate significant difference between 2012 summer–autumn season and 2013 winter–spring season (Fisher's Least Significance Difference Test, $P<0.1$)

Table 4.26 Correlation matrices between agronomic traits and yield, calculated using data from seven field experiments in two cropping seasons (winter–spring 2012, and summer–autumn 2013)

Cropping season	Plot	Plant height	Number of productive tillers	Panicle length	Panicle weight	Number of grains	Number of filled grains	Grain fertility	1,000-grain weight	Total biomass	Harvest index
		(cm)	(plant^{-1})	(cm)	(g)	(panicle^{-1})	(panicle^{-1})	(%)	(g)	(g)	(%)
Winter–spring 2012	B	0.2	0.6***	0.3	0.5***	0.3*	0.4**	0.4**	−0.1	0.8***	0.7***
Summer–autumn 2013	A	0.0	0.6***	0.5**	0.5***	0.3	0.4*	0.3	0.3	0.9***	0.8***
	B	0.2	0.7***	0.4*	0.5***	0.3*	0.4*	0.2	0.2	0.8***	0.6***
	C	0.2	0.5**	0.4**	0.5***	0.3	0.3*	0.2	0.2	0.9***	0.6***
Winter–spring 2013	A	−0.1	0.6***	−0.2	0.5***	0.2	0.5***	0.3	−0.1	0.8***	0.6***
	B	0.4*	0.4*	0.3	0.4**	0.3*	0.4**	0.2	−0.2	0.7***	0.6***
	C	0.6***	0.8***	0.3	0.7***	0.3	0.6***	0.5***	0.2	0.9***	0.7***

Note: *$P<0.05$; **$P<0.01$; ***$P<0.001$

4.5.5 Conclusion: Solutions for Mitigation of Unfavorable Effects of Salinity and Promotion of Rice Cropping

Based on these results, certain solutions are recommended to mitigate the unfavorable effects of soil salinization and promote rice cropping. A dike with sluice gates stretching along the lagoon was constructed from 1994 to 1995 to prevent saltwater intrusion from the lagoon onto land; however, certain sections of the dike have not been constructed because of financial constraints. Therefore, a complete dike system must be constructed along the lagoon to prevent seawater intrusion onto land completely. Second, because there are difficulties associated with drainage in the depression plots, it is necessary to construct adequate facilities to promote rapid drainage of water into the canals during and after flooding and irrigation in these plots. Third, because ECe and ECw increase from May to September, appropriate fresh irrigation water should be supplied frequently for rice fields to lower the ECe and ECw during this period. Farmers must maintain appropriate freshwater levels on the rice fields to compensate for the water lost by evaporation when air temperatures are high. In addition, the irrigation freshwater quantity in the rivers must be provided appropriately to push back the saltwater intrusion from the lagoon; thus, farmers can use fresh water for their fields. Fourth, since soil fertility was poor with low organic matter content and low pH, the application of organic fertilizers and lime should be repeated annually over a long period (Corwin and Lesch 2003). Finally, regarding improving rice breeding: farmers in the Quang Phuoc commune are cultivating a low-yielding local rice variety, KD. Hence, to increase rice yield and income of farmers, this local variety needs to be replaced by the six newly introduced MNR3, MNR4, OM4900, AS996, OM2395, and OM5629 rice varieties. However, the newly introduced OM5629 rice variety may also be suitable for cultivation under the condition of saline soils with ECw ≤ 4.5 dS m^{-1} in the winter–spring cropping season.

4.6 Characteristics of Home Garden and Its Improvement Through Vanilla Introduction in Central Vietnam[6]

4.6.1 Background

4.6.1.1 Farming at Highland: Agroforestry in Uplands

In Vietnam, the mountain areas account for three-quarters of the total natural area. They function as an important means of income for the local people since one-fifth of the population relies on the products from mountains. The area is also home to many ethnic minority groups as well as the majority of poor households in Vietnam. Highland people's livelihood is mainly based on agroforestry activities. Besides, the farming terrain is rather complex—mainly consisting of sloping land. Moreover, highland farming is generally divided by many hills and steams.

Agroforestry, the cultivation of the trees and agricultural crops in intimate combination, is an ancient practice that farmers have used throughout the world. In Vietnam, forms of agroforestry are widely used in many ecological zones as a mode of production (often for subsistence), primarily in the upland areas. Among others, these systems include traditional shifting cultivation typical for ethnic minority groups and home gardens seen in many rural landscapes. It has been developed over many generations and is still common with local people living near or in the forest (Vien 2010: 34–46).

Shifting cultivation regenerates the forest by using slash-and-burn cultivation. This long-standing form of agroforestry overcomes the negative effects of continuous cultivation and creates favorable conditions to restore soil

[6] This section is based on the article by Vu Tuan Minh, one of the authors of this section and has been deposited in the repository of Kyoto University at https://repository.kulib.kyoto-u.ac.jp/dspace/handle/2433/202751. It does not violate any policy of the repository.

nutrients. Farmers rotate swidden fields based on estimated time. This practice has worked sustainably for many years, and the key to its sustainability is the period of time the land lies under fallow. Farmers often collect products from their fallow land such as vegetables, firewood, fruits, foodstuff, and medicinal plants. When population density is sparse and land under this practice is large, the land can remain fallow for a long time. However, under current circumstances, with dense population as well as land allocation policy limits on the area of cultivated land, farmers in many places grow crops or trees during the fallow period as well, which improve the soil nutrients. Additionally, this practice not only reduces the fallow period but also provides firewood to the farmers. The cultivation period is 2 to 5 years depending on upon the rotation plots, total swidden areas, and growth capacity of fertilized soil crops. Besides the shifting cultivation, forest and rice terrace system have existed for many years in central mountainous regions of Vietnam. This system reduces soil erosion and facilitates the water storage for irrigation. It has an important role in preserving irrigation water as it regulates the water supply to rice terraces, prevents landslides, and offers forest products to the farmers (Vien 2010: 34–46).

4.6.1.2 Characteristics of Central Vietnam and Aim of This Section

Central Vietnam is located in a disaster-prone region as floods, typhoons, and other natural disasters often seriously affect agricultural production and the livelihood of local households. Disaster resilience for vulnerable communities in the region can be enhanced through various livelihood activities such as livestock and crop production. Crop production plays an important role in terms of food production and income generation for local farmers. It helps to maintain food security and socio-economic stability. In Central Vietnam, crop production provides more than 67% of all agricultural annual income, with perennial crops accounting for over 7.0%, vegetable production accounting for approximately 16.7%, and annual food crops (rice, cassava, corn, sweet potato) accounting for 76.3% (General Statistical Office of Vietnam 2012). Similarly, in Thua Thien Hue province in Central Vietnam, upland farmers rely mainly on farming for a living and off-farm activities are much less developed. They are dependent on locally available food sources. Generally, food supplies are seasonal and often inadequate to meet the needs of livelihood maintenance throughout the year.

According to TTHPPC (2005: 37–38), 87% of the communities in the mountains of Thua Thien Hue province are dominated by ethnic minorities who have fewer income opportunities than lowlanders. In Vietnam, approximately 83% of the people living in mountainous regions depend on agriculture for their livelihood, and they face increasing difficulties in meeting their daily subsistence needs (Siem and Phien 1999: 9–15). Hillside villages that depend on a combination of paddy and shifting cultivation confront problems of food security, as more than two-thirds of households do not produce sufficient grain to meet their annual needs (Hong and An 2009: 24–30).

As rice production falls short of consumption, maize and cassava are required in order to cover basic nutritional needs (Kono and Rambo 2004). At the same time, the agro-resources on which the economic welfare of the mountain people depends have been severely degraded, with a consequent reduction in productivity per unit area (Hong and An 2009: 24–30). The green landscape in which the villages are located is characterized by large agroforestry sites, including communal village forests, paddy fields, maize fields, and small-scale home gardens (Vien 2003). Combinations of these land uses provide resources that can be used to improve the livelihoods of the local people.

We first analyze the current status of the home garden in terms of crop varieties, crop products, and their consumption, and the contribution of these products to the household's livelihood. Then, we examine the performance of vanilla and its suitability as an additional home garden crop in a mountainous village in Central Vietnam.

4.6.2 Our Study Area

Our study area, Hong Ha commune, is located in the upper Bo River Basin in Thua Thien Hue province, approximately 40 km from both Hue city and the A Luoi district center (Fig. 4.36). The Hong Ha Commune Office (2012) reported a typical example of the livelihood of ethnic minorities in this region. The median monthly income of farmers was 500,000–700,000 Vietnamese Dong (VND) (25–35 USD), providing limited cash for family needs. Although their livelihood is based on self-sufficient production of food, including rice, various vegetables, and forest resources, the products from home gardens offer a valuable source of income as well as various food crops. The commune consists of five villages with a total population of 1,517 and comprises four ethnic groups (Hong Ha Commune Office 2012). Each village harbors different ethnic groups. The Taoi group generally occupies the higher land, the Cotu group occupies the lower land, and several other groups are mixed in with them. All five villages, Can Tom, Can Sam, Pa Hy, Pa Rinh, and A Rom, are located along a valley and surrounded by undulating hills with elevations ranging 100–200 m (Hong and Van 2007: 13–15). These villages are located within 6 km, west-to-east, and 3 km, north-to-south, of each other in the foothills of the mountain region (An et al. 2001: 7–10). Can Tom village is located on top of a hill and has the highest elevation. Can Sam, Pa Hy, and Pa Rinh are located on the hill slopes at intermediate elevations (middle-slope villages). A Rom is located on the lowest part of the slopes (a foothill village).

Hong Ha has a tropical monsoon climate where the dry and wet seasons are quite separate. The dry season includes spring and summer, and runs from January to August, while the rainy season runs from September to January. Since prolonged heavy rainfall events occur from September to November, providing more than 70% of the total annual precipitation, floods are usually observed during this period (Hong and Van 2007: 13–15). The average annual rainfall is approximately 2,700 mm and the average temperature is approximately 23 °C (An et al. 2001: 7–10). The agricultural land is situated along the fluvial area and includes terraces, mainly for lowland rice, cassava, and maize, as well as areas on the hills where slash-and-burn cultivation of upland rice and cassava is carried out, and acacia hybrid plantations (*Acacia mangium* × *A. auriculiformis*) are established. The cultivated area covers approximately 239 ha, of which 28.5 ha are used for lowland and upland rice, 174 ha for cassava, and 19.6 ha for maize (Hong Ha Commune Office 2012).

4.6.3 Home Garden Practice

4.6.3.1 Definition and Potential Contribution to Livelihood

Farming systems in Vietnam comprise four land-use types that are closely linked: home gardens, the area used for annual food crop cultivation, fields that serve as communal farmland, and areas used for shifting cultivation (Hanoi University of Agriculture 2011). The home garden can be defined as a farming system that combines different physical, social, and economic functions on the land around the family home (Nguyen 1992: 2–12). It is part of the land used for cultivation and is usually fenced (VASS 2005: 26–27). Home gardens are sometimes called backyard or kitchen gardens. These gardens have an established tradition and offer great potential for improving household food security and alleviating micronutrient deficiencies. Gardening can enhance food security in several ways, most importantly through direct access to a diverse range of nutritionally rich food sources, an increase in purchasing power by allowing savings to be made on food bills, generating income from sales of garden products, and providing food during seasonally lean periods (FAO 2010). This type of garden has been traditionally used by villagers, and land ownership is transferred by inheritance or sale (Truong and Binh 1985: 73–101). Many types of home garden systems in Asia have been recorded in terms of their structure, combination, and function (Kumar and Nair 2004; Mohri

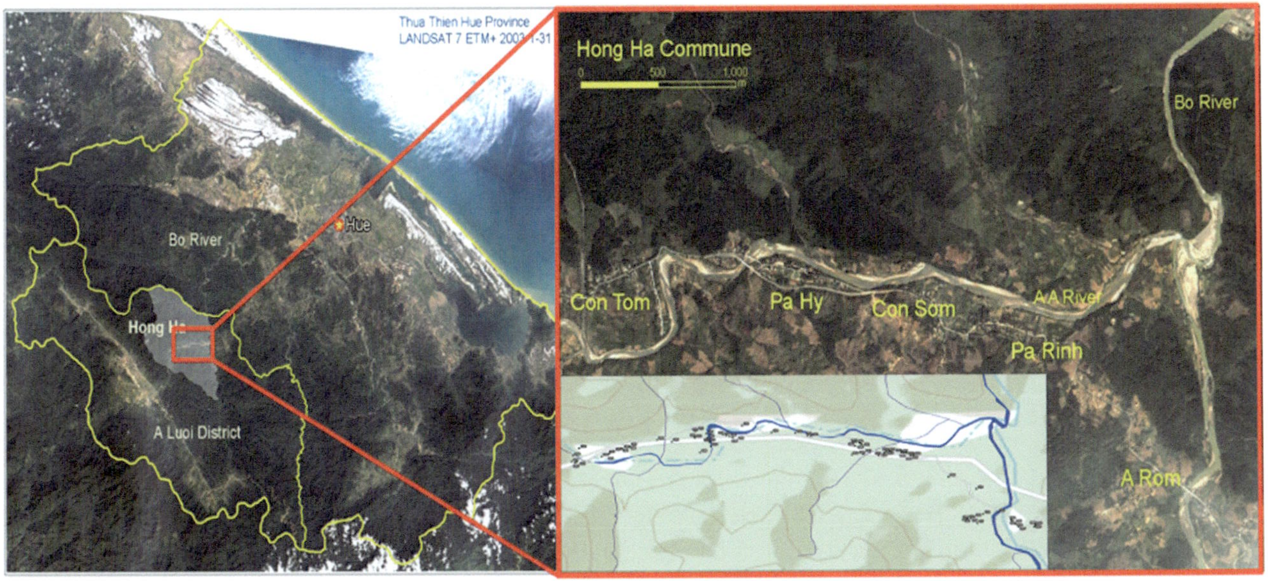

Fig. 4.36 The five village study sites in Hong Ha commune

et al. 2013). Indonesia, Sri Lanka, and Vietnam have home garden systems called Pekarangan, Kandyan, and VAC, respectively. These three systems differ in terms of the structure of their horizontal zoning and vertical stratification. While the Pekarangan and Kandyan represent complex horizontal zoning and vertical stratification systems based on height and a high diversity of species, resulting in intimate plant associations in a virtually closed canopy structure, the VAC system has a simpler vertical structure. VAC stands for Vuon, Ao, and Chuong in Vietnamese, which can be translated as "garden, pond, and livestock pen." It originated in the Red River delta and midlands of northern Vietnam and uses land for various agri-aquacultural activities in a domestic setting (Nguyen 1992: 2–12).

4.6.3.2 Brief Description of the Study Methods Used

We conducted a field survey in September 2012. The information on cultivation was collected from the statistical book provided by the People's Committee Office and the other data were collected from interviews. Among the five villages in the commune, 30% of each village's households were randomly selected for a questionnaire survey, using 95 out of the 309 households to collect detailed information on home garden practices such as location, size, cropping patterns, consumption, and income.

4.6.3.3 General Characteristic of Home Garden

All the households had home gardens around their houses. The home garden size in Hong Ha varied from 130 to 15,000 m^2 with a median size of approximately 1,000 m^2. In 90% of the households, the gardens were less than 5,000 m^2 in size (Table 4.27). Studies of home gardens in different ecological and geographical regions elsewhere in Vietnam have shown that the size ranges from approximately 500–3,200 m^2 in the northernmost areas and from 2,000 to 22,000 m^2 in the southernmost ones (Trinh et al. 2003).

Table 4.27 Distribution of home garden size in Hong Ha (Vu et al. 2015a)

Land size (m^2)	No. of households (n = 95)	Percentage (%)
100–1,000	51	54
1,000–5,000	34	36
5,000–10,000	7	7
10,000–15,000	3	3

Approximately 30% of the households had fishponds in the home garden. These were not only used for raising fish but also provided water for garden irrigation and household activities. The average pond size was approximately 22.4 m^2. The average pond size elsewhere in Vietnam, as calculated by Trinh et al. (2003), was 1,500 m^2 in the southernmost site and 220 m^2 in the northernmost site.

4.6.3.4 Vegetable and Staple Crop Cultivation

The natural and planted vegetation in home gardens consisted of vegetables (Table 4.28), staple crops (Table 4.29), fruit trees, and industrial crops (Table 4.30)—the specific mix being mainly based on their crop physiology and usage. Banana is also included in the category of fruit vegetables, as the villagers mostly use the flowers of planted bananas and obtain banana fruit to eat from naturally occurring plants in the forest, which are diploid banana species. Although bamboo, papaya, and jackfruit are perennial crops and sweet potato is a tuber crop, in this study they were all considered as vegetables since not only the mature shoots but also the young shoots, fruit, and tubers are typically consumed as vegetables. Leafy vegetables and fruit vegetables are relatively common in Hong Ha's gardens. Taro, hot pepper, and pumpkin were the main crops planted in 56%, 54%, and 52% of the gardens, respectively. Taro has also been reported as a major food crop in both Javanese and Kandyan home gardens (Mohri et al. 2013), while hot peppers are also grown in 75% of such gardens in Northeast Thailand (Miyagawa and Kochan 1990).

The farmers in Hong Ha commune have adopted a specific schedule for planting the various kinds of vegetables used for household

Table 4.28 Vegetables grown in home gardens and their percentage occurrence (Vu et al. 2015a)

No	Crop	Scientific name	Crop Calendar												(%)*	Purpose
			J	F	M	A	M	J	Jl	A	S	O	N	D		
Leafy vegetables																
1	Taro	*Colocasia esculenta* L.	■	■	■	■	■	■	■	■	■	■	■	■	56	Home
2	Brassica family	*Cruciferaceae*	■	■	■	■	■	■							46	Home & Sale
3	Amaranth	*Amaranthus* sp.	■	■	■	■	■	■	■						42	Home
4	Sweet leaf	*Sauropus androgynus* L.	■	■	■	■	■	■	■	■	■	■	■	■	36	Home
5	Water convolvulus	*Ipomoea aquatica* F.	■	■	■									■	33	Home & Sale
6	Sweet potato leaf	*Ipomoea batatas*	■	■	■	■	■	■	■	■	■	■	■	■	19	Home
7	Centella	*Centella asiatica* L.	■	■	■	■	■	■	■	■	■	■	■	■	15	Home
8	Ceylon spinach	*Basella alba* L.	■	■	■	■	■	■	■	■	■	■	■	■	9	Home
Fruit vegetables																
9	Hot pepper	*Capsicum annuum* L.; *C. frutescens*	■	■	■	■	■								54	Home
10	Pumpkin	*Cucurbita pepo* L.	■	■	■	■									52	Home
11	Papaya	*Carica papaya* L.	■	■	■	■	■	■	■	■	■	■	■	■	37	Home
12	Gourd	*Lagenaria vulgaris* L.	■	■	■	■	■	■	■						36	Home
13	Angled luffa	*Luffa cylindrical* L.	■	■	■	■	■	■	■	■	■	■			34	Home
14	Eggplant	*Solanum melongena*	■	■	■	■	■								20	Home
15	Banana flower	*Musa paradisiaca* L.	■	■	■	■	■	■	■	■	■	■	■	■	19	Home
16	Bitter gourd	*Momordica charantia* L.	■	■	■	■									12	Home
17	Chayote	*Sechium edule* Sw.				■	■	■	■	■					2	Home
18	Tomato	*Lycopersicum esculentum* Mill	■	■	■	■	■								1	Home
Flavor vegetables																
19	Lemon grass	*Cymbopogon citratus* Stapf.	■	■	■	■	■	■	■	■	■	■	■	■	51	Home
20	Piper lolot	*Piper lolot* L.	■	■	■	■	■	■	■	■	■	■	■	■	47	Home
21	Ginger	*Zingiber offcinale* Rosc.	■	■	■	■	■	■	■	■	■	■	■	■	23	Home
22	Turmeric	*Curcuma domestica* Var.	■	■	■	■	■	■	■	■	■	■	■	■	11	Home
23	Galangal	*Alpinia galangal* L.	■	■	■	■	■	■	■	■	■	■	■	■	2	Home
24	Coriander	*Coriandrum sativum* L.	■	■	■	■									2	Home
25	Basil	*Ocimum* spp.	■	■	■	■	■	■	■	■	■	■	■	■	2	Home
26	Onion, garlic	*Allium* spp.	■	■	■									■	1	Home
27	Water cress	*Enhydra fluctuans*	■	■	■	■	■	■	■	■	■	■	■	■	1	Home
Other vegetables																
28	Sweet potato	*Ipomoea batatas*	■	■	■	■	■	■	■	■	■	■	■	■	43	Home
29	Bamboo shoot	*Bambusa* spp.	■	■	■	■	■	■	■	■	■	■	■	■	39	Home & Sale
30	Green jackfruit	*Artocarpus heterophyllus* Lamk	■	■	■										38	Home
31	Winged yam	*Dioscorea alata*				■	■	■	■						2	Home

*Percentage of households growing each crop

consumption. Sixteen species used as food crops are grown year-round, such as taro, sweet leaf, papaya, lemongrass, *Piper lolot*, sweet potato, and bamboo shoots, and 14 species are cultivated from January till harvest. Thus, crops can be grown in the home gardens over a long

Table 4.29 Staple crops grown in home gardens and their percentage occurrence (Vu et al. 2015a)

No.	Crop	Scientific name	Crop Calendar												(%)*	Purpose
			J	F	M	A	M	J	Jl	A	S	O	N	D		
1	Cassava	*Manihotesculenta*	■	■	■	■	■	■	■	■	■			■	79	Home & Sale
2	Maize	*Zea mays* L.	■	■	■	■	■								36	Home
3	Sweet potato	*Impomoeabatatas* L.	■	■	■	■	■	■	■	■	■	■	■	■	12	Home

*Percentage of households growing each crop

period of time in order to maximize harvest. After harvest, 96% of these vegetables are used for home consumption and 4% for sale, including bamboo shoots, taro, brassicas, and papaya.

Staple crops include cassava, maize, and sweet potato tubers from the farms and home gardens (Table 4.29). Hybrid cassava (KM-94 and KM-98) and hybrid maize remain the most important cash crops in the home gardens of Hong Ha commune (An et al. 2001: 7–10). After these crops were introduced in some villages in 2000, they spread quickly by means of farmer migration throughout the commune. One of the most important benefits of cassava production is that the end-products become trading commodities, due to the presence of tapioca powder factories in the urban area approximately 40 km northeast of the commune. The farmers usually practice mixed planting of maize with sweet potato or *Centella* vegetables in part of the garden that is not shaded by tall trees. They transplant maize in January and February, and harvest it in April and May, while continuing to take care of their other crops. Maize is also commonly grown in the garden under rain-fed conditions with zero or minimal application of fertilizer.

4.6.3.5 Perennial Crop Cultivation

Various perennial crops consisting of 33 species of fruit trees and industrial crops are grown in home gardens (Table 4.30), both for cash sale (13%) and subsistence consumption (87%). Fruit crops such as jackfruit, mango, pomelo, orange, pineapple, and guava are generally grown close to the house on the flatter land, but are also often planted around fishponds. These crops are quite common and were found in over 50% of all home gardens surveyed. In addition, other newly introduced trees, such as *Wrightia*, acacia hybrids, and jatropha, are often planted behind the house. Near the base of these trees, many kinds of flavor vegetables are grown, such as ginger, turmeric, galangal, lemongrass, *P. lolot*, and basil. In approximately 64% of the households, small amounts of industrial crops are planted in the home garden, a little farther from the house on flat land. Seven industrial species for home consumption are grown. Tea and pepper were the most common industrial crops and were grown by 52 and 24% of the households studied, respectively.

Leaf litter is the main soil improvement agent used in the home garden, and the fallen stems and dried branches of many fruit trees are used as fuel. The practice of planting trees such as bamboo and tea around the home to serve as a fence is also common. These trees acted as a windbreak at the beginning of the rainy season when storms are often a problem, and provide good shade in the summer. Customarily, the perennial plants grown close to such fences are mainly used by only one household, although the neighbors can share the benefits provided by them.

4.6.3.6 Agronomic Practices and Structure of Home Gardens

At present, in Hong Ha commune, chemical fertilizers and pesticides are commonly used for crops and all the farmers understand their effectiveness. Approximately 99% of the farmers

Table 4.30 Perennial crops grown in home gardens and their percentage occurrence (Vu et al. 2015a)

No	Crop	Scientific name	Harvesting time												(%)*	Purpose
			J	F	M	A	M	J	Jl	A	S	O	N	D		
	Fruit trees															
1	Jackfruit	*Artocarpus heterophyllus* Lamk.				■	■	■	■						78	Home
2	Mango	*Mangifera indica*				■	■	■							62	Home
3	Pomelo	*Citrus grandis*								■	■	■			60	Home & Sale
4	Orange	*Citrus aurantium* spp.				■	■	■	■						59	Home & Sale
5	Pineapple	*Ananas comous* L.					■	■	■	■					53	Home
6	Guava	*Psidium guyava* L.					■	■	■	■			■	■	51	Home & Sale
7	Longan	*Euphoria longana* Lamk.				■	■								46	Home
8	Lemon	*Citrus limonia*		■	■	■	■	■	■	■	■	■	■		45	Home
9	Banana	*Musa* spp.	■	■	■	■	■	■	■	■	■	■	■	■	45	Home
10	Papaya	*Carica papaya* L.	■	■	■	■	■	■	■	■	■	■	■	■	37	Home
11	Custard-apple	*Annona muricata* L.					■	■	■						23	Home
12	Carambola	*Averrhoa carambola L.*				■	■	■	■	■	■	■	■		22	Home
13	Star-apple	*Chrysophyllum cainito* L.	■	■	■										22	Home & Sale
14	Dragon fruit	*Hylocereus undatus* Britt.							■	■					18	Home
15	Areca nut	*Areca catechu* L.									■	■	■		17	Home
16	Persimmon	*Diospyros kaki* Thun.								■	■	■			14	Home
17	Tangerine	*Citrus reticulata*					■	■							5	Home
18	Sapodila	*Achras sapota* L.					■	■							5	Home
19	Coconut	*Cocos nucifera* L.					■	■	■	■					5	Home
20	Muricate custard-apple	*Annona squamosa*						■	■	■					4	Home
21	Rambutan	*Nephelium lappaceum* L.					■	■							4	Home
22	Avocado	*Persea gratissima* Gaertn.						■	■	■					3	Home
23	Calabura	*Muntigia calabura* L.					■	■	■	■	■	■			2	Home
24	Mulberry	*Morus alba* L.							■	■	■				2	Home
25	Jambose	*Syzygium jambos* L.					■	■							2	Home
26	False mango	*Spondias cytherea*								■	■	■	■	■	1	Home
	Industrial crops															
27	Tea	*Camellia sinesis*	■	■	■	■	■	■	■	■	■	■	■	■	52	Home
28	Pepper	*Piper nigrum* L.							■	■	■				24	Home
29	Coffee	*Coffea* sp.						■	■	■					9	Home
30	Vanilla	*Vanilla planifolia*									■	■	■		6	Home & Sale
31	Rubber	*Hevea brasilliensis*				■	■								5	Home
32	Cinnamon	*Cinnamomum* L.				■	■	■	■						3	Home
33	Sugarcane	*Saccharum* L.			■	■	■	■	■						2	Home

*Percentage of households growing each crop

interviewed expressed a preference for using chemical fertilizer combined with organic fertilizer because this practice led to a higher crop yield. However, the survey results revealed that the organic compost produced in the households is mostly used on paddy fields. It is also applied in home gardens but only in combination with chemical fertilizers, and no farmer used only compost or chemical fertilizers. These survey results covered 15% of the total number of gardens in the commune, including 4% of gardens for vegetables only, 6% for fruit only, and 5% for either vegetables or fruit trees harvested in the dry season. Any chemical fertilizer remaining after application to paddy or maize fields is stored for later use. Approximately 93% of the farmers surveyed do not apply pesticides to crops in their gardens, which may result in lower overall productivity because of insect damage.

The harvest times for perennial crops in the home gardens indicate that their output tends to complement the harvest season for vegetables (Table 4.30). Twenty-two species of fruit trees are harvested in the dry season, when the high drought risk often means that several vegetable crops cannot provide a significant harvest. Therefore, the combination of perennial and annual crops in the home gardens enables families to still have access to fresh foods and supplement their otherwise low income throughout the year.

From an agronomic viewpoint, the advantages of mixed cropping noted in this study agree with the findings of previous studies. The land is used effectively in a multi-layered manner, with the upper layer occupied by large fruit trees and the lower layer by smaller fruit trees and other crops. Furthermore, vine crops such as peppers, dragon fruit, vanilla, and even certain cucurbit vegetables are able to coil up the stems of the taller crops, thereby making effective use of all possible space above ground (Tanaka and Watabe 1981).

Canopy stratification in the home gardens consists of three layers in relation to plant height (Fig. 4.37). The first layer includes plants ranging from 0.5 to 2 m in height, which account for 56% of the garden crops. The second layer includes trees with a height of 2–4 m. These account for the lowest percentage of garden crops, at only 12%. The third layer includes trees taller than 4 m. These accounted for the remaining 32%. The plants comprising the upper layer are fruit-bearing trees such as mango, jackfruit, longan, guava,

Fig. 4.37 Mixed multi-layered cropping in a home garden (Vu et al. 2015a)

Table 4.31 Household income derived from the home garden (Vu et al. 2015a)

Household category*	Percentage of households (%)	Average total income(VND/year)	Average income from garden (VND/year)	Average garden size (m^2)	Contribution of income (%)
Poor	28	4,800,000	2,400,000^a	1,100^a	50.0
Near poor	15	6,000,000	1,836,000^a	1,200^a	30.6
Medium	48	8,400.000	2,311,000^a	2,300^a	27.5
Fair	8	10,000.000	1,438,000^a	3,500^a	14.4

*Based on the Vietnamese government's old standard (2010)
The same letter "a" in a column indicates no statistical difference by Tukey at $p<0.05$

star apple, coconut, avocado, carambola, areca nut, and other tree species such as *Wrightia (muc)*, bamboo, and acacia. The second layer consists of fruit-bearing trees such as citrus (pomelo, lemon, orange, tangerine), papaya, banana, sapodilla, custard apple, muricate custard apple, and persimmon, and industrial crops such as coffee, cinnamon, tea, rubber, sugarcane, pepper, and vanilla (in conjunction with their support trees). Several leafy, fruit and flavor vegetables are grown under the canopies of these fruit trees.

The combination of multi-layered plant canopies and ponds commonly found in the home gardens facilitates a degree of climate regulation on a local scale because the dense vertical plant growth helps determine the microclimatic conditions by influencing ecological factors such as soil moisture and erodibility, solar radiation, wind speed, and ambient temperature (Gebauer 2005; Jose and Shanmugaratnam 1993; Millat-e-Mustafa et al. 1994; Salam et al. 1995).

4.6.3.7 Contribution of Home Garden Products to Household Income

The results of the study indicated that home gardens play a key role in providing nutritional and food security to households by ensuring a steady supply of annual and perennial crops. Annual crops contribute to nutrition in various forms, as both staple crops and seasonal crops (leafy, fruit, or flavor and tuber vegetables), but are less profitable, overall, while perennial crops play a pivotal role in providing nutritious fruit during the dry season and are also more profitable (e.g., jackfruit, mango, pomelo, orange, lemon, and banana).

As mentioned previously, the combination of annual and perennial crops in the home garden serves an important function in providing diversification of daily food products as well as economic commodities. According to the Vietnamese government's old[7] standard for farm income, the total income for farming households can be classified into four categories, according to the income level. The contribution from the home garden to the total income is highest in "poor" households, although the size of the home garden is also large in households defined as "medium" and "fair" (Table 4.31). This is because the latter benefits from having varied sources of income, such as small-scale businesses, pensions, and activities carried out in the forest or shifting cultivation areas. These results indicate that the contribution of home garden is most significant for poor households.

As seen in the commune database, the annual income of farmers in Hong Ha commune is not sufficient to meet family needs. Paddy fields provide food enough for 8 to 10 months of home consumption, shifting cultivation areas and reforestation areas growing foods such as cassava, maize, and non-timber forest products provide food enough for 5 months of home consumption, and almost all the green fresh foods required are supplied by home gardens, especially in the off-season for the aforementioned crops. This type of land use basically resulted from the living and planting needs of individual families, and the actual practices involved vary widely

[7] Vietnam Prime Minister, 2010. Decision 09/2010/QD-TTg on poverty line 2011–2015.

from household to household, depending on their physical and economic situations. Farmers with sufficient resources can make full use of different types of land for diversified production, while others depend mainly on home garden practices because of the limited land area and lack of labor.

4.6.4 Introduction of Vanilla

4.6.4.1 Vanilla as a New Crop in Home Gardens in Central Vietnam

In Central Vietnam, farmers grow mainly fruit trees and annual crops (except cereals) in their home gardens. Many farmers plant crops, but this not done intensively in terms of labor input because they have not yet found suitably valuable crops to plant in their gardens. Some new crops have been introduced to Hong Ha farmers, such as coffee (2002), pepper (2003), and dragon fruit (2003) but they have not been very successful, despite being valuable cash crops in other places such as A Luoi and Nam Dong town, especially in the Central Highland Region.

As part of trials carried out for livelihood improvement in the community, a team from Hue University of Agriculture and Forestry and Kyoto University introduced vanilla (*Vanilla* spp.) to Hong Ha in 2007. Vanilla is easily planted and cultivated. It needs no chemical fertilizers and does not harm the host trees to which it clings for support, which means that it does not need much space to grow and can increase the production potential per unit area. Women, children, and the elderly can care for the plants. Vanilla is the world's third most expensive spice after saffron and cardamom because its unique oil is a valuable component in many processed foods (such as cakes, candies, ice cream, tea, and wine), medicine, and other products (Dexin 2009: 2–3). Approximately 6 kg of green fruit produces 1 kg of processed pods (Parthasarathy et al. 2005: 2–10). The price of processed vanilla pods was 200 USD/kg in 2001. When crops failed in 2003, the price rose to 500 USD/kg in 2004 (Sharon 2008: 7–9). During 2005–2007, prices dropped markedly to 25 USD/kg and fell as low as 7.5 USD/kg before recovering to 12 USD/kg in recent years (Fuavao 2012). Because farmers can get 300–600 kg of processed vanilla fruit per hectare, with careful management, vanilla crops can yield 3,600–7,200 USD/ha.

4.6.4.2 Brief Description of Methods

In 2007, vanilla cuttings were brought to Hong Ha commune 1 week after removal from their parent plants in Central Sulawesi, Indonesia. A group of eight volunteer farmers with experience in home garden management was organized under the supervision of researchers and the leaders from four villages belonging to the commune. These villages represented three different topographies: upper-slope hillside, mid-slope hillside, and lower-slope hillside. In the initial meeting, we described the characteristics of the vanilla plant and how to plant it. A total of 60 cuttings were distributed (5–10 plants per farmer), and their planting was demonstrated and supervised in each farmer's garden (Table 4.32).

4.6.4.3 Number of Vanilla Cuttings Transplanted

During the 5-year study, the number of host trees in the 8 farmers increased from 59 in 2007 to 164 in 2013 (Table 4.32). In Hong Ha, vanilla vines can be easily planted alongside host trees in August and September. The growers choose vanilla plants that were healthy and productive and usually trim them after flowering is complete. They also found that trimming encouraged the growth of new shoots, rejuvenated the plants, and improved overall health. The cuttings were re-planted around host trees and the growers could also sell the cuttings produced by trimming to other farmers from surrounding areas.

The cuttings used were approximately 30–50 cm long to ensure that the propagated vine had 4–6 nodes. The cuttings were left in the shade for several days to allow the wounds to heal and fungal infection was prevented by application of a 0.5% lime solution. These practices were adapted from local techniques used for growing orchid species. Previous recommendations specified that cuttings should be

Table 4.32 Pod grading in vanilla-growing households from 2009 to 2012 (Vu et al. 2015b)

Household	Year	Total weight (g)	Total fruit (pods)	Highest quality (pods)	Intermediate quality (pods)	Lowest quality (pods)
QT	2009	35	21	6 (29%)	15 (71%)	0 (0%)
TO	2010	40	40	31 (78%)	9 (22%)	0 (0%)
DE		55	59	26 (44%)	27 (46%)	6 (10%)
TO	2011	30	17	10 (59%)	7 (41%)	0 (0%)
DE		160	133	38 (29%)	80 (60%)	15 (11%)
NA		40	14	7 (50%)	5 (36%)	2 (18%)
OI		150	89	18 (20%)	71 (80%)	0 (0%)
TA		180	123	31 (25%)	90 (73%)	2 (2%)
TO	2012	140	52	48 (92%)	4 (8%)	0 (0%)
Total		830	548	215	308	25

Highest quality: >15 cm; Intermediate quality: 10–15 cm; Lowest quality: <10 cm

approximately 60–80 cm long and be treated with 3% copper oxychloride to prevent fungal infection (Alexander et al. 2009; Parthasarathy et al. 2005: 2–10). Finally, seedlings were transplanted next to host trees in the garden. Dry trash or dry grass was heaped around the base of the host trees to prevent waterlogging and the plants were watered. Adoption of these practices also allowed the growers to manage the input of organic matter and produce compost for use as a fertilizer in their garden.

4.6.4.4 Vanilla Performance: Growth and Physiology

After the flowering season was complete, the growers employed the following training technique to facilitate further vegetative growth. When the vines attained a height of 1–1.5 m, the shoot was trained downward very close to the ground (within 20–30 cm), then covered with soil mixed with organic matter (trash and litter) to encourage rooting. The remaining shoot length was coiled back up again so that it could reattach to host trees. Support trees were pruned at a convenient height to allow easy training of the vines. In this manner, the vanilla plant was continually cut, transplanted, and encouraged to expand into the space available in the home garden.

In the first year after planting, vanilla plant growth did not show a significant increase in terms of either vine length or the number of nodes because the growth rate was still low. In addition, farmer skill was still limited as this was the first time growing vanilla. By the third year after planting, the growers had learned to choose suitable host trees and their management skills had improved, resulting in improved growth in terms of both in vine length and node number. After 3 years, the vines averaged 398.5 cm in length, with 64.1 leaves and a node length of 6.2 cm. This increase in length was very promising and is comparable with figures reported by Sujatha and Bhat (2010) from India during peak growth.

Sujatha and Bhat (2010) noted that irrigation and rainfall are the most important factors affecting vanilla growth during the summer because vanilla can absorb water through both aerial and terrestrial roots—depending on the ambient humidity and the presence of soil water. During summer, vanilla plants without shade normally dry out and wilt. The canopy trees in the garden are thus important for managing shade and moisture. Growth improved in the second and third years as the growers began to properly manage the amount of shade required.

4.6.4.5 Vanilla Performance: Flowering and Fruiting

In Hong Ha commune, the flowering season lasts from late January to April, with most flowering occurring over a 1–2 month period. During this season, both inflorescence buds and new vegetative buds are produced. An inflorescence bud can only be clearly identified with new vine buds when the tip splits into two or three oval parts approximately 10 days after bud initiation. The vines produced 81% inflorescence buds and 19% vine buds during the 2010 season. This flower percentage greatly surpassed the figure of 27.2% reported by Havkin-Frenkll and Belanger (2010: 16–32), and by the fifth year after planting this figure had reached 97.0%. It was also reported that heavy flowering one year was generally followed by reduced flowering the following year.

In the household gardens studied, the time taken from initiation of the inflorescence bud to the first flower took an average of 14.3 days, with another 3 days to bear the first fruit. Vanilla cannot self-pollinate because the rostellum prevents the stigma from coming into direct contact with pollen grains (Rain 2004: 6–8). Flowering and pollination usually continue for approximately 3–5 weeks, with the best time for pollination being from 9 am to 10:30 am.

Although the leaves yellowed slightly in summer when shading was reduced (20–30%), the plants still yielded fruit. Based on interviews, 30–45 L of water per clump, spread out over three waterings per week, was good practice. In general, vanilla plants require a minimum of approximately 15 L/clump/week, which can be provided by family members on a flexible schedule.

Even though each bunch of flowers produced 14–15 flowers, on average only five flowers bore fruit after artificial pollination. Because each plant required a lot of nutrients and water in the dry season, plant growth was often stunted by the end of the flowering season. These results were similar to those reported by Rain (2004: 6–8), who noted that pollination accounts for approximately one-third of the required labor input per plant. Some growers stated that the investment of skill required for hand pollination was justified when this ratio of pollinated flowers was maintained. Fruit took 8–9 months to mature. The total number of fruits per bunch varied greatly, from 0.5 to 9.2, resulting in fruit set rates of 3.6–61.7%. This variation occurred because some growers lacked the skills to estimate pollen maturation time accurately and to manipulate the rostellum in order to reveal the tip of the stigma.

The green fruit of marketable pods can attain lengths of 10–20 cm and become yellow as it matures. In 2010, marketable green fruit averaged 12.3 cm in length and 0.98 cm in width 2 months after pollination. The fruit grew considerably between 20 and 40 days after pollination. The results observed did not differ significantly among growers.

4.6.4.6 Vanilla Performance: Fruit Harvesting, Curing, and Pod Grading

Vanilla pods were ready for harvest 8–9 months after flowering. Generally, harvesting in the farms studied began in November and ended in December. Because of the variation in flower bud initiation, there was no uniformity in fruit development or pod maturity, even within the same bunch, and each pod was harvested individually based on its color. Immature dark green pods were not harvested. A slight yellow discoloration at the fruit tip indicated maturity. Harvested fruit has no aroma and is worthless in the retail market. To help develop the vanilla aroma and produce a marketable product, curing should commence within a week of harvest.

The farmers used methods based on locally available materials such as water tanks for killing, bamboo baskets for drying, and wooden boxes or cartons for sweating, and they made use of their own experience with regard to water temperature and conditioning time. Green vanilla pods are flavorless and contain little vanillin, earning only 5–10 USD per kilogram, while cured pods are worth as much as 45–50 USD per kilogram. During processing, the fresh pods lose up to 75% of their original weight. The average weight of each cured pod from the farms studied was approximately 1.5 g.

The commercial value of mature fruit is determined by pod length. Due to the relatively limited range of cultivation techniques available to local growers and because vanilla production is still in the early stages of development in Hong Ha commune, in order to evaluate the quality of the fruit after processing, we classified the pods into three categories. If the pod was greater than 15 cm in length and had a uniform dark color, it was categorized as "highest quality", 10–15 cm long brown pods were considered "intermediate quality," and those less than 10 cm in length were considered "lowest quality." The results revealed that both the total amount of cured pods produced and their quality varied from year to year (Table 4.32). Most cured pods were of intermediate quality (308 pods), followed by highest quality (215 pods) and lowest quality (25 pods). The main reason given by growers for this variation was that pollinating *all* the flowers in a bunch would produce variable fruit quality. Therefore, pollinating only 40–50% of flowers per bunch was recommended.

Although there are certain standard methods used for curing, they may need modification in accordance with local and seasonal conditions such as bean size and quality, weather, and available facilities, and while curing may be done on either a household or industrial scale, suitable space, equipment, and human resources are essential. Curing of green fruits should commence within a week of harvest. Matured, light green, or yellow-colored fruits should be harvested individually, avoiding broom harvesting (in which case green fruit and mature fruit are harvested together). The four steps of vanilla pod processing, which were successfully demonstrated by the local farmers, are as follows (Fig. 4.38):

1. Killing: Graded beans are transferred to a bamboo basket and immersed in hot water at a temperature of 65–70 °C for the required period.
2. Sweating: The treated beans are then transferred immediately to a cardboard box/

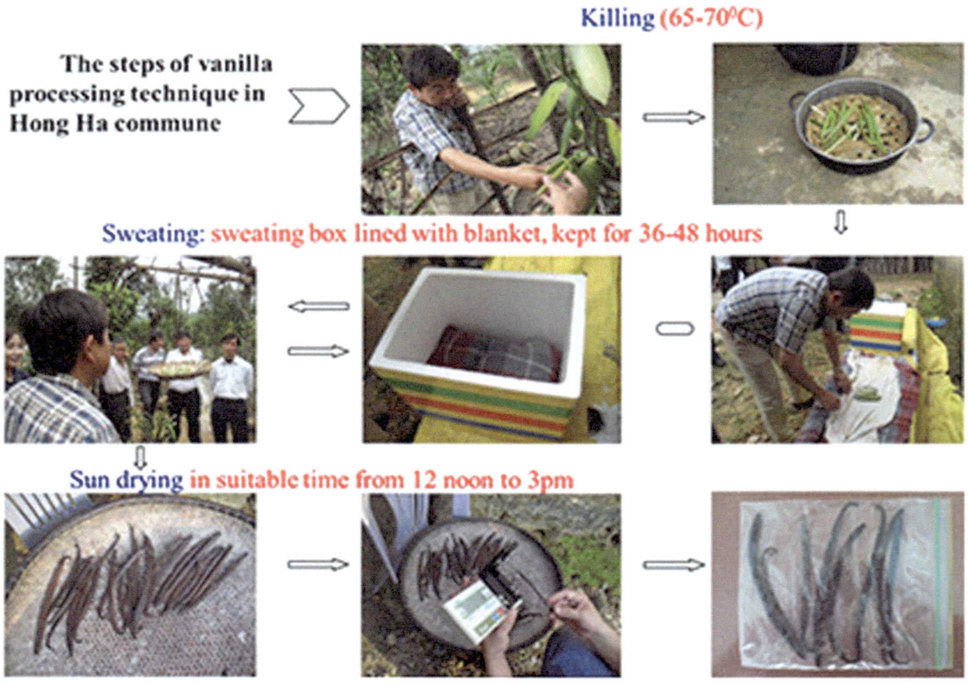

Fig. 4.38 Production process for cured vanilla pods

wooden box lined with a blanket, for sweating, and kept in there for 36–48 h at an initial temperature of 48–50 °C. By then, the beans will have attained a light brown color and started to give off the characteristic vanilla aroma.

3. Sun drying: The beans are then spread out in the hot sun (from 12 noon to 3 pm) on a clean black blanket placed over a wooden frame. The temperature of the pod should then rise to 50 °C. Later on, the bundles are transferred back to the sweating box and further sun drying and sweating is carried out, depending on the grade. The pods should be examined every day during sun drying and conditioning in order to prevent infection.
4. Conditioning (slow drying): The next step involves spreading out the vanilla beans on racks kept in a well-ventilated room maintained at a temperature of approximately 35 °C and relative humidity of 70%. To ensure a good aroma, conditioning should not be carried out where there is smoke from kitchens, ovens, or cooking fires.

4.7 Transition of Human and Livestock Excreta Use and Environmental Impact in Suburban Hanoi

4.7.1 Introduction

The administrative boundary of Hanoi, the capital of Vietnam, was expanded in 2008. This expansion added 15 suburban districts, tripling its size (33,000 km^2) and doubling its population (6.5 million) (HSO 2010). Recently, the suburban areas of the city have faced significant changes, especially owing to rapid modernization, which has affected waste and wastewater management as well.

Human and livestock excreta are rich in nutrients. It has long been a practice to use these valuable nutrients for agriculture. By doing so, people have managed human waste and wastewater as well as livestock waste within their communities through agro-sanitation systems. Agro-sanitation systems managed the waste and wastewater in a closed loop between crop and livestock subsystem. Human and livestock excreta, a natural source of nutrients such as nitrogen (N) and phosphorus (P), were intensively applied for crop yield and soil fertility in the early days of agriculture.

While there are advantages of nutrients in human waste for agriculture, there are certain disadvantages, including environmental pollution such as groundwater pollution by N and eutrophication induced by N and P. With advances in agro-technology, chemical fertilizers, a synthetic source of nutrients, have become common in use. With higher nutrient content, faster nutrient release, and convenience of use, chemical fertilizers have become a more preferable source of nutrients than human and livestock excreta. Chemical fertilizer consumption has been dramatically increasing (Heffer and Prud'homme 2010). As a result, the utilization of human and livestock excreta for agriculture has been decreasing (Sy et al. 2005).

While the use of human and livestock excreta for agriculture decreased (Harada et al. 2010a), some human and livestock excreta have been applied to fishponds for fish farming. With the increasing use of livestock waste for fish farming, the role of fishponds has been found to be a platform for conveyance of waste and wastewater of farming systems to the environment. In this context, the traditional agro-sanitation system has been transformed to an agro-fishery-sanitation system. In this system, a large proportion of nutrients is potentially trapped and accumulated in pond sediment (Boyd 1995; Nhan et al. 2008). However, increasing input of nutrients into fishponds may cause significant discharge of nutrients from fishponds to water bodies (Boyd 1995).

Thus, with economic growth and rapid urbanization, changes in farming systems and their waste and wastewater management have significantly contributed to the contamination of groundwater and rivers. To deal with these issues, it is necessary to understand how waste and wastewater management have changed and how they affect the nutrient balance in the farming system in order to consider a proper nutrient management strategy.

Material flow analysis (i.e., MFA) is a systematic tool used to understand the flows and stocks of materials within a system defined at a particular point in space and time (Brunner and Rechbenger 2004). It has been applied to understand nutrient balance, for example, in Thailand (Briggs and Fvnge-Smith 2008; Buathong et al. 2013), China (Chen et al. 2008), and Vietnam (Harada et al. 2010b; Montangero et al. 2007; Nga et al. 2011). To address the challenges and seek solutions regarding waste and wastewater management in sur-urban Vietnam, this work examined the historical changes on waste and wastewater management manners in suburban Hanoi by developing material flow on N and P flows based on Pham et al. (2015) and Pham et al. (2017).

4.7.2 Study Site

The study site is Trai hamlet in Van Tu commune, Phu Xuyen district, Hanoi, located on the bank of the Nhue river, 40 km south of the center of Hanoi (Fig. 4.39). It has a population of 800 people in an area of 56.1 ha, with a typical land-use pattern of suburban areas. Farming is the major occupation of the residents and rice is the dominant crop. The area of paddy fields, which is located beside the residential area and has not changed from 1980 to 2010, covers more than 90% of the area (Fig. 4.40).

There is no sewerage system or a centralized wastewater treatment facility that receives the local wastewater in the area. Local people in the area have access to three types of sanitation facilities: First, a flush toilet that directly discharges human waste to the environment. Second, a septic tank, which is an underground tank receiving wastewater from water flush toilet, settling solid matter and discharging the supernatant. Third, a dry toilet, which is a toilet without water. Unlike flush toilets, human waste is mixed with wood materials and/or ash so that the waste is decomposed. The residue taken from a dry toilet can be used as compost in agriculture. The dry toilet is mainly installed in places where tap water cannot be made available or where there is no sewerage system.

Like elsewhere in suburban areas of Hanoi, local people use fecal matter from a dry toilet for agriculture (Fig. 4.41). The fecal matter derived from human and livestock excreta used for agriculture is produced and stored in each household. The part beneath the dry toilet is a chamber for storing fecal matter derived from human excreta. Separately, some households store livestock excreta in their house, and when the human fecal chamber is full, the human excreta is mixed with the livestock excreta in a storage tank.

All the fecal matter from a dry toilet is not used for agriculture. Some of it is discharged into fishponds and/or open canals, which then ends up in Nhue river (Fig. 4.42) together with untreated graywater, which is domestic wastewater excluding wastewater from water flush toilets. Despite pollution and eutrophication, the river is the main irrigation source for paddy fields in this area.

4.7.3 Historical Changes in Human Excreta Management[8]

To understand waste and wastewater management and relevant material flows in the hamlet, this study conducted an interview survey of 100 households in this study site. Based on these interviews and previous studies, a material flow of N and P with focus on paddy fields was developed, as the paddy field is an essential platform for considering the impact on the environment by the nutrients.

Let us focus on the changes on the management of nutrients derived from human excreta to understand how much and where they flow. Figure 4.43 shows the historical changes in nutrient amount derived from human excreta. As shown in the figure, the total amount of N and P

[8] The subsections from 4.7.3 to 4.7.8 are based on the article by Pham Huong Giang, one of the authors of this section and have been deposited in the repository of Kyoto University at https://repository.kulib.kyoto-u.ac.jp/dspace/handle/2433/199483. These do not violate any policy of the repository.

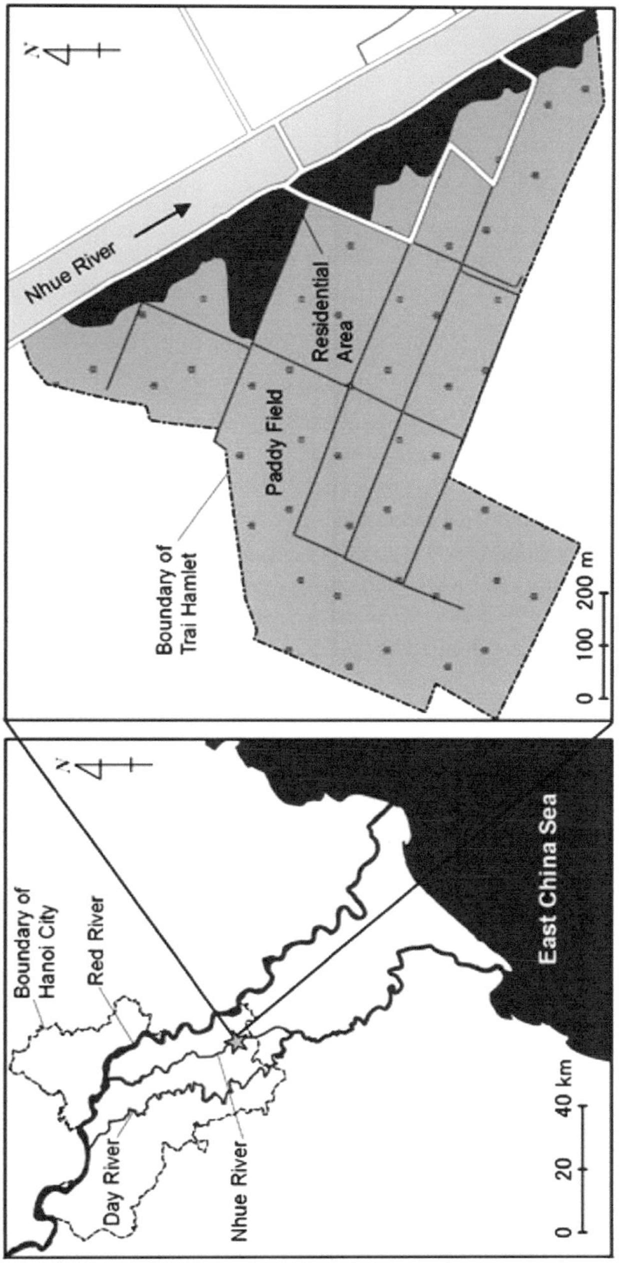

Fig. 4.39 Map of the study site

(a) Paddy fields (b) Residential area

Fig. 4.40 Paddy fields and residential area in Trai hamlet. Photo by Pham Huong Giang

from human excreta increased 1.7 times for both N and P from 1980 to 2010 due to 1.6 times population growth in this study site. However, the proportion of N and P from human excreta entering the paddy fields sharply decreased after 2000. Application of human excreta to paddy fields has a long history in this area. In 1980, 97% of N and P from human excreta were applied to paddy fields. These ratios decreased to 50% in case of N and 49% in case of P in 2010. Instead of being used to intensively apply to paddy fields, human excreta flowed to septic tanks and to fishponds.

These transitions of N and P load may be owing to economic modernization. Dry toilets, from which human excreta can be used for agriculture, have been gradually replaced by flush toilets. In 2010, 44% of households using flush toilets stopped applying human excreta to agriculture; instead, they discharged excreta to septic tanks or fishponds. The amount of N and P from human excreta that flowed to fishponds were 2.1 and 0.3 kg/ha/year in 2010, while no excreta flowing to fishponds in 1980.

In summary, owing to gradual replacement of dry toilets with flush toilets, the nutrient use of human excreta to the paddy fields decreased. Subsequently, a significant portion of the nutrients were transferred to septic tanks, and more nutrients were directly discharged to fishponds as feed and to the environment directly.

4.7.4 Historical Changes in Livestock Excreta Management

Let us focus on the changes in the management of nutrients derived from livestock excreta to understand how much and where they have flown. Figure 4.44 shows the historical changes in N and P amounts derived from livestock excreta from 1980 to 2010. Corresponding to the change in livestock numbers, the total amount of N and P derived from livestock excreta was 2.8 and 2.9 times, respectively, from 1980 to 2000 since the government had a policy of increasing livestock production (ACI 2002). It then decreased by 14 and 16% respectively from 2000 to 2010 due to livestock disease owing to H5N1 outbreak. The recycling practices of livestock excreta have also changed remarkably. The ratio of N and P flowing from livestock to paddy field decreased from 54 and 80% in 1980 to 30 and 42% in 2010. Instead, that flowing to biogas systems or fishponds increased. N and P amounts from livestock to fishponds dramatically increased by 24 and 30 times, respectively, from 1980 to 2010. In 2010, livestock contributed 17.27 kg N/ha and 3.9 kg P/ha to fishponds, which is equivalent to 31 and 37% of the total amount of N and P derived from livestock excreta, respectively.

(a) Dry toilet

(b) Fecal matter from a dry toilet

(c) Fecal matter collection from a fecal chamber

(d) Livestock excreta in a fecal chamber

Fig. 4.41 Dry toilet and fecal matter stored and used in Trai hamlet. Photo by Hidenori Harada

(a) Fish pond (b) Open canal (c) Nhue river

Fig. 4.42 Fish pond, open canal, and Nhue river, where the excreta is discharged. Photo by Ryo Sakaguchi

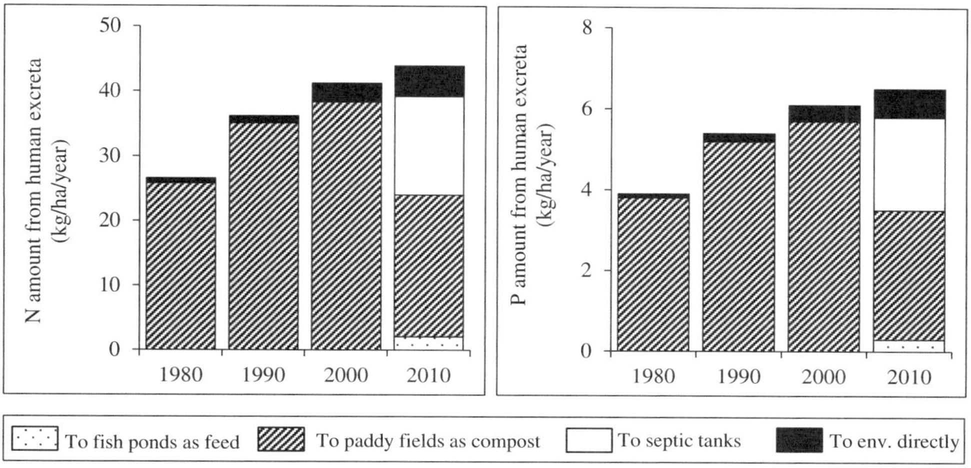

Fig. 4.43 Nitrogen and phosphorus derived from human excreta (kg/ha/year). Modified from Pham et al. (2015)

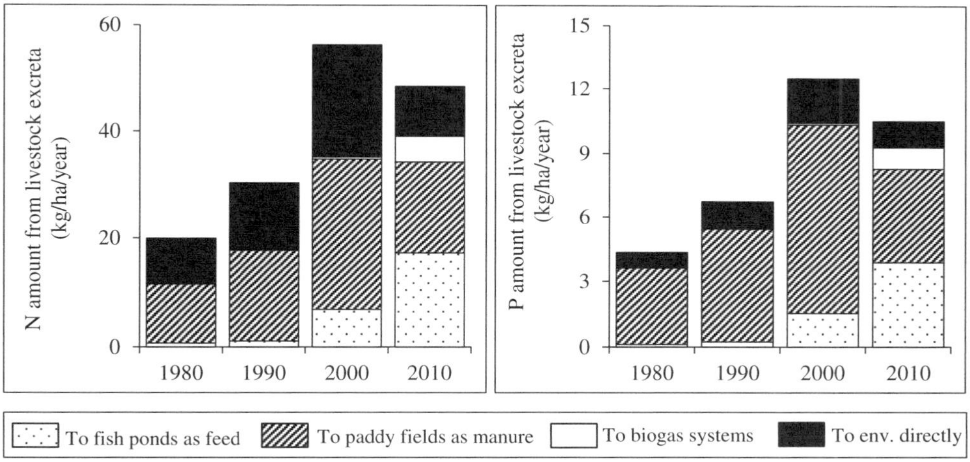

Fig. 4.44 Nitrogen and phosphorus derived from livestock excreta (kg/ha/year). Modified from Pham et al. (2015)

A biogas program for Vietnam's livestock sector was implemented in 2003 in suburban and rural Vietnam to help increase farmers' income and reduce environmental pollution (Dung et al. 2009). The project was introduced to the study area in 2006; however, only 16% of households raising pig constructed biogas plants to treat pig excreta; 53% of the households discharged pig excreta into fishponds. According to Vu et al. (2007), pig farmers in northern Vietnam raise pigs solely to provide excreta as a feed for their fishponds because they believe that fish in ponds fertilized with pig excreta grow significantly faster than when other feeds are applied. Thus, fishponds have gradually become a platform for receiving nutrients from livestock rather than from paddy fields in the study area.

4.7.5 Historical Changes in Chemical Fertilizer Consumption

Although the study area has a custom to use human and livestock excreta as agricultural inputs, chemical fertilizers are now widely

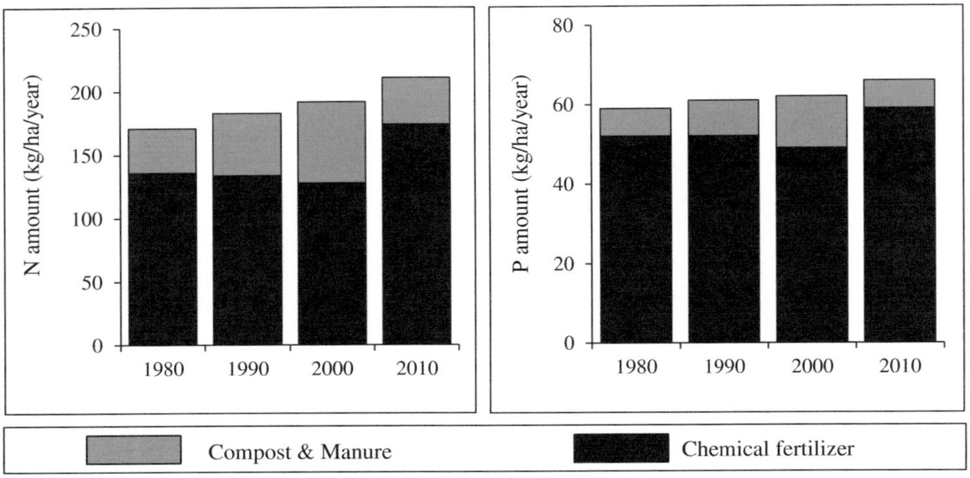

Fig. 4.45 Nutrient derived from compost and manure and chemical fertilizer. Reprinted from Pham et al. (2015) with permission from IWA Publishing

applied to paddy fields. As mentioned above, the application of both human and livestock excreta to the paddy fields is decreasing. Let us now focus on the changes in chemical fertilizer consumption.

Figure 4.45 illustrates time transition of chemical fertilizer consumption compared to that of the application of compost/manure (human and livestock) application. As can be seen in the figure, the nutrient inputs derived from chemical fertilizer decreased by 6% from 1980 to 2000 for both N and P and then increased 1.4 times for N and 1.2 times for P from 2000 to 2010. Nutrient inputs derived from compost/manure peaked in 2000 but those derived from chemical fertilizer contracted. It is indicated that a large amount of compost/manure applied to paddy fields could reduce the amount of chemical fertilizer. Although the government has policies to increase livestock production (ACI 2002), the nutrient inputs from manure (livestock) have been gradually decreasing, as mentioned in the previous section. Due to economic development, chemical fertilizers have become a more preferable source of nutrients rather than compost/manure and their adoption is expected to increase in the near future.

4.7.6 Historical Changes in Nutrient Balance of Paddy Fields

In order to consider sound nutrient management, we focused on changes in the balance of nutrients that flow in and out of paddy fields. N and P balances in paddy fields in 1980 and 2010 are shown in Fig. 4.46. The dominant flows for N were chemical fertilizer, irrigation, production, and runoff, while those for P were chemical fertilizer, production, and soil accumulation and discharge. The total inputs to paddy fields were 435 kg N/ha/year and 90 kg P/ha/year in 2010, which are 1.5 and 1.3 times higher than those in 1980, respectively. The contribution of chemical fertilizers to the 1980 and 2010s total inputs was 47 and 40% in the case of N, respectively, and 74 and 65% in the case of P, respectively. Especially, the inputs from irrigation water represented 29% of the total amount of N (125 kg N/ha/year) and 12% of the total amount of P (11 kg P/ha/year) in 2010 since the irrigation source, Nhue river, was considerably contaminated with N and P by this time (VEA 2012). The total inputs to paddy fields were estimated to exceed the recommended level, 200 kg N/ha/year and 52% kg P/ha/year

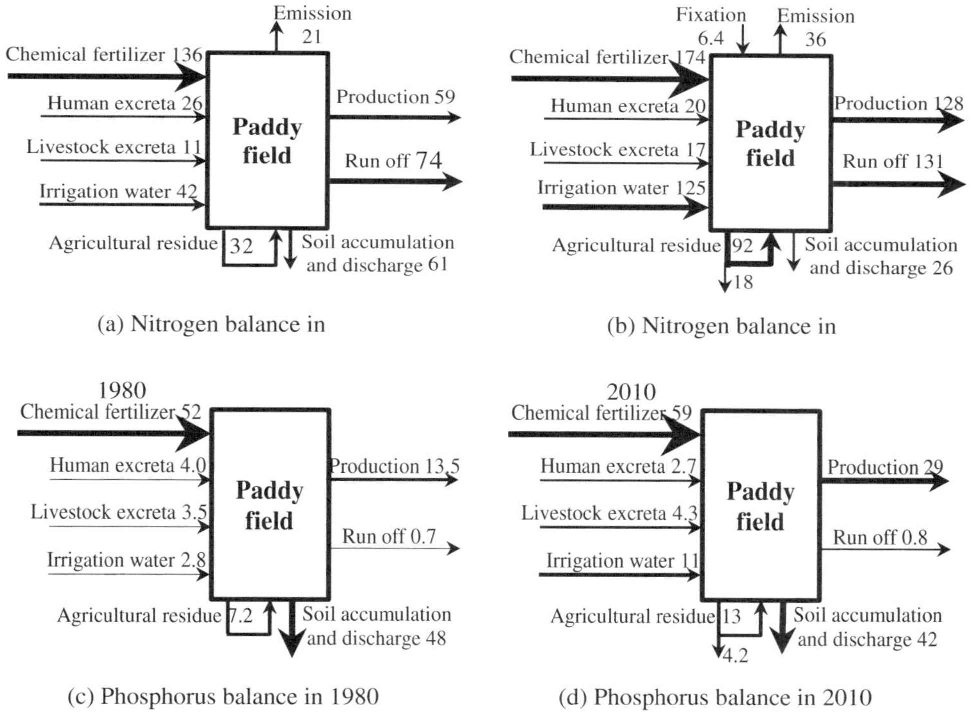

Fig. 4.46 Nitrogen and phosphorus balance of paddy fields in 1980 and 2010 (kg/ha/year). Reprinted from Pham et al. (2015) with permission from IWA Publishing

(Bo et al. 2003) in both the years (1980 and 2010). The excessive application of N and P inputs to paddy fields, especially because of large input of chemical fertilizers, has caused a greater burden of N in the water bodies and P in the soil. The differences between N and P loads can be partly explained by the higher runoff coefficient of N to surface water and the larger fraction of P accumulating in soil (Carpenter et al. 1998).

Therefore, for a sound material cycle, chemical fertilizer consumption needs to be reduced and human and livestock excreta should be promoted for reuse in agriculture. Currently, 33 kg N/ha/year and 5 kg P/ha/year of human/livestock excreta is discharged directly or via septic tank/biogas systems to the environment. If these amounts are applied to paddy fields, use of chemical fertilizers could be reduced by 19% for N and 8% for P. This may not only contribute to better nutrient management but also help achieve environment-friendly sanitation.

4.7.7 Historical Changes in Nutrient Loads to the Water Bodies

In order to analyze how much and which nutrients have impacted the nutrient load on water bodies, let us focus on changes in nutrient loads to the water bodies.

Figure 4.47 describes N and P loads to the water bodies and their changes from 1980 to 2010. The total amount of N and P loads to the water bodies gradually increased 1.9 and 3.2 times from 1980 to 2010, respectively. As can be seen in the figure, the N load from paddy fields contributed a big proportion to the total N load, 131.3 kg/ha/year in 2010, which is 1.8 times as high as the load in 1980. In contrast, fishponds were the main contributor, which accounted for 71% of total P load to the water bodies. P load from fishponds discharged to the water bodies increased by 6.2 times between 1980 and 2010. The differences between N and P loads can be partly explained by the higher runoff of N to

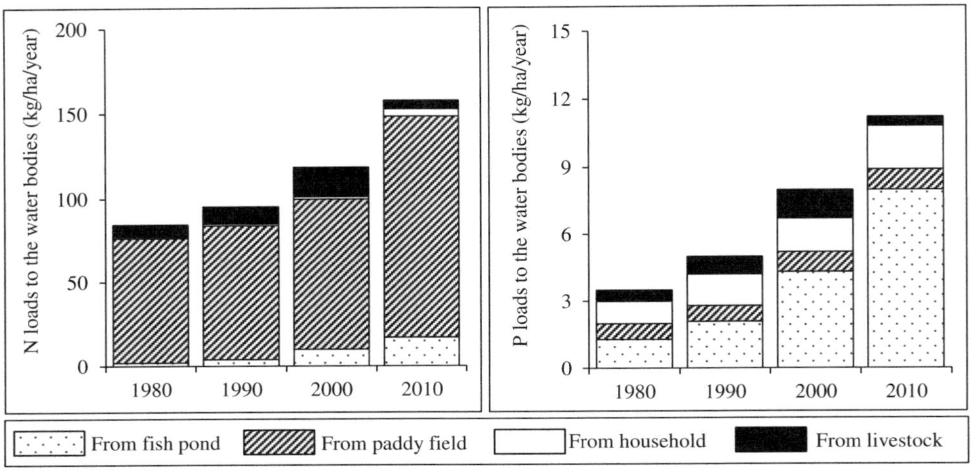

Fig. 4.47 Nutrient loads to the water bodies (kg/ha/year). Modified from Pham et al. (2017)

surface water and the larger fraction of P accumulating in pond sediments (Carpenter et al. 1998). The increase in N load from paddy fields to the water bodies through runoff resulted from the application of excessive N to paddy fields. In modern times, chemical fertilizers have replaced human and livestock excreta as the preferred nutrient sources for paddy fields. As mentioned in the previous section, the N input from chemical fertilizers to paddy fields increased by 1.3 times from 1980 to 2010 and reached 174 kg/ha in 2010, exceeding the optimum amount (100–140 kg N/ha/year) recommended by Huan et al. (2000).

The dramatic increase in nutrient outputs from fishponds to the water bodies could be explained by the increase in inputs to fishponds from livestock, as mentioned in the previous section. Although fish cultivation has generally been only developed at small scales (household level), the excessive nutrient inputs to fishponds have resulted in severe eutrophication in fishponds of the study area. Montangero et al. (2007) also reported water pollution and eutrophication in Viet Tri city because of large nitrogen input generated from households and livestock excreta to fishponds. Fishponds have thus gradually become a platform for conveyance of domestic and livestock wastes from this area to the environment.

4.7.8 Conclusions

The study examined the recent changes in waste and wastewater management as well as nutrient flows in suburban Hanoi, where human excreta and livestock excreta were traditionally used for agriculture.

By developing a material flow model of N and P, this study exhibited the historical changes in domestic waste management, nutrient usage, and the pollution loads to the environment in the study area. Paddy fields were the key component of nutrient cycles in this study area. Owing to economic development, the traditional waste recycling manners have been decreasing and chemical fertilizers are becoming a preferable source of nutrients compared to human and livestock excreta. The total inputs to paddy fields in 2010 were 425 kg N/ha/year and 90 kg P/ha/year, which are 1.5 and 1.3 times higher than those in 1980. As of 2010, the largest input flow to paddy fields is chemical fertilizer, contributing 40% N and 65% P of the total input. The total inputs of N and P to paddy fields are estimated to exceed the recommended level by 2.0 and 3.5 times, respectively. Excessive application of N results in a large N burden on the water bodies through runoff and excess accumulation of P in soil and/or contamination in ground water.

The total of P loads to the water bodies gradually increased by 1.9 and 3.2 times respectively from 1980 to 2010. While paddy fields accounted for a large proportion of the total amount of N loads, fishponds were considered as the biggest contributor, contributing 71% of the total of P load in the water bodies. The total amount of N and P loads from fishponds greatly increased between 1980 and 2010. As a result, fishponds have had an important role in the nutrient cycle of the study area.

It is necessary to create a nutrient cycle system based on the ever-changing reality on-ground, such as the spread of flush toilets, as well as based on the good recycling habit of local people, such as the use of human excreta in agriculture. Sanitary as well as environment-friendly nutrient management will prove to be a key future waste management challenge as cities and suburbs grow.

4.8 Livelihood and Its Challenges of a Mountainous Area in Central Vietnam

4.8.1 Economic Growth and Imbalanced Regional Development

Vietnam got the status of "lower middle-income" country in 2009 going through two decades of rapid economic growth (World Bank 2011). The Doi Moi policy which has enabled the country to shift from a centrally planned to a market-oriented economy contributed to this achievement. In other words, the implementation of the Doi Moi policy transformed Vietnam into a market-driven economy. Since then, Vietnam has witnessed a steady economic expansion, reaching 6.68% in 2015 and achieved a rapid growth of disposable income as well as consumers' expenditures, which reached up to 150 million USD in 2016 (Nguyen et al. 2017). This high economic growth rate (average 6.78%) had been kept until 2019. Under the pandemic condition due to COVID-19, economic growth rate has been kept at nearly 2.91% in 2020 although most ASEAN countries experienced a negative growth rate. This shift has brought about a process of decentralization in which some decision-making powers have been transferred from the center to provincial governments as well as a process of devolution of forest- and land-use rights to communities and households (Clement 2010).

Unprecedented economic growth and improvement of access to basic social services have reduced the percentage of the population living in extreme poverty to almost zero and the national income poverty rate from 21% in 2010 to 10% in 2016 (Pimhidzai 2018).

However, some aspects of the government's socio-economic and environmental accomplishments have been criticized because achievements in poverty alleviation have not been evenly distributed across the country. Although the population living in acute poverty rate declined, most of these improvements in living standards have been found in lowland areas while mountainous areas have been left behind with correspondingly higher poverty rates. This is a matter of great national concern because 75% of Vietnam's land surface is classified as sloping land and this is the highest percentage in any Asian country (Vien et al. 2006).

4.8.2 Policies and Programs for Forestry and Development

Nearly 75% of the land in Vietnam is comprised of hilly, highland, or high mountains. This particular natural geography causes the constitution of limited farmland, which accounts for only 15% of the area (Jong et al. 2006). Three types of forests can be found in mountainous areas, which are natural, semi-natural, and planted forests. The forest cover in 1943 was approximately 45% of the area and was reduced to approximately 24% in 1983. It, however, rebounded during the 1980s and increased to approximately 28% in 1993 (De Koninck 1999). Reforestation projects by the central government have exclusively enhanced forest plantations and not focused restoring natural forests with native species. Reforested plants consist of foreign species such as acacia and eucalyptus (McElwee 2009). From the early 1990s, the area of natural forests decreased at a slower rate than before, and then started to increase gradually (Jong et al. 2006). Forest management regimes differ region by region across Vietnam. Thulstrup et al. (2013) reported forest cover changes at two communes in Central Vietnam from 2000 to 2010 due to effects of the reforestation program being started from 2000. It showed that many of the dense forests had been replaced by forest plantations. Many studies have pointed out how local level social inequality was expanded through the implementation of government induced programs (Thulstrup et al. 2013), which were reforestation and development ones. These programs have been conducted at mountainous areas for ethnic minorities living there over the last four decades. The government classified forest land in accordance with three major functions as production, protection, and special use, as defined by the Forest Protection and Development Law in 1991 (Meyfroidt et al. 2013; World Bank 2011). Production forests have been designated to ensure a country level forest product supply from timber and non-timber forests and have a function of environmental protection, which enable to prevent natural disasters and soil erosion through reservation of soils and water streams (Jong et al. 2006; Thulstrup et al. 2013). Production forest occupies sixty thousand km^2 of the country while protection forest occupies fifty thousand km^2 and special-use forests almost twenty thousand km^2 (World Bank 2011). Special-use forests are designated for nature conservation, environmental protection, protection of historical and cultural relics, and tourism (Thulstrup 2015). Since 1993, the central government has strengthened its special-use forest system.

The central government has tried to remedy the environmental condition and socio-economic situation in mountainous areas by accomplishing the two objectives, such as forestry-based economic development and reforestation (Jong et al. 2006; Meyfroidt and Lambin 2008). Program 327 was launched in 1993 and worked until 1998. This program aimed at achieving the forest restoration, forest protection, forest plantations, agriculture, and aquaculture development (Jong et al. 2006) and was finally replaced by Program 661 (Jong et al. 2006; McElwee 2009). The program was aiming to reforest fifty thousand km^2 of land, which comprised twenty thousand km^2 of special-use and protection forest land and thirty thousand km^2 of production forest (Jong et al. 2006; McElwee 2009). During the execution of the forestry programs, Programs 135 and 134 as infrastructure and socio-economic programs had been continued as well. From 1998 to 2005 as the first phase of Program 135, there were strong needs of the program which were no access via all-weather roads to commune center in approximately 600 communes, no existence of health clinics in approximately 800 communes, no primary school or not enough classrooms in 47% of all communes, no access to market in approximately 900 communes, temporary structure in 70% of houses, access to clean water among less than 40% of the population, and no connection to the electricity network in more than 50% of communes (Sunderlin and Ba 2005). Program 135 was the most consequential poverty alleviation program for residents living at remote and difficult areas, which is officially named "Socio-economic Development for the Communes Facing Greatest Hardships in the Ethnic Minority and Mountainous Areas." This program was designed to have ethnic minorities in mountainous areas alleviate poverty, bridge the income gap with the majority group and exclude risks for social instability (Ha 2009; Kyeyune and Turner 2016). Program 134 was implemented after Decision 134/2004/QD-TTg in July 20, 2004 (Chronic Poverty Research Centre 2010). Several of the policies aiming at providing land of agriculture, houses, residential lands, and water to poor ethnic minorities are found in Decision 134. In some cases, these programs mentioned above have been implemented in the same area for the achievement of synergy effects by all of them. For example, activities under Program 135 have endorsed Program 661 in order to develop forest-processing industries as well as the reforestation of mountainous areas. Thus, several policies and programs were aiming at not only increasing forest areas but also alleviating poverty of ethnic minorities by relating forest to their livelihood, which has been affecting livelihood of those who have been living in mountainous areas and changing its composition.

4.8.3 A Case Study Site

4.8.3.1 Overview of Doi Village

Doi village is one of the typical villages in mountainous areas in Vietnam and is located in the south-eastern part of Nam Dong district, Thua Thien Hue province, in Central Vietnam (Fig. 4.48). This village and its surrounding areas have been affected by socio-economic changes due to the Doi Moi reform program and it experienced drastic land-use changes through the forest land allocation and the settlement policy for local residents from the 1970s. In addition, this village is also affected by urbanization in near cities such as Hue and Danag cities. The Cotu people, one of the ethnic minorities living primarily in the mountains inland along the border with Laos, consist of 100% of the residents in this village. They used to be mainly engaged in slash-and-burn cultivation, but paddy-field rice cultivation and continuous upland farming have become popular with the local communities of mountainous regions since slash-and-burn cultivation was prohibited when all the forest areas were nationalized. A semi-structured interview survey with all 151 households was conducted from September to October 2014, to investigate the actual situation regarding people's livelihoods, and a GIS analysis was conducted to understand land use related to their livelihood (Fig. 4.49).

4.8.3.2 Cotu People

The Cotu people is one of 53 ethnic minority groups in Vietnam living mainly in the districts of Quang Nam province (74.2%) and Thua Thien Hue province (23.8%). The traditional livelihood of Cotu people is agriculture, with the cultivation of dry rice, cassava, and corn by the slash-and-burn method (cutting and burning plants on the land). Beside this, their traditional livelihoods also consist of hunting, gathering forest products, and fishing (from natural springs in forests) using self-produced tools (Århem 2009; Nguyen 2016).

The Cotu people's culture has developed based on their life of self-sufficiency and spiritual belief in the power of local gods (Århem 2009; Tran 2009). Their tangible culture consists of traditional architecture, handicraft, and cuisine, for which forest resources are used as main materials/ingredients (Tran 2009). The Cotu people have a strong belief in local gods, such as forest,

Fig. 4.48 The location of Doi village

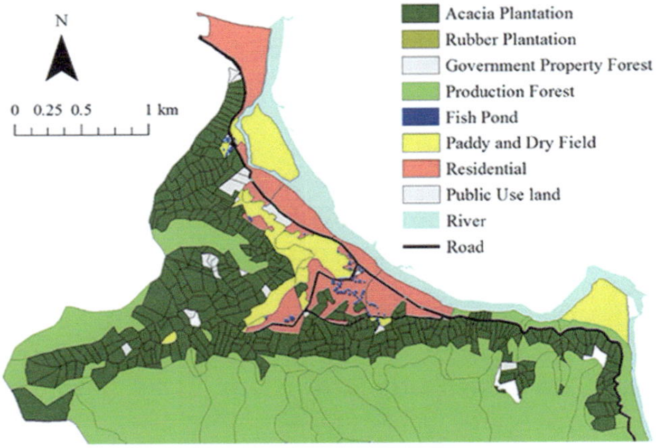

Fig. 4.49 Land-use map in Doi village

land, mountain, stream, village, and house gods, who are believed to heavily influence human well-being and crops. This belief can be found clearly during performances at festivals, rituals, and ceremonial occasions (Århem 2009; Tran 2009). The language of the Cotu people is Katuic, a branch of the Mon-Khmer linguistic family in Vietnam, which has been used well for conversation in Cotu villages. Meanwhile, its character became extinct (Århem 2009). Villages are the most fundamental social unit for Cotu people. In the past, a village was a self-managed social unit and was governed by a village patriarch (male person). Nowadays, the power of the village patriarch is shared with a village leader who is appointed by the Commune People's Committee (Århem 2009). The village patriarch influences significantly the Cotu villagers because he has considerable experiences in traditional worship, customs, cultivation, and plays an important role in maintaining community solidarity. Nowadays in Central Vietnam, they are supposed to be moved to flatland in mountainous areas and change their main livelihood from slash-and-burn cultivation due to Vietnam central government policies, which aimed to increase forest areas.

4.8.3.3 Livelihood in Doi Village and Its Trend

The semi-structured interview survey revealed types of livelihood, which can be categorized into three groups. Livelihoods in the first group are agricultural products cultivated at agricultural lands, which are acacia, rubber, rice, and cassava (Fig. 4.50). Acacia can be harvested and sold every 5–7 years so that acacia plantation became the main resource of income while rubber can be harvested around 10 years after rubber tree planting. Wet-paddy rice agriculture has been conducted since their moving to current flatland. Appropriate lands for wet-paddy rice agriculture are limited in mountainous areas so their own lands areas are relatively small, which are around 0.5–2.0 ha, which cannot be the main income resource. Cassava plays a key role in food security in the world, especially in developing countries. In Doi village, cassava is cultivated widely in their forestry lands when height of acacia trees is low. Acacia and rubber are sold at a town near the village as a cash crop while rice and cassava are mainly for self-consumption in this village. Those in the second group are agricultural products cultivated at their home garden, which are vegetables, fruits, grain crops, and potato (Fig. 4.51). These livelihood in the two groups are conducted based on their own lands. That means those who does not possess their own lands cannot do these livelihoods. Finally, livelihoods which are not related with land use are in the third group, which are day-labor job (cutting and transportation of acacia), hunting, public officer/office workers, livestock production, and military pension (Fig. 4.52). While this survey detected a variety of their livelihoods (Table 4.33), many of the households were conducting acacia forestry, and

Fig. 4.50 Livelihoods categorized in the first group

(a) Vegetables

(b) Banana

Fig. 4.51 Livelihoods categorized in the second group

(a) Day-labor job (cutting and transportation of acacia)

(b) Livestock production

Fig. 4.52 Livelihoods categorized in the third group

Table 4.33 The number of households by the type of livelihood (the result of multiple answers)

	Livelihood	The number of households
First group	Acacia plantation	119
	Rubber plantation	74
	Rice	116
	Dry-field crop (cassava, vegetable, fruits)	111
Second group	Crop production at home garden for self-consumption	63
	Crop production at home garden for selling	86
	Crop production at home garden for selling when abundant crops are harvested	9
Third group	Livestock production	
	Pig	51
	Buffalo	12
	Beef cattle	51
	Chicken	110
	Duck	8
	Goat	1
	Day-labor job (acacia cutting)	63
	Public officer and other office workers	22
	Real estate business	4
	Hunting and collection of non-timber forest products	28
	Military pension	13

villagers said that these numbers and the areas were increasing recently. Accordingly, acacia forests are dominant in the landscape at Doi village (Fig. 4.53). Tokito et al. (2015) pointed out that the increasing number of households in which the main income relies on acacia plantation might weaken community resilience toward volatile global market price in acacia.

This survey revealed that acacia forestry had become more and more dominant and the potential risks from this mono-culture had been getting worse. In addition, we found a lack of crops for personal consumption and a situation of non-land use due to people's reliance on acacia forestry. Also, younger generations, even with a higher education, tend to stay in the village due to difficulties in finding a job in the urban areas. This causes many households to inherit parts of their relatives' limited farmland and conduct agricultural activities on tiny plots, which is gradually leading to poverty in the area (Tokito et al. 2015).

4.8.4 Conclusion

In the village, loss of land-use diversity, the mono-culture of acacia plantations, the poverty induced by the increasing population were found to be the main issues which would be losing their sustainability. Considering this, it is almost impossible for this area to depart from the money economy that is encroaching upon from the cities. Accordingly, the following can be suggested to enhance resilience. In the mountainous areas, acacia plantations are now the main source of income for villagers so they are unable to quit these activities, that is, irreversible trend. They would, however, sink into poverty in the case of bad fluctuations in market prices for acacia products, so residents and the local government should be aware of the high vulnerability of acacia forestry if they wish to extend their acacia plantation activities in this area. Enhancement of livelihood diversity can be one solution to mitigate the current potential risk. In particular, some crops for personal consumption would be helpful

Fig. 4.53 Acacia plantation at Doi village

for residents who can cultivate them in their own gardens or on small vacant plots. Thus, land-use management to maintain the area for these crops and political or personal efforts to enhance the cultivation of these crops are required. In addition, the residents who move to the cities should be assisted. Their education level is getting higher by the support by the central government but this is not reflected in their rate of employment at cities. This includes complicated matters and social problems so the local/central government should recognize these issues and prepare appropriate policies or systems for local minorities living there.

This section introduced the fact that people's livelihoods in mountainous area are faced with various difficult problems. Especially, people's livelihoods, which were mostly decided by the government a few decades ago, do not have a long history so enough resilience has not been acquired. Local people inherently possess their own adequate resilience to economic changes as well as hazardous events such as droughts, storms, flooding, heavy rains, and landslides because most of their capabilities have been improved in small steps by adapting to these events over many generations. This discipline, however, collapses when anthropogenic pressures rapidly invade and encroach upon people's lives. Recently, drastic economic changes in Vietnam have affected people's livelihoods, and residents' ways of life have been forcibly changed leading to a loss in their powers of resilience. This discipline could be found in Doi village through investigation of their livelihood. Therefore, a future sustainable strategy based on our suggestions should be carried out by taking this low resilience into account.

4.9 Community Resilience of Dam-Displaced Ethnic Minorities in Central Vietnam

4.9.1 Introduction

Vietnam's rapid economic growth is generating an ever-growing demand for energy. Electricity demand, in particular, has increased by 10%

per annum from 2015 to 2019 (Rogers 2019). Although the country is currently developing offshore oil and other energy sources and the share of hydropower in the national energy mix is expected to decline rapidly, as of 2020 Vietnam still relied on hydropower generation for 30% of total electricity supplies (US International Trade Administration 2021). The electricity, totaling 17.50 MW, is generated by more than 70 large hydropower dams, which blanket the country's ten major river systems (Dao 2010). While these dams have helped fuel industrial growth, mitigate seasonal flooding, provide irrigation and drinking water, and further the goal of universal electrification, dam and reservoir construction has displaced an estimated 240,000 residents, many of them being members of Vietnam's 53 economically and socially marginalized ethnic minority groups (Bui and Schreinemachers 2011).

Those displaced by dam construction in Vietnam are legally required to be compensated for homes, land, and productive assets and resettled to another site, with the goal of, at the least, restoring original levels of income and living conditions. However, in Vietnam as in many other nations, reality has fallen short: a 2010 survey found that 82% of those displaced by dams in Vietnam were worse off than before displacement (Institute for Development Consultation 2010).

Sociologist Cernea (2000) developed the influential Impoverishment Risk and Reconstruction framework to identify the poverty risks associated with development-forced displacement (Table 4.34). For dam projects, it is clear that the global trend of high impoverishment risk applies in the case of Vietnam as well.

Researchers criticize dam-forced displacement from a human rights as well as from a risk perspective, citing international declarations such as the Guiding Principles on Internal Displacement of the United Nations (2004), which posit the right to be protected from arbitrary displacement. The UN Declaration on the Right to Development from 1986 states that development should benefit all individuals, who should be allowed "active, free, and meaningful participation." Yet, dam-displaced populations rarely benefit from or participate in decision-making for large-scale dam projects.

Several studies have identified loss of productive land and constrained access to forests and other common resources as being the most detrimental outcomes of resettlement in Vietnam (Singer and Hoang 2014). Other adverse

Table 4.34 Likelihood of each type of risk for dam projects in Vietnam applying Cernea's Impoverishment Risk and Reconstruction framework (2000)

Landlessness	Very high	Most AP farmland in inundated area is permanently lost
Joblessness	Medium/High	Traditional jobs often disappeared after resettlement, little job opportunity at the new places
Homelessness	Low/Medium	New houses are built but poor quality
Marginalization	Medium/High	Social and cultural impacts particularly adverse for ethnic minorities
Morbidity	High	Disruption of access to medical facilities may increase risks
Food Insecurity	High	Poor land provision and difficulty of establishing new plots will be felt most in early days
Loss of access to common property resources	Very high	Restricted access to forests and fisheries will disrupt current practices
Social disarticulation	Medium/High	Depends on resettled community and relations with host communities

Adapted from Strategic Environmental Assessment of the Hydropower Master Plan in the Context of the Power Development Plan VI Final Report 2009, Ty (2008)

impacts include insufficient land and cash compensation (Neef and Singer 2015) and conflict with host communities over access to common resources (Dao 2010). As Leslie (2005: 203) has noted, dams create need "where before none existed," transforming self-sufficient farmers into residents requiring some kind of development initiative to satisfy their basic needs.

The research described below, conducted from 2011 to 2018, is divided into three sections according to the overall research focus of each study: exploring post-resettlement adaptation, the roles of various stakeholders in the process, and community resilience; comparing outcomes for rural and peri-urban resettlement; and assessing prospects for a benefit-sharing scheme to improve outcomes for affected people (APs). The projects discussed here were funded by the Japanese government and carried out by multidisciplinary research teams from Kyoto University Graduate School of Global Environmental Studies, Hue University of Agriculture and Forestry, and Danang University of Technology.

4.9.2 Understanding the Process of Post-Resettlement Adaptation and Community Resilience

4.9.2.1 Research on Dam-Forced Community Resettlement in Central Vietnam

We investigated conditions for two resettled hamlets in a mountainous area of Dong Giang district, Quang Nam province, Central Vietnam (Fig. 4.54). The two adjacent hamlets, Aden and Tro Gung, together comprise a village called Cutchrun. Approximately 95% of the 120 households are ethnic Cotu, an indigenous ethnic minority group that traditionally practices shifting cultivation and hunting. The villagers were displaced in 2006 by construction of the 83 m-high 210 MW A Vuong dam on a tributary of the Vu Gia Thu Bon river and resettled to a site approximately 16 km from the dam reservoir (Fig. 4.55).

Using household surveys, semi-structured interviews, key informant interviews, and focus group discussions, we assessed village conditions by applying the Sustainable Livelihoods Framework (Scoones 1998), a popular analytical tool in development studies. We first investigated the community capitals, or resources that each household could call upon for their livelihoods and improved living conditions: natural, physical, financial, human, social, and cultural capital. We also investigated the operant policy and legal framework, including development assistance, and the strategies that households have adopted to improve livelihoods. Although no survey was conducted before displacement, secondary research provided income levels before the move and the villagers were asked to assess changes in their living conditions (Singer and Watanabe 2014).

4.9.2.2 Results

Based principally on results from household surveys and focus groups, we identified a decline in living conditions as well as income after resettlement (Table 4.35). Annual income averaged 8.79 million Vietnam Dong (VND) per household in 2012, compared to 10–12 million VND before displacement, with 92–94% of residents in the two hamlets under the rural poverty line, according to village officials. APs suffered a 50% decline in food production, as replacement land was insufficient and nonproductive, and access was constrained to forests and the river, traditional sources of supplementary protein. In addition, livestock suffered from high levels of morbidity. This has resulted in food insecurity and child malnutrition rates of 32–45%. Due to poor harvests, residents began cutting and burning nearby protected forest to convert it for swidden cultivation, resulting in decreasing forest cover.

Residents were provided with 40 m^2 poorly constructed housing on 1.85 m-high piles, which was hot in summer and cold in winter. The steep front steps made the houses dangerous for young children and elderly family members, and the staircases were damaged during storms. Toilets were provided, but more than 70% were

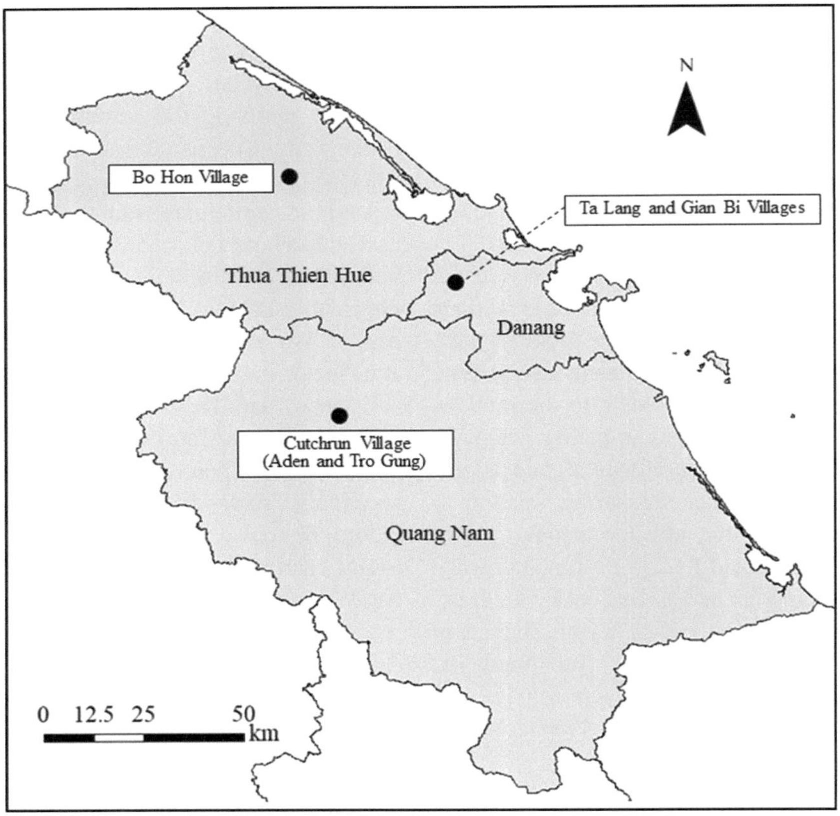

Fig. 4.54 Research site

no longer functional after a few years. Due to the poor housing stock, residents were forced to repair and renovate their homes themselves. They applied traditionally acquired construction skills to reconstruct staircases, erected siding to convert the ground floor space into living space, and built adjacent traditional "kitchen houses" of thatch or bamboo (Fig. 4.56).

Because the villagers had traditionally practiced residential mobility, they were able to reconstruct their homes and could also build and maintain a traditional community house, or *guol,* which helped in fostering stronger community ties and ethnic identity. However, the need to expend energy and secure funds and materials for constructing additional structures on their housing plots meant that less land was available for a home garden, thereby reducing income opportunities. A district official recognized that providing shoddily constructed houses to skilled builders was a mistake, saying, "We should have given more rights to the people and we should have done more to prepare the new sites. The government doesn't need to build houses for people; let them build them by themselves (personal communication, 2011)."

When asked about their greatest challenges during a focus group session, villagers cited the poor quality and quantity of productive land, lack of irrigation for rice paddies, high livestock morbidity, and poor water supplies during the dry season.

4.9.2.3 Discussion

An analysis of the resources, or community capitals, that were available to the resettled households (Fig. 4.57), revealed that physical capital has improved since resettlement, due

Fig. 4.55 Resettled hamlet of Aden, Ma Cooih commune, Dong Giang district, Quang Nam province

Table 4.35 Comparison of pre-and post-resettlement conditions: Aden and Tro Gung hamlets

	Prior to move (2006)	Post resettlement (2012)	Notes
Ave. HH income	10–12 million VND	8.79 million VND	After PFES program started, income rose to avg. 14.8 million VND
Physical capital	Poor	Improved	Road, electricity installed. School built in village. Homes built but shoddy construction
Financial capital	Poor	Poor	Insecure income from harvests, manual labor; little diversification of livelihoods; 93% poor
Natural capital	Excellent	Degraded	Lost access to fisheries, hunting; received land of poor quality so food insecure; high livestock morbidity
Human capital	Poor	Slightly improved	Better education, some skills and agricultural training
Social capital	Good	Good	Communities intact; strong community cohesion
Cultural capital	Good	Slightly degraded	Fewer festivals than before; built community house and traditional homes

to provision of infrastructure such as electricity, roads, a primary school, and a small health clinic. Human capital is unchanged but financial capital has decreased, due to reduced agricultural productivity. Villagers have tried to diversify crops and introduce new livelihoods such as aquaculture, but low educational levels and a lack of bridging capital has made it

Fig. 4.56 Original house and adaptations: from left, original construction, modified staircase and bottom floor, traditional "kitchen house"

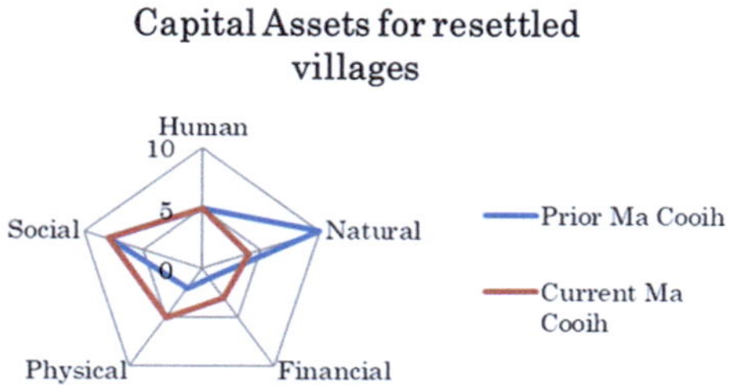

Fig. 4.57 Capital assets among resettled households

difficult for them to try new ventures or migrate to urban areas for employment.

The most negative change has been in natural capital, due to poor productive land, loss of access to river fisheries and non-timber forest products, and loss of forest due to the dam project and the perceived need to destroy forest to convert the land for cultivation. As concerns community resilience to cope with future vulnerabilities, the picture appears mixed. Without more education, job skills, and financial resources, it will be difficult to improve livelihood prospects. However, since the communities were resettled mainly intact and there is little social or financial inequality, and since Cotu identity remains strong, social networks and community cohesion appear to be maintained.

4.9.3 Does Location Matter? Comparing Rural and Peri-Urban Resettlement of Dam-Displaced Populations

4.9.3.1 Resettlement and Geographical Considerations

The Vietnam government's approach to improving post-resettlement livelihoods and living conditions for dam-displaced populations prioritizes financial compensation and agricultural support, but it has not proven successful for residents with poor endowments of financial or human capital who are resettled in remote rural areas where the soil is poor and supplementary livelihoods are limited. Dams are typically located in remote upland areas, where there is a severe lack of level, arable land, making replacement land difficult for local governments or project authorities to acquire. The remote location and limited education, skills, and financial capital available to APs make it especially difficult for them to diversify livelihoods by cultivating new crops, learning new occupations like training as mechanics or tailors, or migrating to distant cities to work as manual labor or factory workers.

However, there have been a small number of APs who have been resettled close to the fringe of urban areas, where they can continue to practice agriculture as their primary occupation but can benefit from the following:

- Improved access to arable, well-irrigated land
- Access to urban markets and employment
- Better educational opportunities
- Health and welfare services
- Proximity to Kinh residents, the majority ethnic group, providing exposure to new values, agricultural techniques, and technologies.

This research explored the hypothesis that peri-urban resettlement can contribute significantly to the livelihood adaptation of displaced households, allowing them to diversify beyond livelihoods based upon natural resources to a range of market- and migration-based strategies due to interaction with majority Kinh residents as well as urban proximity. However, we also hypothesized that the potential increase in financial and human capital of peri-urban resettlement in areas dominated by Kinh residents may also be accompanied by strained social networks, increased economic inequality, vulnerability to discrimination by the Kinh, and weakened traditional practices and ethnic identity among resettled ethnic minority communities, which may result in a decline in overall community cohesion and resilience. In addition, high land prices in urban areas may make it difficult for resettlers to expand land holdings and access to forests, fisheries, and other natural resources may be poor.

To test the hypothesis, the research team selected two peri-urban resettlement

Table 4.36 Survey responses related to social and human capital

Village	Site 1—peri-urban, Hue		Site 2—peri-urban, Danang		Site 3—rural Quang Nam	
Children's educational level	Elem./Below	69.4	Elem./Below	54.9	Elem./Below	62.6
	Jr. High School	10.9	Jr. High School	13.7	Jr. High School	23.2
	High School	13.0	High School	21.6	High School	10.7
	University	6.7	University	9.8	University	3.6
Strong mutual assistance	73.9% agree		100% agree		98.2% agree	
Would borrow money from neighbors	6.5%		11.8%		14.3%	
Ask neighbors to look after homes	41.3%		49%		60.7%	
Cantril life satisfaction median score (1–10)	4.9		5.8		6.2	

communities (Fig. 4.54) and one rural community and carried out household surveys (50–60 for each village) and key informant interviews with local officials to learn more about the relative conditions of each community. The three communities were as follows:

1. **Bo Hon village**, Binh Thanh commune, Hung Ta district, Thien Thua Hue province: 54 households (HHs) (278 people); lowland, peri-urban; located 25 km from Hue city; resettled in 2006 due to dam construction.
2. **Ta Lang and Gian Bi villages**, Hoa Bac commune, Hoa Vang district, Danang province: 50 HHs of total 241 HHs (440 people) are poor/near-poor; lowland, peri-urban; located 15 km from Danang industrial zone; resettled in 1997 for poverty alleviation.
3. **Aden hamlet**, Cutchrun village, Ma Cooih commune, Dong Giang district, Quang Nam province: 95 HHs (450 people); 97% of HHs are poor, remaining are near-poor; upland, rural; located 3 hours' drive from Danang; resettled in 2006 due to dam construction.

4.9.3.2 Results and Discussion

According to the household survey results, average monthly income varied widely among the three communities, with the two villages near urban Danang recording the highest income of 2.783 million VND as compared with 2.188 million VND for Bo Hon, near Hue, and only 910,000 per month for Cutchrun village in rural Quang Nam (Table 4.36). However, it should be noted that monthly expenditures are much higher in the peri-urban sites, from 1.6 to 1.7 million VND per month compared to 640,000 VND for rural Cutchrun. The sources of income are more varied in the peri-urban areas as well. Residents in Hue are mainly farmers but they also engage in labor for Kinh neighbors, who own acacia plantations, while those in Danang commute to a nearby industrial park for work, or engage in construction. Many younger residents from both communities have migrated to Ho Chi Minh, Hanoi, and other large urban areas for employment as well. In Cutchrun, however, farming and gathering of non-timber forest products continue to be the main income sources, with very few residents able to obtain industrial work or migrate to large urban centers. When asked why not, the residents explain that they lack urban connections and the skills and high school education needed to secure factory jobs.

We found differences in educational levels as well, with more residents of Danang able to continue to secondary school and university than the other two areas. Both Danang and Cutchrun show high levels of mutual assistance, although Cutchrun residents are more likely to borrow money from neighbors and entrust their homes to them. Interestingly, Cutchrun residents, although they have lower incomes, record higher average scores when asked to indicate their satisfaction

Table 4.37 Cultural capital

Village	Site 1—peri-urban Hue	Site 2—peri-urban Danang	Site 3—rural Quang Nam
Influence from Kinh neighbors	Yes 30 No 13 Don't know 4	Yes 45 No 6	Yes 37 No 19
Often wear Cotu attire	Yes 26 No 20	Yes 44 No 7	Yes 51 No 5
Religious practices	31 Cotu 19 Non-religious 1 Buddhist	32 Cotu 13 Non-religious	52 Cotu 4 Non-religious
Children's names	48 Kinh 3 Cotu	44 Kinh 1 Cotu 2 Unsure	27 Kinh 29 Cotu
Effects from Kinh proximity	45 Yes 6 No	30 Yes 13 No	37 Yes 19 No
Cotu practices	Lack funds for Cotu festivals; rarely use Guol	Rarely sponsor Cotu festivals; speak Vietnamese at home	Continue buffalo sacrifice festival, frequent use of Guol

with their lives. Research has shown that income tends to be loosely correlated with life satisfaction levels; thus, it is possible that non-monetary factors may also contribute to these responses.

Finally, when assessing cultural capital (Table 4.37), some clear differences can be distinguished.

Residents of Cutchrun in Quang Nam continue to practice traditional Cotu rituals such as the buffalo sacrifice at harvest time and they make frequent use of the community house, while the peri-urban communities no longer engage in ritual practices. Cotu language is less spoken in Danang than in Quang Nam, and almost all children are given Kinh first names. In Cutchrun, residents still wear traditional attire and speak of less direct influence from Kinh neighbors.

In conclusion, the hypothesis of improved livelihood opportunities but reduced social and cultural capital for peri-urban resettlement seemed to be supported (Singer et al. 2014). The study suggests that peri-urban resettlement can promote expanded labor opportunities and diversification of livelihoods to ameliorate the impact of loss of common resources and scarce productive land. It also demonstrates that rural upland villages are not yet integrated into the migration networks that are important livelihood strategies for young peri-urban residents. It suggests that the proximity of Kinh residents may hasten adoption of the majority culture and language for ethnic minorities. Finally, it seems probable that social capital may remain intact if a community is resettled together and there is limited economic inequality.

4.9.4 Payment for Forest Environmental Services: Can APs Become Beneficiaries of Development?

4.9.4.1 Introduction

Many of Vietnam's impoverished dam-displaced communities are now receiving increased income due to nationwide implementation of a benefit-sharing scheme, the Payment for Forest Environmental Services (PFES) program. PFES was first piloted in 2009 and scaled up nationwide following enactment in 2010 of the 99/2010/ND-CP Law on Payment for Environmental Services. This law stipulates that three of the beneficiaries of dam construction—hydroelectric utilities, public water utilities, and ecotourism providers—pay a tax on hydropower generation, irrigation supplies, and tourism revenues, respectively, that accrues to provincial governments. Part of the proceeds are disbursed to a district-level Forest Protection and Development Fund that provides a regular stipend to residents living in the dam watershed, including resettled residents, to monitor forests for illegal logging and maintain forest health. The program's premise is that hydropower and water suppliers benefit from a steady flow of clean water, erosion mitigation, and other ecosystem provisioning services, so they should contribute to financing these services (Singer et al. 2014). The scheme not only alleviates poverty by generating a long-term, sustainable income stream for local residents and by allowing participants to cultivate acacia and other tree crops on part of the forest land, it also educates residents about the importance of resource conservation and helps to reduce deforestation by illegal logging or for cropland conversion.

As of 2018, the PFES program was being implemented in 41 of Vietnam's 58 provinces, covering 40% of all forest land and 500,000 households, according to a German consultant with the Ministry of Agriculture and Rural Development (personal communication, 2018). It has been backed by several prime ministerial proclamations and decrees that stipulate payment levels and beneficiaries. According to one NGO official, as of 2017 the project was generating 1 billion USD per year from taxes on hydropower and irrigation. Provinces with more dams can receive more revenues, and the roster of taxpayers is expanding to include forest owners, plantation companies, and households with land leases for forest land (personal communication, 2018).

4.9.4.2 Case Study in Quang Nam Province

Examination of one pilot PFES program helps to illuminate both the promise and the challenges of the scheme. After a successful trial run in two adjacent communities, the PFES program was implemented in the resettlement village of Cutchrun in Central Vietnam in January 2014. Every household in the two hamlets making up the village participated in the PFES program, initially by joining workshops led by the international NGO Winrock to learn about forest ecosystem services and to acquire skills for monitoring and maintaining forests, including holding patrols to prevent illegal logging, trimming branches, and tree-planting in degraded forest areas. The training course and initial baseline surveys and mapping of allocated forest areas was funded by the Asian Development Bank.

Each household was allocated 22 ha of forest to monitor and allowed to conduct commercial forestry in a portion of the land. Village residents were divided into groups of 13 households, with one member of each family expected to participate in weekly forest patrols and forest maintenance. Monitoring of forest quality was conducted by staff of the district-level Forest Protection Management Unit. Each household received the same payment of 274,000 VND per hectare per year, or a total of 6.02 million VND per year.

The program yielded clear benefits in furthering the twin objectives of forest protection and poverty alleviation. Local officials claimed that the scheme has reduced illegal logging by 50%, and villagers spoke of increased desire to maintain forest cover and biodiversity, despite potential benefits from conversion of forest for agricultural production. As for economic benefits, the PFES payments provided an average boost of 40% to household income, which increased from 8.79 million VND in 2012 per household to an estimated 14.8 million VND. However, it should be noted that for Aden and Tro Gung, this would mean that the average household would remain below Vietnam's rural poverty threshold of 400,000 VND per capita per month. In addition, many villagers noted that payments by the hydropower utility, Vietnam Electricity (EVN), have been greatly delayed and are often non-existent, eroding expectations of future financial security.

4.9.4.3 Discussion

The findings above were augmented by later field visits to two PFES schemes being implemented in dam-displaced communities in Tuong Duong and Que Phong districts in Nghe An province in northern Vietnam. Nghe An's PFES program includes a total of 28,000 households, 14 forest protection groups, and 46 communities monitoring 300,000 hectares of forest land. The participants in Nghe An welcomed the extra income and the opportunity to plant acacia trees on degraded forest land in their allocated patrol area, but they complained about delayed PFES payments and lower payment rates due to a decline in hydroelectric generation.

Based on the results of visits to all three sites, we have identified several common flaws with Vietnam's PFES program:

1. High transaction costs, particularly for baseline assessments, training of residents, and periodic monitoring.
2. Irregular and inaccurate forest monitoring.
3. Irregular payments in many communities.
4. Poor monitoring of local implementation and little transparency.
5. Lack of capacity by local government for coordination and training.
6. Little implementation of a K-coefficient to differentiate payment amounts based on such variables as forest type, forest cover, and ease of access or of performance-based rewards, possibly affecting motivation for continued compliance. This is often the preference of residents, who prefer equal payments despite differences in forest quality or residents' contributions.
7. Many analysts have also noted that the PFES program is at variance with conventional PES schemes in that participation by both service providers and beneficiaries is mandatory and payment terms have not been negotiated by these parties.

Nevertheless, this first nationwide PES scheme in Asia has been successful in improving forest cover and alleviating poverty for most resettled communities. With improved administration, it can be regarded as a model for other countries dealing with similar environmental and economic challenges.

The research described here has deepened the understanding of the complex factors that contribute to the successful resettlement of dam-displaced populations (Singer 2019). It is suggested that multiple stakeholders, including civil society organizations, be given a larger role, and that APs be allowed greater participation in decision-making. Recognizing that most outcomes are extremely adverse, the Vietnamese government has decided to cease plans for new large hydropower projects and is seeking to apply benefit-sharing and additional livelihood assistance to existing communities; yet, more research on their effectiveness is needed.

Column 5

Intestinal parasite and sanitation in Vietnam

Are you aware of parasitic diseases in your daily life? Apparently, over 200 species of parasites live in humans worldwide. They range in size from those too small to be seen with the naked eye to large enough to be grasped with fingers. In most cases, parasitic infections may cause only mild symptoms, such as fatigue and mild abdominal pain; occasionally, however, these infections may lead to severe abdominal pain and vomiting, necessitating medical attention. In fact, these infestations can be fatal in some cases. Intestinal parasitic diseases are common in Vietnam. For instance, over 17 million people are infected by intestinal nematodes in the country (IHME 2019). The most common parasites in Vietnam are roundworms and whipworms, which are relatively large helminths that parasitize humans and pigs.

A typical mode of intestinal parasitic infection is the fecal–oral route. The parasites are introduced into the body through the ingestion of untreated water or food contaminated by fecal matter. The risk of infection exists even through the daily diet, as raw vegetables (Fig. 4.58), such as herbs, and semi-raw pork are often consumed. In our surveys in Hanoi in 2015, parasitic eggs (Fig. 4.59) were detected in raw herbs as well as in rivers and ponds. Even in urban areas, the risk of oral infection from dietary

Fig. 4.58 Raw vegetables in rural area in Hanoi. Photo by Ryo Sakaguchi, October 21, 2013

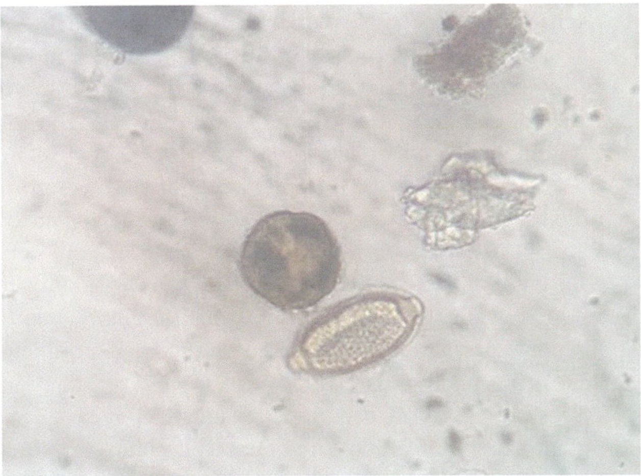

Fig. 4.59 Parasite eggs observed under a microscope. Photo by Ryo Sakaguchi, November 12, 2015

sources is inevitable, and this risk may be even higher in rural areas because of frequent exposure to contaminated water and soil in daily life.

Measures that each individual can exercise include avoiding the consumption of raw foods and drinking untreated water. Regular intake of anthelmintics is also important. These practices are expected to be effective as parasite control measures. However, raw herbs are important in Vietnamese food culture, and changing peoples' habit of eating raw foods is challenging. We need to consider other approaches. Specifically, sanitary food production and consumption, as well as the provision of good sanitation, including sanitary management of wastewater, human waste, and fecal sludge, are crucial to ensure a hygienic living and agricultural environment, which would prevent the contamination of food and water.

As Vietnam's economy continues to grow, many people will aspire even more hygienic living conditions and safe enjoyment of the local food culture. There will also be great demand from overseas tourists, who wish to travel with less risk of getting parasitic infections. In response to these demands, we hope that further efforts will be sought to develop good sanitation in Vietnam for creating a more hygienic living and agricultural environment and for allowing people to safely enjoy the Vietnamese food culture.

Fig. 4.60 A urine-diverting dry toilet in a suburban community of Hanoi used for a number of decades. Adopted from Harada and Fujii (2020)

Column 6

Urine-diverting Dry Toilets.

Have you ever heard of a urine-diverting dry toilet (UDDT)—a unique resource-oriented sanitation system in Vietnam? UDDTs were once widely used in the rural areas of northern Vietnam and are used even today in some parts (Fig. 4.60). Like other East Asian countries, Vietnam follows the tradition of using human waste in agriculture. The unique feature of a UDDT is that the shape of the toilet bowl is modified such that the feces falling directly down and the urine falling forward are collected and handled separately.

Feces and urine are different in appearance and composition. Clearly, majority of the pathogenic microorganisms in human waste are present in feces; surprisingly, however, the majority of the nutrients useful in agriculture, such as nitrogen, phosphorus, and potassium, are present in urine. The fact that urine smells like ammonia sometimes after its release is an indication that it is rich in nitrogen. However, many people believe that majority of the nutrients are contained in feces, due perhaps to the image of compost. Utilizing this difference, feces can be efficiently dried and sanitized without being mixed with urine by spreading ash or other alkaline agents. Meanwhile, urine, which is relatively safe to handle and rich in nutrients, can be used for agriculture as a liquid fertilizer more effectively and securely without mixing with other human waste. To safely use human waste through these mechanisms, the health authority in northern Vietnam recommended UDDTs as sanitary latrines and launched a campaign in 1956 to build double-vault dry toilets, including those that do not divert urine and feces (Winblad and Simpson-Hebert 2004). Thanks to this campaign, Vietnam became one of the few successful countries in the world where UDDTs were widely used.

Fig. 4.61 A urine-diverting dry toilet in rural Malawi, newly introduced by a non-governmental organization. Photo by Hidenori Harada

Nowadays, however, UDDTs in Vietnam are gradually being replaced by water-flush toilets, although 12% of the population in a suburban community of Hanoi, for example, continued using UDDTs in 2010 (Giang et al. 2015). At present, UDDTs are once again being promoted as environmentally friendly sanitary latrines in South Asia and Africa, where a significant proportion of the population does not have access to basic sanitation (Fig. 4.61). In addition, UDDTs can be utilized as disaster-response toilets, exploiting their simple mechanism for sanitizing fecal matter by separating urine and feces (Harada et al. 2012). Scenarios wherein water is not available and sewerage is not working during a disaster are common in many areas where UDDTs are being promoted. In emergencies, such as natural disasters, containment of fecal matter is critical from a hygienic standpoint, and this can be achieved simply by separating feces from urine. Although research is ongoing on how UDDTs can achieve long-term acceptability and sustainability in newly introduced areas (Harada and Fujii 2020), the potential of UDDTs has been receiving renewed attention. UDDTs that were once widely used in rural Vietnam may one day become a solution to sanitation problems in South Asia and Africa.

Column 7

Daily Meals of the Cotu

When I first visited Cotu village in Nam Dong district of Thua Thien Hue province in 2014, acacia forestry dominated the landscape as far as I could see, with a few croplands and paddy fields nestled between in mountain ravines. This made me wonder what the locals usually eat. When I asked this question to a Vietnamese researcher who was accompanying me on the trip, he responded that "Cassava is a staple of their diet."

Over the next few months, I stayed near the village and visited many houses for an interview survey. One day, one of the villagers invited me to their lunch. Although it was not special, they prepared a variety of dishes, which clearly depicted a rich history and culture (Fig. 4.62). This was completely different from what I had heard earlier.

Common ingredients in Cotu's cuisine include home garden crops and semi-domesticated vegetables. Especially in the Cotu village, we often see the Taro. People prefer to plant it because it is easy to take care of, grows fast, and reproduces naturally (vegetative reproduction). Thus, locals can harvest Taro when they need it and as much as they want. According to the villagers, they get Taro cuttings from their parents and relatives when they move due to resettlement or marriage. This shows how deeply Taro is embedded in Cotu's life. Taro is also preferred because it is filling. According to village elders, this was especially useful in staving off hunger during a disaster or food shortage. Villagers of Cotu always keep Taro in their home gardens, which almost acts as a fail-safe crop. When I joined their lunch, the dishes, of course, abundantly contained Taro.

Although Cotu has been facing land-use change and biodiversity loss in recent times, locals have cleverly used the forest resources to create a rich dietary culture based on their long experience. Such a type of regional custom has a high affinity for the local environment, and this fact is almost unrecognized even among other Vietnamese people.

Fig. 4.62 Left: Salt-massaged Taro (leaf stalk of *Xanthosoma violaceum*). Right: Baked catfish with vegetables in leaf of *Alpinia zerumbet*

References

ACI (Agrifood Consulting International) (2002) Livestock policy briefs for Vietnam. Project Brief Series. ACI

Alexandre D, Rajan S, Rajamony L, Ushakumary K, Kurien S (2009) The ad hoc package practices recommendations for organic farming. Kerala Agricultural University

An LV, Bao LQ, Ngoan LD, Hieu NM, Cach NT, Sen HT, Phuong VTM, Hong NX, Hoa HH, Quan TT, Tri TM, Nam NP, Toan NH, My NK, Van NTM (2001) Research on resources management based on communities in Hong Ha Commune, A Luoi, Thua Thien Hue. Summary Report Period 1998–2001. Hue University of Agriculture and Forestry (in Vietnamese)

Anh TN, Harada H, Fujii S, Anh PN, Lieu PK, Tanaka S (2016) Preliminary analysis of phosphorus flow in Hue Citadel. Water Sci Technol 73(1):69–77. https://doi.org/10.2166/wst.2015.463

Århem N (2009) In the sacred forest: landscape, livelihood and spirit beliefs among the Katu of Vietnam. Göteborg University. Accessed 12 Dec 2022 from https://www.academia.edu/10640092/In_the_Sacred_Forest_Landscape_livelihood_and_spirit_beliefs_among_the_Katu_of_Vietnam

Babatunde A, Truong T, Pradeep K, Robert M, Marie C, Robert D (2016) Economics of climate change adaptation: agriculture sector analysis for Vietnam. UNDP. https://doi.org/10.13140/RG.2.2.36707.66083

Baldock JA, Nelson PN (2000) Soil organic matter. In: Sumner ME (ed) Handbook of soil science. CRC Press, Boca Raton, Florida, pp B25–84. https://doi.org/10.1038/194324b0

Baulch B, Nguyen HTM, Phuong PTT, Pham HT (2010) Ethnic minority poverty in Vietnam, Working paper. Chronic Poverty Research Centre. Accessed 12 Dec 2022 from https://assets.publishing.service.gov.uk/media/57a08b0040f0b649740008e0/WP169.pdf

Bell RW, Rerkasem B, Keerati-Kasikorn P, Phetchawee S, Hiranburana N, Ratanarat S, Pongsakul P, Loneragan JF (1990) Mineral nutrition of food legumes in Thailand with particular reference to micronutrients. ACIAR Technical Report Series Reports No 16

Bo NV, Mutert E, Sat CD (2003) Balanced fertilization for a better crop in Vietnam. Potash and Phosphate Institute of Canada (Southeast Asia Program)

Boyd CE (1995) Bottom soils, sediment and pond aquaculture. Chapman and Hall, New York

Briggs MRP, Fvnge-Smith SJ (2008) A nutrient budget of some intensive marine shrimp ponds in Thailand. Aquac Res 25:789–811. https://doi.org/10.1111/j.1365-2109.1994.tb00744.x

Brunner PH, Rechbenger H (2004) Practical handbook of material flow analysis. Lewis Publishers

Brzeski VJ, Newkirk GF (2002) Lessons in resource management from the Tam Giang Lagoon. Dalhousie University, Coastal Resources Research Network

Buathong T, Boontanon SK, Boontanon N, Surinkul N, Harada H, Fujii S (2013) Nitrogen flow analysis in Bangkok City, Thailand: area zoning and questionnaire investigation approach. Procedia Environ Sci 17:586–595. https://doi.org/10.1016/j.proenv.2013.02.074

Bui TMH, Schreinemachers P (2011) Resettling farm households in northwestern Vietnam: livelihood change and adaptation. Int J Water Resour Dev 27(4):769–785. https://doi.org/10.1080/07900627.2011.593116

Ca VT, Hieu PV, Luong CV, Tien DD (2011) An experimental application of a quantitative method for the conservation of seagrass ecosystems in the coastal areas of Ly Son Island, Quang Ngai. Vietnam J Marine Sci Technol 11(4):47–56 (in Vietnamese). https://doi.org/10.15625/1859-3097/385

Carpenter SR, Caraco NF, Correll DL, Howarth RW, Shrapley AN, Smith VH (1998) Nonpoint pollution of surface water with phosphorus and nitrogen. Ecological Appl 8(3):559–568. https://doi.org/10.2307/2641247

Cernea MM (2000) Risks, safeguards and reconstruction: a model for population displacement and resettlement. Econ Polit Weekly 35(41):3659–3678. https://www.jstor.org/stable/4409836

Chen M, Chen J, Sun F (2008) Agricultural phosphorus flow and its environmental impacts in China. Sci Total Environ 405:140–152. https://doi.org/10.1016/j.scitotenv.2008.06.031

Chieu TT (1994) Soil mapping in Vietnam and the application of FAO-UNESCO and soil taxonomy soil classification systems. Vietnam Soil Sci J 4:7–9 (in Vietnamese, with English abstract)

Clement F (2010) Analysing decentralised natural resource governance: proposition for a "politicised" institutional analysis and development framework. Policy Sci 43(2):129–156. https://doi.org/10.1007/s11077-009-9100-8

Congalton RG, Green K (1999) Assessing the accuracy of remotely sensed data: principles and practices. Lewis Publishers, FL, Boca Raton

Corwin DL, Lesch SM (2003) Application of soil electrical conductivity to precision agriculture: theory, principles, and guidelines. Agron J 95:455–471. https://doi.org/10.2134/agronj2003.0455

Dai NH (2010) Management and restoration of seagrass beds in Nui Thanh District, Quang Nam Province. Vietnam J Marine Sci Technol 10(2):63–75 (in Vietnamese) https://doi.org/10.15625/1859-3097/10/2/913

Dan NV, Hai HT, An NH (2006) Results of livelihood analysis by participatory rural appraisal in Quang Phuoc Commune 2006: integrated management of lagoon activities in Tam Giang Cau Hai Lagoon (IMOLA) Project. 1–8. People's Committee of Thua Thien Hue Province, Hue City

Dao N (2010) Dam development in Vietnam: the evolution of dam-induced resettlement policy. Water Altern 3(2):324

De Koninck R (1999) Deforestation in Vietnam. International Development Research Center (IDRC), Ottawa
Dexin Ch (2009) Bright vanilla. Hainan Bright Industry Co., Ltd.
Dierolf T, Fairhurst T, Mutert E (2001) Soil Fertility Kit. GTZ-GmbH, FAO, PT Jasa Katom, and PPI and PPIC. Oxford Graphic Printer
Dung TV, Hung HV, Hoa HTL (2009) Biogas user survey 2007–2008. Biogas development Programme for Livestock Sector in Vietnam 2007–2011
ENVI (2009) FLAASH User's Guide, Version 4.7. ENVI, CO: Boulder
FAO (2005) The state of the World's land and water resources for food and agriculture (SOLAW). Accessed 25 May 2007 from https://www.fao.org/3/i1688e/i1688e08.pdf
FAO (2006) Study on livelihood, aquaculture source and aquaculture raising in Tam Giang—Cau Hai Lagoon, Thua Thien Hue. FAO, Rome
FAO (2010) Improving nutrition through home gardening. Accessed 18 July 2012 from https://www.echocommunity.org/en/resources/c7e44987-37e1-4ef1-bbd5-8b3e3b507032
FAO (2017) FAO yearbook, Fisheries and aquaculture statistics 2015. FAO
FAO-ISRIC-ISSS (1998) World reference base for soil resources. Acco Press, Leuven
Fuavao J (2012) Pacific Islands exporting vanilla to New Zealand. Pacific Island Trade and Invest (PT&I), USA
Funakawa S, Kosaki T (2007) Potential risk of soil salinization in different regions of Central Asia with special reference to salt reserves in deep layers of soils. Soil Sci Plant Nutr 53:634–649. https://doi.org/10.1111/j.1747-0765.2007.00186.x
Funakawa S, Suzuki R, Karbozova E, Kosaki T, Ishida N (2000) Salt-affected soils under rice-based irrigation agriculture in southern Kazakhstan. Geoderma 97:61–85. https://doi.org/10.1016/S0016-7061(00)00026-4
Garcia RA, Hedley JD, Hoang CT, Fearns PR (2015) A method to analyze the potential of optical remote sensing for benthic habitat mapping. Remote Sens 7(10):13157–13189. https://doi.org/10.3390/rs71013157
Gebauer J (2005) Plant species diversity of home gardens in EL Obeid, Central Sudan. J Agric Rural Dev Trop Subtrop 106:97–103
Gee GW, Bauder JW (1986) Chapter 15. Particle-size Analysis. In: Jackson ML, Lim CH, Zelazny LW (eds) Part 1. Physical and mineralogical methods. Method of soil analysis, 2nd edn. American Society of Agronomy and Soil Science Society of America, WI, Madison, pp 383–411
Giang PH, Harada H, Fujii S, Lien NPH, Hai HT, Anh PN, Tanaka S (2015) Transition of fertilizer application and agricultural pollution loads: a case study in the Nhue-Day River basin. Water Sci Technol 72(7):1072. https://doi.org/10.2166/wst.2015.312
General Statistical Office of Vietnam (2012) Part I Result of the 2011 rural—agriculture and fishery census. Statistical Publishing House, Hanoi (in Vietnamese), In Statistical Year Book of Vietnam
General Statistics Office of Vietnam (2014a) Yield of main annual crops. Accessed 16 Feb 2023 from https://pxweb.gso.gov.vn:443/sq/f72aac50-4408-420a-8879-085185af8322
General Statistics Office of Vietnam (2014b). Planted area of main annual crops. Accessed 16 Feb 2023 from https://pxweb.gso.gov.vn:443/sq/d2428fa0-c5e0-4fd8-86cd-91a6b87cb7b7
Ghafoor AG, Qudir M, Murtaza G (2004) Salt affected soil: principles of management. Allied Book Centre, Lahore
Grattan SR, Zeng L, Shannon MC, Roberts SR (2002) Rice is more sensitive to salinity than previously thought. Calif Agric 56(6):189–198. https://doi.org/10.3733/ca.v056n06p189
Ha PQ, Hien BH, Hoa HTT, Tu PK, Ninh HT, Loan BTP, Quynh VD, Dufey JE (2005) Overview of sandy soils management in Vietnam. In: Proceedings of the international management of tropical sandy soils for sustainable agriculture symposium. Thailand, pp 348–352
Ha VQ (2009) Programme 135—Sharing lessons on poverty reduction and development schemes for ethnic minorities in Vietnam. United Nations
Hanoi Statistical Office (2010) Hanoi statistical yearbook 2009. Statistical Publishing House, Hanoi
Harada H, Fujii S (2020) Challenges and potentials of ecological sanitation: experiences from the cases in Vietnam and Malawi. Sanitation Value Chain 4(1):3–16. https://doi.org/10.34416/svc.00015
Hanoi University of Agriculture (2011) Vietnam's encyclopedic dictionary of agriculture. Agriculture Publishing House, Hanoi
Harada H, Dong NT, Matsui S, Fujii S (2010a) Traditional urine-diversion practices in sub-urban Hanoi, Vietnam: Replaced or promoted? Water Practice Technol 5(3):1–6. https://doi.org/10.2166/WPT.2010.058
Harada H, Adachi T, Fujii S, Nguyen PHL, Huynh TH (2010b) Phosphorus flow analysis in Hanoi focusing on wastewater, agriculture and stockbreeding. Environ Eng Res 47:465–474. http://library.jsce.or.jp/jsce/open/00516/2010/47-0465.pdf
Harada H, Kobayashi H, Fujieda A, Kusakabe T, Shimizu Y (2012) Urine-diverting system for securing sanitation in disaster and emergency situations. Leadersh Manag Eng 12(4):309–314. https://doi.org/10.1061/(ASCE)LM.1943-5630.0000180
Haruyama S, Dinh HT, Le VT (2006) Problem soils of the paddy fields in the eastern coastal plain of the Red River Delta, Vietnam. J Geogr (Chigaku Zasshi) 115(6):737–749. https://doi.org/10.5026/jgeography.115.6_737
Havkin-Frenkll D, Belanger F (2010) Handbook of vanilla science and technology. Wiley

Heffer P, Prud'homme M (2010) Fertilizer Outlook 2010–2014. In: 78th IFA annual conference Paris (France). International Fertilizer Industry Association (IFA), Paris

Hoang CT, Phap TT, Tuan NQ, Son TPH (2011) Application of remote sensing and geographical information system (GIS) techniques in assessing the existing situation of seagrass meadow at coastal wetland in Huong Phong Commune, Huong Tra District, Thua Thien Hue Province. Hue Univ J Sci 65(2):231–239 (in Vietnamese)

Hoang CT, Cole A, Fotedar R, O' Leary M (2015) Seasonality and distributions of macro-algae *Sargassum* beds at Point Peron, Shoalwater Islands Marine Park, Western Australia. J R Soc Western Australia 98:97–98. https://doi.org/10.23919/oceans.2015.7404388

Hoang CT, Cole AJ, Fotedar RK, O'Leary M, Lomas MW, Roy S (2016a) Seasonal changes in water quality and *Sargassum* biomass in southwest Australia. Mar Ecol Prog Ser 551:63–79. https://doi.org/10.3354/meps11735

Hoang CT, Lomas MW, Ishizaka J (2016b) Satellite-derived estimates of primary production during the Sargasso Sea winter/spring bloom: Integration of in-situ time-series data and ocean color remote sensing observations. Reg Stud Marine Sci 3:131–143. https://doi.org/10.1016/j.rsma.2015.07.002

Hoang CT, O'Leary M, Fotedar R (2016c. Remote-sensed mapping of *Sargassum* spp. distribution around Rottnest Island, Western Australia using high-spatial resolution WorldView-2 satellite data. J Coastal Res 32(6):1310–1321. https://doi.org/10.2112/jcoastres-d-15-00077.1

Hoang CT, Garcia R, O'Leary M, Fotedar R (2016d) Identification and mapping of marine submerged aquatic vegetation in shallow coastal waters with WorldView-2 satellite data. J Coastal Res 75(sp1):1287–1291. https://doi.org/10.2112/SI75-258.1

Hoang TTH, Phan TC, Hoang MT, Chen W, Bell R (2010) Sandy soils in South Central Coastal Vietnam: Their origin, constraints and management. In: Proceedings of 19th world congress of soil science, soil solutions for a changing world. Brisbane, Australia, pp 251–254

Hong NX, Van NTM (2007) Sustainable livelihoods for people of the upland: the problem of land tenure and bargaining strategy (case study in Hong Ha, A Luoi, Thua Thien Hue). Science Report. Hue University of Agriculture and Forestry (in Vietnamese)

Hong TTT, An LV (2009) Management ago-forestry land used and livelihoods for A Luoi District people, Thua Thien Hue: Result of survey in 2009, Community-Based Natural Resource Management in the upland of Vietnam, period 2000–2011. Hue University of Agriculture and Forestry (in Vietnamese), CARD Office

Hong Ha Commune Office (2012) Statistics of Hong Ha People's committee. Hong Ha Commune Office

Hu J, Kawamura H, Hong H, Qi Y (2000) A review on the currents in the South China Sea: seasonal circulation, South China sea warm current and Kuroshio intrusion. J Oceanogr 56:607–624. https://doi.org/10.1023/A:1011117531252

Huan TN, Tan PS, Hiraoka H (2000) Optimum fertilizer nitrogen rate for high yielding rice based on growth diagnosis in wet seeded culture of rice. In: Proceedings of the 2000 annual workshop of JIRCAS Mekong Delta Project, pp 60–68

Hung L, Tien PC (1986) Upper paleozoic stratigraphy in Indochina. In: Geology of Campuchia, Laos and Vietnam. Science Publisher, Hanoi, pp 53–64 (in Vietnamese)

IHME (Institute for Health Metrics and Evaluation) (2019) 2019 Global Burden of Disease (GBD) study. Accessed 14 May 2022 from https://vizhub.healthdata.org/gbd-results/

IMF (International Monetary Fund) (2018) Global price of Shrimp [PSHRIUSDM]. Accessed 22 April 2018 from FRED, Federal Reserve Bank of St. Louis: https://fred.stlouisfed.org/series/PSHRIUSDM

Inoue K, Saizen I, Tokito M (2021) Asian Rural Future 2030, The Association of Rural Planning, Japan. Accessed 13 Dec 2022 from http://rural-planning.jp/wordpress/wp-content/uploads/2022/02/AsianRuralFuture2030.pdf

Institute for Development Consultation (CODE) (2010. Displacement, resettlement, living stability and environmental and resources protection in hydropower dam projects. CODE, Hanoi (in Vietnamese)

Japan Fisheries Resource Conservation Association (2005) Water standard for fisheries (in Japanese)

JICA (2013) Agricultural transformation and food security 2040: ASEAN region with a focus on Vietnam, Indonesia, and Philippines—Vietnam Country Report. Accessed 13 Dec 2022 from https://openjicareport.jica.go.jp/pdf/12145546.pdf

Jong WD, San DD, Hung TV (2006) Forest rehabilitation in Vietnam: Histories, realities and future. CIFOR (Center for International Forestry Research), Jakarta

Jose D, Shanmugaratnam N (1993) Traditional home gardens of kerala: a suitable human ecosystem. Agrofor Syst 24:203–213. https://doi.org/10.1007/BF00706892

Kirk GDJ (2004) Biogeochemistry of submerged soils. John Wiley and Sons, Chichester

Koedsin W, Intararuang W, Ritchie RJ, Huete A (2016) An integrated field and remote sensing method for mapping seagrass species, cover, and biomass in southern Thailand. Remote Sens 8(4):292. https://doi.org/10.3390/rs8040292

Kono Y, Rambo AT (2004) Some key issues relating to sustainable agro-resources management in the mountainous region of mainland Southeast Asia. Southeast Asian Stud 41(4):550

Kumar BM, Nair PKR (2004) The enigma of tropical home gardens agroforestry system. Agrofor Syst 61:135–152. https://doi.org/10.1023/B:AGFO.0000028995.13227.ca

Kyeyune V, Turner S (2016) Yielding to high yields? Critiquing food security definitions and policy implications for ethnic minority livelihoods in upland Vietnam. Geoforum 71:33–43. https://doi.org/10.1016/j.geoforum.2016.03.001

Lang NT, Buu BC, Viet NV, Ismail AM (2010) Part III: agriculture. Strategies for improving and stabilizing rice productivity in the Coastal Zones of the Mekong Delta, Vietnam. In: Hoanh CT, Szuster BW, Suan-Pheng K, Ismail AM, Noble AD (eds) Tropical deltas and coastal zones: food production, communities and environment at the land-water interface. CPI Antony Rowe Ltd, Wallingford, pp 209–222

Le BT (1998) Vietnam-territories and geographical areas. The Gioi Publishing House, Hanoi (in Vietnamese)

Le BT (2009) Vietnam-Nature. Viet Nam Education Publishing House, Hanoi (in Vietnamese)

Leslie J (2005) Deep water: the epic struggle over dams, displaced people and the environment. Picador, New York

Li Y, Shi Z, Li F (2007) Delineation of site-specific management zones based on temporal and spatial variability of soil electrical conductivity. Pedosphere 17(2):156–164. https://doi.org/10.1016/S1002-0160(07)60021-6

Lyzenga DR (1981) Remote sensing of bottom reflectance and water attenuation parameters in shallow water using aircraft and Landsat data. Int J Remote Sens 2(1):71–82. https://doi.org/10.1080/01431168108948342

Manuputty A, Lumban-Gaol J, Agus SB (2015) Seagrass mapping based on satellite image WorldView-2 by using depth invariant index method. Sci./ILMU Kelatun: Indonesian J Marine Sci 21(1):37–44. https://doi.org/10.14710/ik.ijms.21.1.37-44

Martins D (2009) Vanilla—an emerging crop in western Uganda, tool for conservation and use of pollination services initial survey of good pollination practices. FAO, Italy

McElwee P (2009) Reforesting "Bare Hills" in Vietnam: Social and environmental consequences of the 5 million hectare reforestation program. AMBIO J Hum Environ 38(6):325–333. https://doi.org/10.1579/08-R-520.1

Mekong River Commission (2002) Land resource inventory for agricultural development (basin wide) project. Part III soil database final report June 2002. Mekong River Commission, Phnom Penh

Meteorological Station of Thua Thien Hue Province (2012, 2013) Climatic conditions from January 2012 to May 2013

Meyfroidt P, Lambin EF (2008) The causes of the reforestation in Vietnam. Land Use Policy 25:182–197. https://doi.org/10.1016/j.landusepol.2007.06.001

Meyfroidt P, Vu TP, Hoang VA (2013) Trajectories of deforestation, coffee expansion and displacement of shifting cultivation in the Central Highlands of Vietnam. Glob Environ Chang 23(5):1187–1198. https://doi.org/10.1016/j.gloenvcha.2013.04.005

Millat-e-Mustafa MD, Hall JB, Tekhlchaimanot Z (1994) Structure and floristics of Bangladesh Homegarden. Agrofor Syst 33:263–280. https://doi.org/10.1007/BF00055427

Miller JJ, Curtin D (2006) Chapter 15. Electrical conductivity and soluble ions. In: Carter MR, Gregorich EG (eds) Soil sampling and methods of analysis, 2nd edn. CRC Press, pp 161–171

Miyagawa S, Kochan S (1990) Village homegarden cultivation in Northeast Thailand. Jpn J Tropic Agric 34:235–242

Mohri H, Lahoti S, Saito O, Mahalingama A, Gunatilleke N, Irham Hoang VT, Hitinayake G, Takeuchi K, Herth S (2013) Assessment of ecosystem services in home garden system in Indonesia, Sri Lanka and Vietnam. Ecosyst Serv 5:124–136. https://doi.org/10.1016/j.ecoser.2013.07.006

Montangero A, Belevi H (2007) Assessing nutrient flows in septic tanks by eliciting expert judgment: a promising method in the context of developing countries. Water Res 41(5):1052–1064. https://doi.org/10.1016/j.watres.2006.10.036

Montangero A, Cau LN, Anh NV, Tucan VD, Nga PT, Belevi H (2007) Optimizing water and phosphorus management in the urban environmental sanitation system of Hanoi, Vietnam. Sci Total Environ 384:55–66. https://doi.org/10.1016/j.scitotenv.2007.05.032

Moral FJ, Terrón JM, Marques da Silva JR (2010) Delineation of management zones using mobile measurements of soil apparent electrical conductivity and multivariate geostatistical techniques. Soil Tillage Res 106:335–343. https://doi.org/10.1016/j.still.2009.12.002

Moran MS, Jackson RD, Slater PN, Teillet PM (1992) Evaluation of simplified procedures for retrieval of land surface reflectance factors from satellite sensor output. Remote Sens Environ 41:169–184. https://doi.org/10.1016/0034-4257(92)90076-v

Motohiro C (2014) Lagoon and water surface management commune in Thua Thien Hue, Central Vietnam in 19–20 century. Hiroshima J Asian Res 19:1–23 (in Japanese)

Naklang K, Whitbread A, Lefroy R, Blair G, Wonprasaid S, Konboon Y, Suriya-arunroj D (1999) The management of rice straw, fertilizers and leaf litters in rice cropping systems in Northeast Thailand. Plant Soil 209:21–28. https://doi.org/10.1023/A:1004571015620

Neef A, Singer J (2015) Development-induced displacement in Asia: conflicts, risks, and resilience. Dev Pract 25(5):601–611. https://doi.org/10.1080/09614524.2015.1052374

Nga DT, Antoine M, Hung MV, Phuc PD, Kei N, Thamarat K (2011) Assessing nutrient fluxes in a Vietnamese rural area despite limited and highly uncertain data. Resour Conserv Recycl 55:849–856. https://doi.org/10.1016/j.resconrec.2011.04.008

Ngan PK, Huoc LH, Long PD (1986) Lower Paleozoic stratigraphy in Indochina. In: Geology of Campuchia, Laos and Vietnam. Science Publisher, Hanoi, pp 30–46 (in Vietnamese)

Nguyen KN (1992) VAC's handbook technique. Agriculture Publishing House, Hanoi (in Vietnamese)

Nguyen NN, Özçaglar-Toulouse N, Kjeldgaard D (2017) Toward an understanding of young consumers' daily consumption practices in post-Doi Moi Vietnam. J Bus Res 86:490–500. https://doi.org/10.1016/j.jbusres.2017.10.002

Nguyen TMA (2016) Transformation of ethnic minorities' society in Central Vietnam: Livelihood negotiation during the transition from moral economy. [Doctoral dissertation, Okayama University]. Okayama University Scientific achievement Repository. https://ousar.lib.okayama-u.ac.jp/54345

Nguyen VT, Dang NT, Nguyen HD (2001) Seagrass of Vietnam. Publishing House of Science and Technology, Hanoi (in Vietnamese)

Nhan DK, Verdengem MCJ, Milstein A, Verreth JAV (2008) Water and nutrient budgets of ponds in integrated agriculture-aquaculture systems in the Mekong Delta, Vietnam. Aquaculture 39(11):1216–1228. https://doi.org/10.1111/j.1365-2109.2008.01986.x

NIFS (National Institute for Soil and Fertilizer) (2002) General information on main soils of Vietnam. The World Publishing, Hanoi

Okamoto Y, Tanaka U, Mizuno K, Nguyen PN (2008) Seasonal change of sediment condition in aquaculture area and fishery resource management in Sam-An Truyen Lagoon, Central Vietnam. J Jpn Agric Syst Soc 25(1):71–78. https://doi.org/10.14962/jass.25.1_71 (in Japanese, with English abstract)

Okamoto Y, Tanaka U, Mizuno K, Le VA (2012a) Spatial transition of the fishery activities in Sam-An Truyen Lagoon, Central Vietnam. J Jpn Agric Syst Soc 28(2):63–71. https://doi.org/10.14962/jass.28.2_63

Okamoto Y, Tanaka U, Mizuno K, Nguyen PN, Le VA (2012b) Changes on characteristics of sediment and fishery environment in Tam Giang Lagoon, Central Vietnam. J Jpn Agric Syst Soc 28(1):19–27. https://doi.org/10.14962/jass.28.1_19

Parthasarathy VA, Anandarj M, Rema J, Sasikumar B, Suseela Bhai R (2005) Vanilla (extension pamphlet). Indian Institute of Species Research

Pham HG, Harada H, Fujii S, Nguyen PHL, Huynh TH, Pham NA, Tanaka S (2015) Transition of fertilizer application and agricultural pollution loads: a case study in the Nhue-Day River basin. Water Sci Technol 72(7):1072–1081. https://doi.org/10.2166/wst.2015.312

Pham HG, Harada H, Fujii S, Nguyen PHL, Huynh TH (2017) Transition of human and livestock waste management in rural Hanoi: a material flow analysis of nitrogen and phosphorus during 1980–2010. J Mater Cycles Waste Manag 19:827–839. https://doi.org/10.1007/s10163-016-0484-1

Pimhidzai O (2018) Climbing the ladder: poverty reduction and shared prosperity in Vietnam. The World Bank, Washington DC

Power JF, Peterson GA (1998) Nitrogen transformations, utilization, and conservation as affected by fallow tillage method. Soil Tillage Res 49:37–47. https://doi.org/10.1016/S0167-1987(98)00153-6

Qadir M, Ghafoor A, Murtaza G (2000) Amelioration strategies for saline soils: a review. Land Degrad Dev 11(6):501–521. https://doi.org/10.1002/1099-145X(200011/12)11:6<501::AID-LDR405>3.0.CO;2-S

Rain P (2004) Vanilla: The culture history of the world's favorite flavor and fragrance. J. P. Tarcher/Penguin, New York

Rogers M (2019) Vietnam's renewable energy future. McKinsey Sustain. https://www.mckinsey.com/business-functions/sustainability/our-insights/sustainability-blog/vietnams-renewable-energy-future

Rowell DL (1994) Chapter 14. The preparation of saturation extracts and the analysis of soil salinity and sodicity. In: Rowell DL (ed) Soil science: methods and applications. Longman Group, Essex, pp 277–302

Sahrawat KL (2005) Fertility and organic matter in submergence rice soils. Int Crop Res Inst Semi-arid Trop India 88(5):735–739

Salam MA, Babu KS, Mohanakumaran N (1995) Home garden agriculture in kerala revisited. Food Nutr Bull 16(3):220–223

Schroeder TA, Cohen WB, Song C, Canty MJ, Yang Z (2006) Radiometric correction of multi-temporal Landsat data for characterization of early successional forest patterns in western Oregon. Remote Sens Environ 103(1):16–26. https://doi.org/10.1016/j.rse.2006.03.008

Scoones I (1998) Sustainable rural livelihoods: a framework for analysis. IDS Working Paper no 72, Institute of Development Studies, Sussex

SEAFDEC (Southeast Asian Fisheries Development Center) (2002) Fishing gear and methods in Southeast Asia, Vietnam. SEAFDEC

SEAFDEC (Southeast Asian Fisheries Development Center) (2018) Fisheries Country Profile: Viet Nam. Accessed 23 April 2018 from http://www.seafdec.org/fisheries-country-profile-viet-nam/

Senanayake SGJN, Wijerathen V (1988) Heritability and genotypic and phenotypic correlations of yield, yield components, and protein content in Cowpea (*Vigna unguiculata* (L.) Walp.). Beiträge zur Tropischen Landwirtschaft und Veterinärmedizin 26(3):279–283

Sharon A (2008) A report on vanilla cost benefit analysis. Sustainable Agriculture Trainers' Network (SANET), Uganda

Shirato Y, Paisancharoen K, Sangtong P, Nakviro C, Yokozawa M, Matsumoto N (2005) Testing the Rothamsted carbon model against data from long term experiments on upland soils in Thailand. Eur J Soil Sci 56(2):179–188. https://doi.org/10.1111/j.1365-2389.2004.00659.x

Siem NT, Phien T (1999) Upland soils in Vietnam degradation and rehabilitation. Agricultural Publishing House, Hanoi (in Vietnamese)

Singer J (2019) Paying resettled communities for environmental services: legally mandated benefit-sharing for Vietnam's dam-displaced. In: Price S, Singer J (eds) Country frameworks for development displacement and resettlement. Routledge, pp 179–194

Singer J, Hoang H (2014) Participation in resettlement decision-making by dam-displaced villagers in central Vietnam. Asia J Global Stud 6(1):16–29

Singer J, Watanabe T (2014) Reducing reservoir impacts and improving outcomes for dam-forced resettlement: experiences in central Vietnam. Lakes Reserv Res Manag 19(3):225–235

Singer J, Pham HT, Hoang H (2014) Broadening stakeholder participation to improve outcomes for dam-forced resettlement in Vietnam. Water Resour Rural Dev 4:85–103. https://doi.org/10.1016/j.wrr.2014.07.001

Socialist Republic of Viet Nam (2004) National report on disaster reduction in Vietnam—for the World Conference on Disaster Reduction, Kobe-Hyogo, Japan, January 18–22, 2005. Hanoi

Son TPH (2010) Application of remote sensing techniques for coral mapping in coastal waters of Vietnam seas. The scientific report in the Joint Russian-Vietnamese project of the Institute of Marine Biology RAS and Institute of oceanography VAST, Biodiversity and reproduction ability in coral ecosystems of Vietnam

Sugimori Y, Funakawa S, Pachikin KM, Ishida N, Kosaki T (2008) Soil salinity dynamics in irrigated fields and its effects on paddy-based rotation systems in southern Kazakhstan. Land Degrad Dev 19(3):305–320. https://doi.org/10.1002/ldr.843

Sujatha S, Bhat R (2010) Response of vanilla (*Vanilla planifolia*) intercropped in arecanut to irrigation and nutrition in humid tropics of India. Agric Water Manag 97(7):988–994. https://doi.org/10.1016/j.agwat.2010.01.031

Sunderlin WD, Ba HT (2005) Poverty alleviation and forests in Vietnam. CIFOR (Center for International Forestry Research), Bogor

Sy DT, Vien TD, Quang NV (2005) Environment and food safety in peri-urban Hanoi. Centre for Agricultural Research and Ecological Studies (CARES), Hanoi Agricultural University, Wageningen University and Research Centre

Tanaka K, Watabe T (1981) Traditional cropping systems of small farmers in the central and southern Deccan plateau Area. Southeast Asian Stud 19(2):205–221. https://doi.org/10.20495/tak.19.2_205

Thang HN, Son TPH (2012) A comparison of bottom reflectance index and depth invariant index method for detecting seagrass distribution in Lap An Lagoon, Thua Thien Hue. Project: seagrass in Thua Thien Hue Province, Vietnam (in Vietnamese)

Thang HN, Trang HT (2015) Seagrass canopies situation assessment by using Landsat 8 multi-spectral imagery in Lap An Lagoon, Thua Thien Hue, Vietnam. Hue Univ J Sci Agric Rural Dev 104(5). https://doi.org/10.26459/jard.v104i5.2955 (in Vietnamese)

Thanh TDz, Hoe ND, Hong ND (1986) A generalying on Devonian system in Indochina. In: Geology of Campuchia, Laos and Vietnam. Science Publisher, Hanoi, pp 46–53 (in Vietnamese)

Thanh NV, Thang LV, Canh HH, Lanh NK, Loi TV (2005) Thua Thien Hue monograph—natural part, 1st edn. Scientific and Society Publishing House, Hanoi (in Vietnamese)

Thi PT (1985) Metamorphic complexes of the socialist republic of Vietnam. In: Lee SM (ed) Explanatory text of the metamorphic map of South and East Asia (1:10,000,000), pp 90–97

Thua Thien Hue Statistical Office (2010) Gross domestic product at current price by economic sector, productivity of employed population by kind of economic activity and output value of aquaculture at constant 1994 price by kind of activity. Stat Yearb 2009 103(230) (in Vietnamese)

Thulstrup AW (2014) Plantation livelihoods in central Vietnam: implications for household vulnerability and community resilience. Norwegian J Geogr 68(1):1–9. https://doi.org/10.1080/00291951.2013.870928

Thulstrup AW (2015) Livelihood resilience and adaptive capacity: Tracing changes in household access to capital in Central Vietnam. World Dev 74:352–362. https://doi.org/10.1016/j.worlddev.2015.05.019

Thulstrup AW, Casse T, Nielsen TT (2013) The push for plantation: drivers, rationales and social vulnerability in Quang Nam Province, Vietnam. In: Bruun O, Casse T (eds) On the frontiers of climate and environmental change. Springer, Berlin, pp 71–89

To XP, Chan LH, Cao TC (2021. Supply of raw acacia wood in Vietnam—status and trends (in Vietnamese)

Toan NV (2004) Characteristics of coastal arenosol soils and present utilization in Northern Central region. Vietnam Soil Sci J 68:25–29 (in Vietnamese, with English abstract)

Tokito M, Saizen I, Asano S (2015) Analysis of the present situation and the vulnerability of ethnic minorities' livelihood structure at a rural village of Central Vietnam. Environ Inf Sci 29:123–128 (in Japanese, with English abstract)

Tomczak M, Godfrey JS (1994) Adjacent seas of the Pacific Ocean. In: Regional oceanography: an introduction. Pergamon Press, Oxford, pp 173–191

Tra TD (1989) The culture of Gracilaria verrucosa in Tam Giang Lagoon, Thua Thien Hue Province, S. R.

Vietnam. Seminar papers: Session 1. Status of culture of tropical agarophyte, pp 20–24. Report of seminar on *Gracilaria* production and utilization the bay of Bengalregion, Songkhai, Thailand, 23–27 October 1989. FAO, Bay of Bengal Programme
Tran NN (1995) The geology of Vietnam: a brief summary and problems. Geoscience Reports of Shizuoka University, pp 1–10
Tran TV (2009) The cotu in Vietnam. Vietnam News Agency Publishing House, Hanoi
Trinh LN, Watson JW, Hue NN, De NN, Minh NV, Chu P, Sthapit BR, Eyzaguirre PB (2003) Agrobiodiversity conservation and development in Vietnamese homegardens. Agr Ecosyst Environ 97(1–3):317–344. https://doi.org/10.1016/S0167-8809(02)00228-1
Truong NV, Binh NN (1985) Business agro-forestry promotion effective to potential labor, land and natural resource. Agricultural Publishing House, Hanoi (in Vietnamese), Ministry of Forestry
TTHPPC (Thua Thien Hue Provincial People's Committee) (2005) Geography book of Thua Thien Hue province—natural. Social Sciences Publishing House, Hanoi (in Vietnamese)
Tu PK, Hoa HTT, Duyet HN, Huong LD, Hao ND, Dung NT, Hieu NM, Ngoan LD, Ha PQ, Lebailly F, Francis F, Haubruge E, Bragard C, Dufey JE (2005) Farming systems in the sandy area of the Thua Thien Hue Province, Central Vietnam. Survey of socio-economic situation and constraints identified by farmers. In: Proceedings of the international management of tropical sandy soils for sustainable agriculture symposium. Thailand, pp 75–80
Tuya F, Haroun R, Espino F (2014) Economic assessment of ecosystem services: monetary value of seagrass meadows for coastal fisheries. Ocean Coast Manag 96:181–187. https://doi.org/10.1016/j.ocecoaman.2014.04.032
Ty PH (2008) Assessment of impoverishment risks, livelihood restoration strategies and land use planning within involuntary displacement and resettlement plan of hydropower development projects. Agriculture Publishing House, Hanoi (in Vietnamese)
Uede T (2008) Validity of acid volatile sulfide as environmental index and experiment for fixing its standard value in aquaculture farms along the coast of Wakayama Prefecture, Japan. Nippon Suisan Gakkaishi (The Japanese Society of Fisheries Science) 74(3):402–411 (in Japanese, with English abstract). https://doi.org/10.2331/suisan.74.402
United Nations (1986). Declaration on the right to development. UN Human Rights, Office of the High Commissioner. Accessed 4 Jan 2021, from https://www.ohchr.org/en/instruments-mechanisms/instruments/declaration-right-development
United Nations (2004) Guiding principles on internal displacement. UN, New York
United States Salinity Laboratory Staff (1954) Diagnosis and improvement of saline and alkali soils. In: Richards LA (ed) Agriculture handbook, vol 60. U. S. Government Printing Office, Washington DC, pp 69–81
US International Trade Administration (2021) Vietnam country commercial guide. Accessed 12 Nov 2021 from https://www.trade.gov/country-commercial-guides/vietnam-power-generation-transmission-and-distribution#:~:text=In%20the%20capacity%20structure%20of,percent%2C%20wind%20power%201%20percent%2C
Van Wambeke, A. 1992. Chapter 13 *Entisols*. In Soils of the tropics: Properties and appraisal (pp. 253–266). McGraw-Hill, New York
VASS (Vietnamese Academy of Social Sciences) (2005) Encyclopedic dictionary of Vietnam, Part 3 (N-S). Vietnam's Encyclopedia Publishing House, Hanoi (in Vietnamese)
VEA (Vietnam Environment Administration) (2012) Monitoring results of project on environmental protection of river basins in Vietnam. Centre for Environmental Monitoring, VEA
Verma TS, Bhagat RM (1992) Impact of rice straw management practices on yield, nitrogen uptake, and soil properties in a wheat-rice rotation in northern India. Fertilizer Res 33:97–106. https://doi.org/10.1007/BF01051164
Vien TD (2003) Culture, environment, and farming systems in Vietnam's northern mountain region. Southeast Asian Stud 41(2):180–205. https://doi.org/10.20495/tak.41.2_180
Vien TD (2010) Marketing and agroforestry development in Vietnam's upland. Agriculture Publishing House, Hanoi
Vien TD, Leisz SJ, Lam NT, Rambo AT (2006) Using traditional swidden agriculture to enhance rural livelihoods in Vietnam's upland. Mt Res Dev 26(3):192–196. https://doi.org/10.1659/0276-4741(2006)26[192:UTSATE]2.0.CO;2
Vietnam Ministry of Agriculture and Rural Development (2021) Status of special-use and protection forests in Vietnam 2019 (in Vietnamese)
Vietnam Ministry of Natural Resources and Environment (2021) Report on the state of the national environment for the period 2016–2020. Dan Tri Publishing House, Hanoi (in Vietnamese)
Vietnam Society of Soil Science (VSSS) (1996) The soils of Vietnam. Agricultural Publishing House, Hanoi (in Vietnamese)
Vu TKV, Tran MT, Dang TTS (2007) A survey of manure management on pig farms in Northern Vietnam. Livest Sci 112(3):288–297. https://doi.org/10.1016/j.livsci.2007.09.008
Vu TM, Mizuno K, Funakawa S, Shinjo H, Tanaka U, Le VA (2015a) Home garden practices and crop contribution to livelihood in mountainous village of central Vietnam. Trop Agric Dev 59(3):118–126. https://doi.org/10.11248/jsta.59.118

Vu TM, Mizuno K, Funakawa S, Shinjo H, Tanaka U, Le VA (2015b) Introduction of vanilla to mountainous villages of central Vietnam. Trop Agric Dev 59(4):199–206. https://doi.org/10.11248/jsta.59.199

Whitbread A, Blair G, Konboon Y, Lefroy R, Naklang K (2003) Managing crop residues, fertilizers and leaf litters to improve soil C, nutrient balances, and the grain yield of rice and wheat cropping systems in Thailand and Australia. Agric Ecosyst Environ 100(2–3):251–263. https://doi.org/10.1016/S0167-8809(03)00189-0

White PF, Oberthür T, Pheav S (1997) The soils used for rice production in Cambodia, A manual for their recognition and management. International Rice Research Institute, Manila

Winblad U, Simpson-Hebert M (eds) (2004) Ecological sanitation (revised and enlarged edition). SEI

World Bank (2011) Vietnam development report 2011: Natural resources management. Joint development partner report to the Vietnam Consultative Group Meeting, Hanoi, December 7–8, 2010. The World Bank, Washington D.C

Zeng L, Shannon MC (2000) Salinity effects on seedling growth and yield components of rice. Crop Sci 40:996–1003. https://doi.org/10.2135/cropsci2000.404996x

Zheng Z, Zhang F, Ma F, Chai X, Zhu Z, Shi J, Zhang S (2009) Spatiotemporal changes in soil salinity in a dripirrigated field. Geoderma 149:243–248. https://doi.org/10.1016/j.geoderma.2008.12.002

Livelihood and Environmental Impact in Urban Areas

5

Hidenori Harada, Hirohide Kobayashi, Miki Yoshizumi, Tung Ngoc Nguyen, An Thi Tran, Izuru Saizen, Minh Phuoc Truong, Bich-Thuy Ly, Yosuke Sakamoto, Ngoc Thi Nhu Do, Yoshizumi Kajii, Dieu-Anh Van, Shigeo Fujii, Shuhei Tanaka, Van Quang Tran, Nguyet-Anh Pham, Quynh-Anh Tran-Nguyen, Lan Chau Nguyen, Lien Pham Hong Nguyen, Atsushi Takai and Toru Inui

Abstract

Vietnam has been witnessing rapid urbanization, which has altered urban livelihood and caused various challenges. In this chapter, after overviewing the cultural and environmental impacts of urbanization in Vietnamese cities, we illustrate the current status and challenges of these issues through case studies in Hanoi, Danang, Hue, and their surrounding areas, where we have been conducting research for a long time. The contents include various urban issues related to community linkage, housing culture, climate change, air pollution, water environment, water use, wastewater and fecal sludge management, and solid waste management.

H. Harada (✉)
Graduate School of Asian and African Area Studies, Kyoto University, Kyoto, Japan
e-mail: harada.hidenori.8v@kyoto-u.ac.jp

H. Kobayashi · I. Saizen · Y. Sakamoto · Y. Kajii · S. Fujii · S. Tanaka · A. Takai
Graduate School of Global Environmental Studies, Kyoto University, Kyoto, Japan

M. Yoshizumi
College of Gastronomy Management, Ritsumeikan University, Kyoto, Japan

A. T. Tran
Faculty of Management Science, Thu Dau Mot University, Thu Dau Mot, Vietnam

M. P. Truong · Q.-A. Tran-Nguyen
The University of Danang - University of Science and Education, Danang, Vietnam

B.-T. Ly · N. T. N. Do · D.-A. Van · L. P. H. Nguyen
School of Chemistry and Life Sciences, Hanoi University of Science and Technology, Hanoi, Vietnam

V. Q. Tran
University of Science and Technology, The University of Danang, Danang, Vietnam

N.-A. Pham
Faculty of Chemistry and Environment, Thuyloi University, Hanoi, Vietnam

L. C. Nguyen
Faculty of Civil Engineering, University of Transport and Communications, Hanoi, Vietnam

T. Inui
Graduate School of Engineering, Osaka University, Suita, Japan

T. N. Nguyen
Faculty of Architecture, University of Sciences, Hue University, Hue, Vietnam

S. Fujii et al. (eds.), *Livelihood and the Environment in Vietnam*, Sustainable Development Goals Series,
https://doi.org/10.1007/978-981-97-2289-1_5

Keywords

Urbanization · Cultural impact · Urban livelihood · Urban challenges · Environmental impact · Environmental pollution

5.1 Introduction

5.1.1 Urbanization in Vietnam

As in many other countries in Southeast Asia, Vietnam has been witnessing rapid urbanization. Cities are the major component of urban areas. The organization of administrative units differs across the country; in Vietnam, it consists of three levels (Law No. 77/2015/QH1). The highest level, the provincial level, constitutes municipalities and provinces: Hanoi, Ho Chi Minh City, Haiphong, Danang, and Can Tho as municipalities and 58 provinces. The municipalities are cities controlled by the central government and have equivalent status to provinces. The number of cities under provinces has increased from 54 in 2010 to 81 in 2021 (General Statistics Office n.d.), and accordingly, the number of cities (municipalities and provincial cities) in Vietnam has increased from 59 in 2010 to 86 in 2021.

The urban population in Vietnam has been rapidly growing. From 20% in 1990, the share of urban population rose to 24% in 2000 and 37% in 2020 (World Bank n.d.). Although the annual growth rate of urban population has been declining, the urban population continues to grow, and by 2040, it is expected to exceed the rural population (United Nations 2018). Thus, as urbanization progresses, changes in livelihood and the urban environment are becoming more apparent.

5.1.2 Change in Urban Livelihood in Vietnam

According to the national living standard survey, with the urban growth described above, the urban livelihood in Vietnam is changing (General Statistics Office 2021). The national average household size in Vietnam remains large, compared to, for instance, the Japanese average of 2.2 in 2020 (Statistics Bureau of Japan n.d.); however, it has declined from 3.9 to 3.6 between 2010 and 2020, and urban household size in 2020 was smaller than the rural one, with an average of 3.5. The proportion of urban population in 2020 was 37%, as mentioned above, but the highest proportion of urban population by age group was 39.5% for the 40–49 years age group, followed by 39.1% for the 35–39 years age group and 38.3% for the 30–34 years age group, indicating that the proportion of urban population is particularly high in the working-age population.

In 2020, 25.2% of the urban population worked in the industry, followed by 19.3% in wholesale and retail sales, and the percentage of the non-farm population has increased from 54.1% in 2010 to 70.4% in 2020. Monthly income per capita was 2,129,500 VND in urban areas and 1,070,400 VND in rural areas in 2010, indicating a difference of 1.99 times between urban and rural areas. With an increase in monthly income per capita to 5,590,200 VND in urban and 3,481,500 VND in rural areas in 2020, the gap between urban and rural areas has narrowed to 1.6 times, albeit still large. Regarding the education level, which is closely related to income, the college and university enrollment rate in urban areas has increased from 14.4% in 2010 to 21.0% in 2020, and the difference between urban and rural areas is significant, with the enrollment rate in rural areas at only 7.8% in 2020.

Furthermore, the urban living environment is changing. According to Vietnamese statistics (General Statistics Office 2021), houses are classified into four types: permanent houses, semi-permanent houses, less-temporary houses, and temporary/other houses. Between 2010 and 2020, while the percentage of permanent houses remained almost constant at around 50%, the percentage of semi-permanent houses increased from 37.8 to 45.9%. In contrast, the proportion of less-temporary houses decreased from 7.5 to 3.2% and that of temporary/other houses from 5.5 to 1.2%, indicating a decrease in the proportion of lower categories. Regarding living space per capita, an increase was observed in all four types of housing between 2010 and 2020, with an overall increase in living space per capita from 17.9 to 25.2 m^2. Specifically, the per capita living area increased from 20.7 to 26.6 m^2 in urban areas and from 16.7 to 24.5 m^2 in rural areas. Interestingly, on average, the per capita living area is slightly larger in urban areas.

5.1.3 Environmental Impact in Urban Vietnam

The urban population is growing while the urban environmental infrastructure is not sufficiently developed, and various environmental problems are emerging. The Viet Nam Environmental Performance Assessment Report MONRE and UNEP-RRC.AP (2010) has identified seven priority environmental concerns: inland water pollution, air pollution from mobile sources, inadequate solid waste management, inadequate forest resources, threats to biodiversity, threats to coastal zones, and climate change. Although all these are affected by urban activities, inland water pollution, air pollution from mobile sources, and inadequate solid waste management are typical environmental problems emerging in cities, particularly with urbanization.

Economic development and population growth have increased the number of motorcycles, which in turn has caused severe air pollution. The growth rate of motorcycles in 2000 was 14.2%. For instance, the number of motorcycles in Hanoi increased from approximately 200,000 in 1990 to 800,000 in 2000, causing severe air pollution, particularly in the major cities of Hanoi and Ho Chi Minh City (MONRE and UNEP-RRC.AP 2010). SO_2, NO_2, PM, and CO emissions must be controlled to prevent air pollution. Regarding PM2.5, for instance, the annual average concentration has been decreasing from 39 $\mu g\ m^{-3}$ in 2010 to 30 $\mu g\ m^{-3}$ in 2017 (World Bank n.d.). However, 34.5% of the population in 2017 was estimated to be exposed to levels exceeding 35 $\mu g\ m^{-3}$ on average annually, which is defined by the World Health Organization (WHO) as Interim Target 1 (IT-1) and is associated with a 15% increase in long-term mortality compared with the Air Quality Guideline (AQG), indicating a major urban environmental issue.

While basic water and sanitation have progressed in urban areas, water pollution in rivers and lakes due to the discharge of untreated or limitedly treated wastewater has become apparent. According to the World Bank (2019), 71% of industrial wastewater is treated in some centralized wastewater treatment plants, while only 12.5% of domestic wastewater is treated. With the increase in population and domestic water use, in 15 years, urban domestic wastewater is estimated to account for nearly 60% of all wastewater, including wastewater from other sectors. Although many cities are planning to develop sewerage, it is expected to require a considerable time before wastewater treatment is widely available across all cities in Vietnam. Further, as of 2005, over 110 industrial and processing zones had been developed. However, there are many small-scale production areas, such as craft villages, where many cases of insufficient wastewater treatment exist, leading to serious pollution problems in areas of wastewater discharge. Thus, industrial wastewater management is a major challenge to creating a sound inland water environment in Vietnam.

The amount of municipal solid waste generated is determined by lifestyle. According to MONRE and UNEP-RRC.AP (2010), the amount of waste generated per person per day in Vietnam was 0.3 kg in rural areas and 0.7 kg in urban areas as of 2010, with urban residents generating more than twice as much waste per capita as rural residents. The collection rate of waste by cart or truck has increased from 39.2 to 68.5% between 2010 and 2020 for the country as a whole, particularly in urban areas, from 79.6 to 91.0% (General Statistics Office 2021). Thus, the collection of municipal solid waste is becoming more widespread in most cities. Simultaneously, many households in Vietnam are conducting small-scale waste segregation, collection, and recycling, which is estimated to account for nearly 8–12% of the collected waste (MONRE and UNEP-RRC.AP 2010). In recent years, Ho Chi Minh City and Hanoi have aimed to achieve larger-scale resource utilization through urban-scale solid waste segregation and recycling; however, many challenges remain to effectively achieve this. The challenge of solid waste management in urban Vietnam is shifting from waste collection to proper disposal and establishment of a sound circular economy.

5.1.4 Contents of This Chapter

With the above background, the present chapter discusses livelihood and environmental issues in urban Vietnam, where urbanization is progressing. Section 5.1 provides an overview of urbanization and environmental issues in Vietnam. The following sections are divided into two sub-chapters. Sections 5.2–5.4 focuses on livelihood in urban areas, including socio-cultural aspects. Section 5.2 describes traditional community activities and linkages and highlights the impact of rapid urbanization on the traditional living environment based on a case study in an old historical district of Hue city. Section 5.3 illustrates the conservation status of traditional houses in cities from the example of a traditional garden house in Hue as well as the historical value and unique characteristics of the living environment. Section 5.4 provides a practical framework for climate change vulnerability assessment and its effect on urban life in Danang city—a leading socio-economic region on the Central Coast of Vietnam. Sections 5.5–5.9 focuses on environmental impact in urban areas. Section 5.5 describes the current status and challenges of air pollution control in Vietnam, with a special focus on Hanoi, where we have conducted research on the impact of motorcycles. In Sect. 5.6, the water environment in Vietnam is overviewed and the river water environment is described based on a case study in the Nhue-Day river basin. In Sect. 5.7, the characteristics of domestic water use in Vietnam are described based on case studies in Hanoi and Danang, and in Sect. 5.8, domestic wastewater and fecal sludge management in Vietnam is summarized briefly, followed by the characteristics of wastewater and sewage sludge management in Hanoi, Hue, and Danang. Finally, in Sect. 5.9, the current status, challenges, and technical prospects of waste treatment, final disposal, and recycling in Vietnam are described, focusing on municipal solid wastes and construction demolition waste.

5.2 Impact of Urbanization on Traditional Community Linkages in the Historic Old Quarters of Hue City, Central Vietnam

5.2.1 Introduction

Since the introduction of Doi Moi (Đổi Mới) in 1987, Vietnam's economic trajectory has been marked by remarkable growth. The country's average annual gross domestic product (GDP) growth rate for the period from 2000 to 2019 stood at roughly 6.48%. In 2003, the urban population constituted 25% of Vietnam's total population, and this proportion escalated to 37.34% by 2020 (World Bank, n.d.). This swift urbanization has discernibly influenced the traditional living environment in Vietnam.

The objective of this section is to delineate traditional community activities and connections, as well as to assess the repercussions of rapid urbanization on the conventional living environment in Vietnam. To this end, a case study from the Gia Hoi area, an ancient district of Hue city located in central Vietnam, is presented.

The case study encompasses the city blocks around Chi Lang Street, Nguyen Du Street, Hien Thanh Street, and the Huong River in the historic old quarters of Hue city, situated to the east of the Hue Citadel (Fig. 5.1). Chi Lang Street, a major thoroughfare, bisects the area. Interviews were conducted from 2009 to 2015 with households in the area to gather basic information about the living environment and to perform

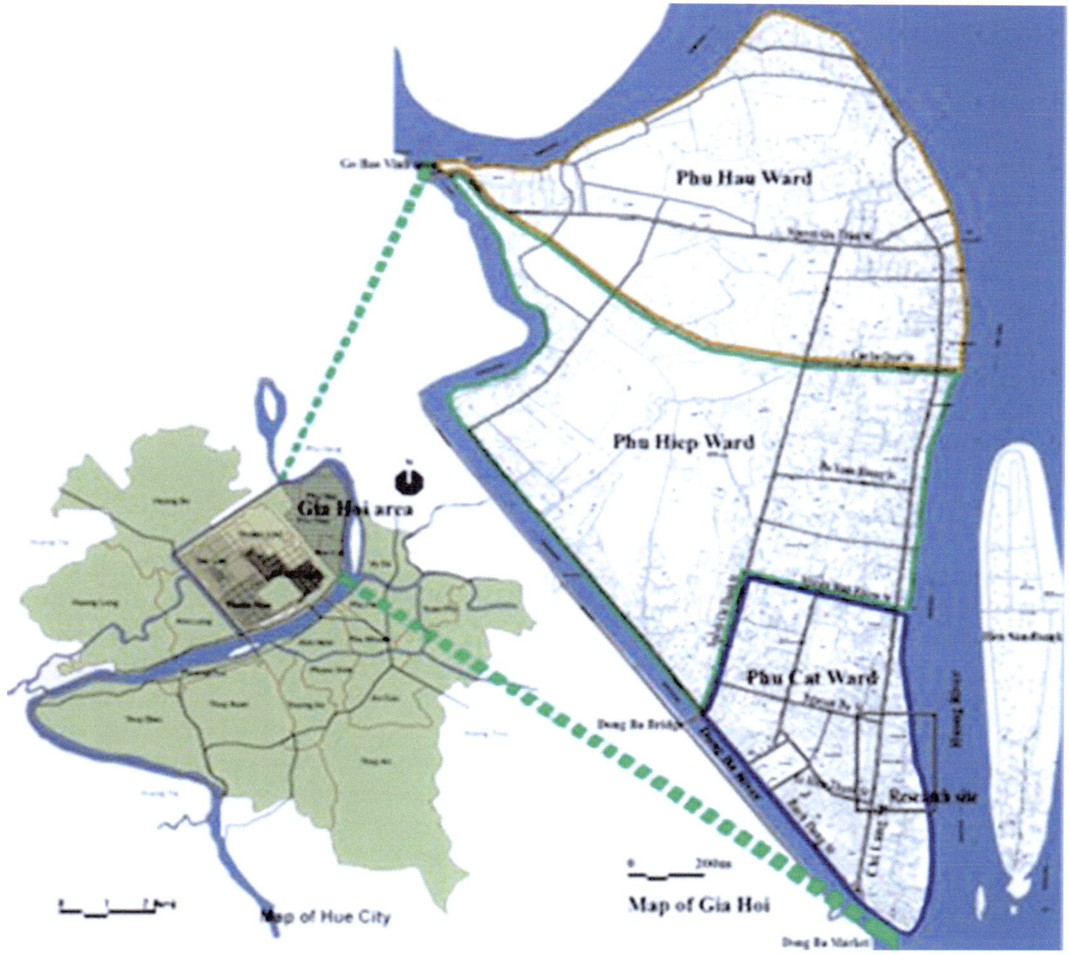

Fig. 5.1 Map of Vietnam, Hue city, and Gia Hoi area

measurements. To glean more detailed insights about the living environment from the residents, a workshop was conducted on July 19, 2009.[1]

5.2.2 Urbanization in Vietnam

5.2.2.1 Urbanization in Hue

The Doi Moi policies, initiated by the Vietnamese government in 1986, were instrumental in fostering a liberal market economy and advocating for private-sector involvement. As a consequence of these policies, the process of urbanization in Vietnam experienced substantial acceleration. Notably, there was a robust increase in the nation's economic prosperity during this same timeline, as evidenced by the steady rise in per capita GDP. This economic surge coincided with a swift shift in the structural composition of the nation's economy, moving from an agrarian foundation toward an increased focus on industrial production and export-driven operations (World Bank 2011).

The phenomenon of urbanization in Vietnam has witnessed a significant surge in recent times. Data indicates an upward trend in the urban

[1] The field surveys, interviews, and the workshop were conducted by a research team of Kyoto University and Hue University.

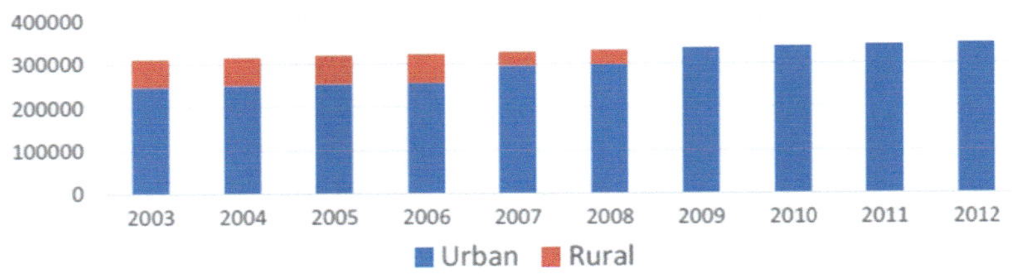

Fig. 5.2 Population in Hue city from 2003 to 2012 (Hue Development of Statics 2008)

population of the country, with urban residents constituting 31% of the total population in 2011, a marked increase from 15% in 1960 and 20% in 1990 (World Bank 2011). Projections from the United Nations suggest that the urban populace will outnumber the rural populace by the year 2040 (World Bank 2011).

Figure 5.2 provides a demographic depiction of Hue city, juxtaposing its urban and rural population from 2003 to 2012. An intriguing observation is the disappearance of rural areas within Hue city since 2009, thus transforming all regions of the city into urban territories. Concurrently, the city has seen a rise in both its population and population density. However, birth rates have depicted a marginal decline, while mortality rates have shown a slight ascension.

5.2.2.2 Urban Development Policies in Hue

Following the inception of the Doi Moi policies, the Vietnamese administration has enforced an array of policies aimed at stimulating evenly distributed economic growth and urban development. Government Decision No. 10 in 1998, concerning the urban system and development strategy until 2020, underscored the necessity of nurturing medium and small-sized cities while constraining the growth of larger cities. The government subsequently acknowledged the potential emergence of megacities, with a population exceeding 10 million, in Government Decision No. 445 in 2009. The Socio-Economic Development Strategy for 2011–2020 further corroborates that urbanization will be pivotal in achieving the nation's objectives of industrialization and modernization (World Bank 2011).

Hue city, the capital of Thua Thien Hue province, is situated in Central Vietnam. As of 2012, Hue city had a population of 346,070, ranking it the sixth-largest city in Vietnam. Hue served as the historical capital of the Nguyen Dynasty, spanning from the dawn of the nineteenth century until the mid-twentieth century. The Complex of Hue Monuments, situated within and around the old city, received recognition as a world heritage site by UNESCO in 1993. Since this acknowledgment, Hue city has drawn millions of visitors, both national and international. An urban development plan aimed at transforming Hue into a metropolis is currently being enacted. In anticipation of further population and tourist growth, city reorganization plans have been proposed, which involve relocating residents from the city center to suburban locales.

Figure 5.3 delineates four regions intended for relocation within the projects. Site A, consisting of the “Eo Bau” and “Thuong Thanh” regions, has been officially marked as displacement site by the Hue People’s Committee. The project was originally planned for execution between 2006 and 2010, but implementation has yet to occur. For Site B, situated along Tran Huy Lieu St. in Phu Hoa ward, a relocation plan has been devised by the Hue People’s Committee, yet no official documentation is publicly available. Site C, involving the relocation of 545 households from the area between Chi Lang street and the Huong river in Phu Cat ward, has resulted in some individuals moving to Bai Da (Site D). This relocation project was conducted during 2007, 2008, and late 2009. The sites vacated following the relocation projects will come under the ownership of the respective local people’s committees and wards, with plans for the development of new tourist routes in these areas.

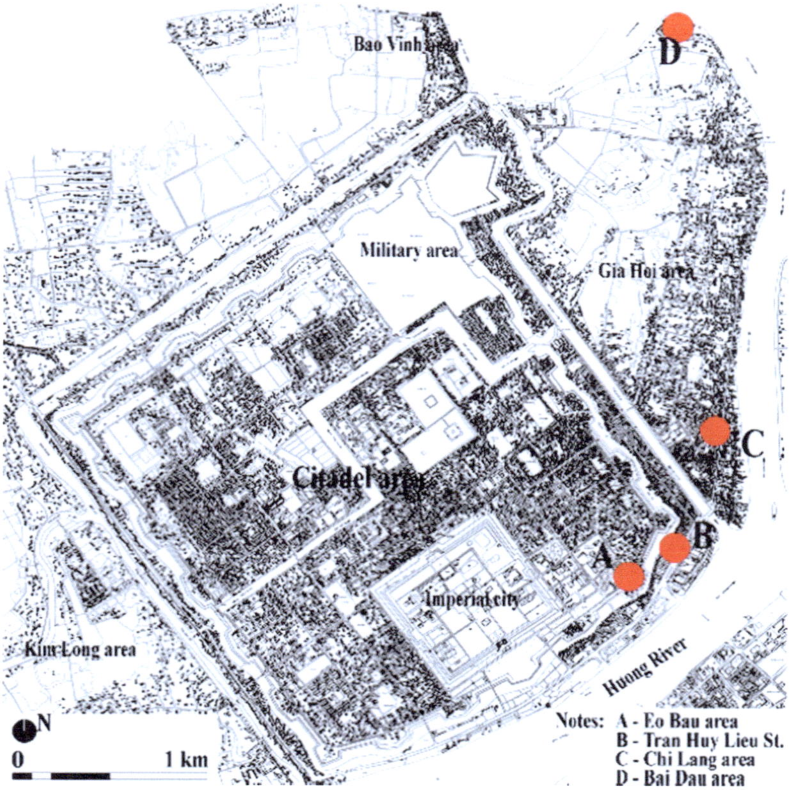

Fig. 5.3 Areas planned to be relocated in the project. *Source* The map was provided by the Department of Construction of Thua Thien Hue Province

5.2.3 Overview of Living Environment in the Case-Study Area

5.2.3.1 Development Process of the Gial Hoi Area

The first Chinese settlers reportedly disembarked at Thanh Ha port in 1794 and subsequently established their residence in the Gia Hoi area of Hue, where they initiated their commercial activities (Hy et al. 1999). Figure 5.4 illustrates the transformative journey of the Gia Hoi area.

In 1836, eight structures, collectively known as "Duyen Giang Bat Trang," were constructed, catalyzing the development of new thoroughfares and burgeoning commercial activities. Two years later, in 1838, King Ming Mang sanctioned the erection of a fresh market, dubbed Gia Hoi, which comprised 89 kiosks and communal buildings featuring two levels. Along the street, extending from the Gia Hoi bridge to Tran Binh Dai, perpendicular to Cho Dinh street, a total of 399 residential houses were constructed, with residents volunteering to erect an additional 149 chambers. All column structures were brick-based, with three chambers segregated by a singular brick wall. Also, a handrail was installed along the riverbank in the Dinh market, which successively relocated from the intersection of Nguyen Du street and Chi Lang street to the crossroads of Cao Ba Quat street and Chi Lang street, before finally settling at its current location near the junction of Nguyen Gia Thieu street and Chi Lang street.

In 1899, when the Dong Gia market (currently known as Dong Ba) was relocated to its present location from within the citadel

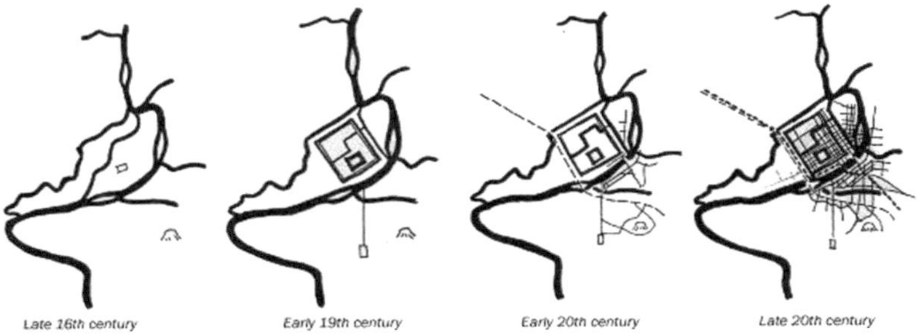

Fig. 5.4 Transformation of Gia Hoi area (Adamson and Ejdeholm 1999)

near Dong Ba bridge, and the establishment of Truong Tien bridge and Truong Tien street (now known as Tran Hung Dao), the commercial epicenter of Hue gradually transitioned from Gia Hoi street to Tran Hung Dao street. Gia Hoi remains as the nomenclature for the ancient quarter within the new urban city. Even though the street is officially named Chi Lang, the local population continues to refer to it as "Gia Hoi," a reminder of the area's past commercial prosperity. With the relocation of the primary markets away from the Gia Hoi area, the district has seen a transition toward urbanization, serving not only as a commercial hub, but also as a residential area.

Data from the Hue Department of Statistics (2008) indicates that the total population of Hue experienced a steady rise from 289,428 in 1998 to 335,747 in 2007, with the average density increasing from 4,077 individuals per km^2 in 1998 to 4,729 individuals per km^2 in 2007. Notably, the urban population has shown a significant increase, with the 2007 urban population being approximately six times larger than the rural population. This demographic shift is indicative of the rapid pace of urbanization in Hue. Within this urban expanse, Phu Cat ward, which includes part of the Gia Hoi area and forms a segment of the case-study area, boasts the highest population density in Hue, recording 25,849 persons per km^2.

5.2.3.2 Types of Houses in the Case-Study Area

According to a November 1997 report by the Institute for Construction Economics, Ministry of Construction, the average size of the majority of houses does not exceed 50 m^2. Multigenerational living, where two to four generations share the same dwelling, is a common characteristic of Vietnamese households (JBIC 1999). Home renovations and room additions occur in response to family expansion. There are instances of rural-to-urban immigrants occupying unused land for unauthorized construction. In urban contexts, homes serve not only as residential spaces, but also as venues for income generation. Frequently, the first floors of roadside houses are converted into shops, a structure commonly referred to as "shophouses." Shophouses typically span two to three floors, with the ground floor devoted to commercial activity and the upper floors serving as residences.

The 1997 Living Standard Report by the Ministry of Construction identifies semi-permanent houses (41.56%) and temporary houses (19%) as the primary types of dwellings in Vietnam. Prior to the implementation of the Doi Moi policies, only a minority of households had access to borrowing options to finance their construction costs, and the majority of the population could not afford to build or refurbish their homes. Presently, however, half of these households have managed to build or own their homes using their capital (JBIC 1999).

The historic main thoroughfare, Chi Lang Street, bisects the area, with various small alleyways branching out. The community's characteristics vary based on the street configuration. Seventy-two percent of all buildings along the main street are shophouses. The

Fig. 5.5 Wooden house

architectural styles in the area encompass traditional Vietnamese wooden houses (Fig. 5.5) and European-style buildings (Fig. 5.6). A field survey conducted in 2009 revealed that 56% of all buildings surveyed are modern concrete houses (Fig. 5.7), 31% are wooden houses, and 13% are European-style houses, making for a diverse architectural landscape along the main street. Conversely, modern concrete housing makes up 86% of all buildings along small alleys, as most of these structures were built after 1945 (71% of housing along small alleys).

Fig. 5.6 European-style house

The population of this area increased significantly following the Vietnam War, particularly after the inception of the Doi Moi policy. Further, after the 1980s floods, some individuals settled illegally in the space behind the main street houses. Survey results indicate that 81% of households along small alleys lack a land certificate, known as the "Red Book" in Vietnam.[2] The average income of households along Chi Lang Street surpasses that of households in small alleys. Chi Lang Street households earn approximately 210 USD per month, while small alley households average just over half that amount, at approximately 110 USD per month.

Given its economic attractiveness (Fig. 5.8), the area surrounding Chi Lang Street is more urbanized. Over the past decade, the street has attracted a substantial influx of new residents from outside Hue. Further, the street has seen the emergence of new establishments such as internet cafes and shops catering to the younger demographic. In line with this development, the local government has sought to redevelop the area along the Huong River into parks or hotels, with plans to relocate current residents to other areas.

5.2.4 Traditional Community Linkage

5.2.4.1 Traditional Community Boundary

Figure 5.9 is a map of the case-study area in the Gia Hoi area. For administrative management, each ward is divided into several basic local administrative zones called "To dan pho" in Vietnamese (Urban and Rural Planning Institute 2004). In the research site, Zones 1, 2, and 11 are included (Fig. 5.2.8). On the contrary, traditionally, each resident belongs to a local community, which is called "Xom.

[2]2 Some people lost the Red Book during the war and flooding; others simply settled down illegally without it.

Fig. 5.7 Concrete modern house

Fig. 5.8 Chi Lang street

The traditional urban communities locally (*Xom*) have been maintained for generations behind the main streets, although it is difficult to identify the cohesion of the community. This is an informally developed neighborhood association, thus a municipality may not always perceive Xom; moreover, the areas covered by Xom and an administrative neighborhood association are sometimes different. Xom is said to originally consist of relative residents living along a small alley.

5.2.4.2 Traditional Community Activities

Figure 5.10 shows various types of community activities in a Xom. The residents in a Xom cooperate in community activities such as cleaning up an alley and organizing traditional festivals such as New Year Festival (*Tết Nguyên Đán*), Mid Year Festival (*Tết Trung Nguyên*), and Mid Autumn Festival (*Tết Trung Thu*), which have contributed in developing the city culture of Hue. One of the most important activities of a Xom is worshiping ancestors, locally called "Cung Xom," which is organized by each Xom. Cung Xom was originally introduced by Chinese immigrants. Cung Xom is organized two times a year (worshiping the Heaven and the Earth on May 23 and welcoming *Tết* holidays). All households contribute money to the wife of a community leader every three months. Before the ceremony, the wife of a community leader and other wives together go to the market and buy items for preparation. The place of the ceremony is in the middle of the alley near the Plantae tree. A community leader is the first person who comes to pray for health, prosperity, and so on. Then, other husbands gradually come and pray. After the offering, food is divided and offered to each household, followed by a social gathering at the community leader's house that is visited by a representative from each household. Such activities have made a significant contribution in developing the culture of Hue city.

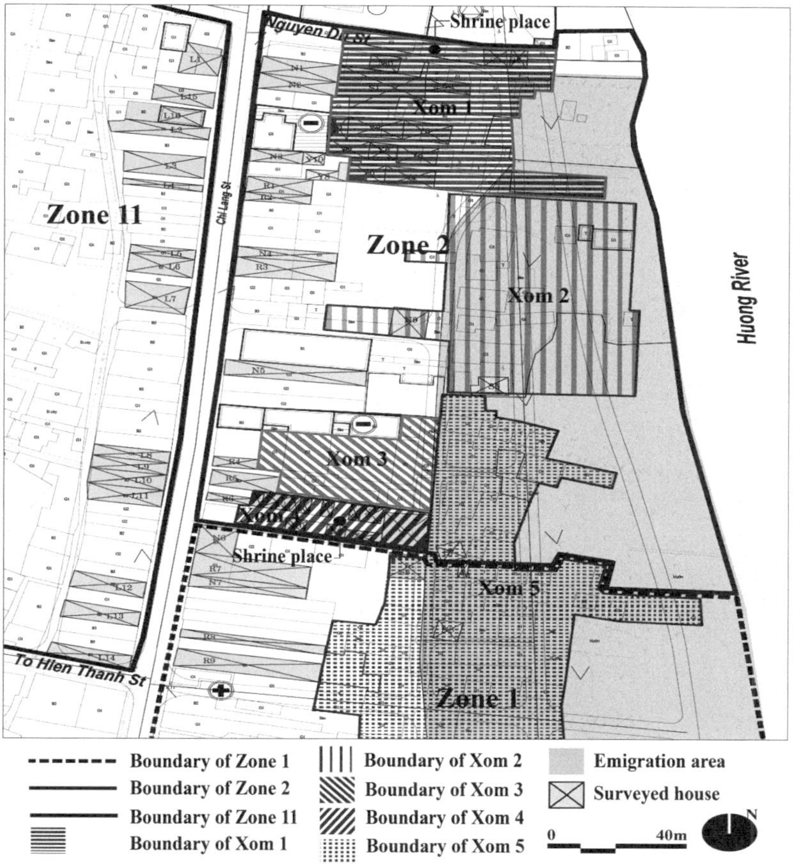

Fig. 5.9 Map of the case-study area in Gia Hoi area, and boundaries of local community "Xom" and present

5.2.5 Urbanization Impacts on Living Environment

5.2.5.1 Living Environment in a Main Street and Small Alleys

Figure 5.11 depicts the extent to which households along Chi Lang Street and those residing in small alleys express satisfaction with their local living environment.[3] Residents along Chi Lang Street exhibit a high level of satisfaction in relation to their housing conditions and the convenience of shopping, work, and educational facilities. However, they display dissatisfaction regarding the availability of open spaces for community interactions and children's playgrounds.

In contrast, households located along the small alleys tend to express greater overall satisfaction with their living environment when compared to those along the main street. Specifically, residents in these smaller thoroughfares display contentment with their respective communities and the level of interaction with neighbors.

5.2.5.2 Factors that Impact Living Environment

Figure 5.12 illustrates the responses to a query about what factors severely impact the living environment. Based on the survey results,

[3] Figures 5.4 and 5.5 and Table 5.1 are the survey results from 69 households of which 32 households belong to the main street community, 28 to the small alley, and the remaining 9 households did not answer the survey.

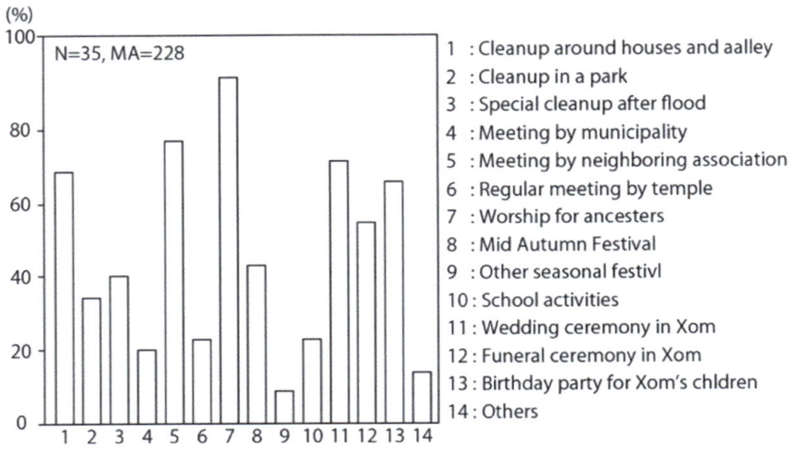

Fig. 5.10 Various community works in Xom. Created by Satomi Shirasaki

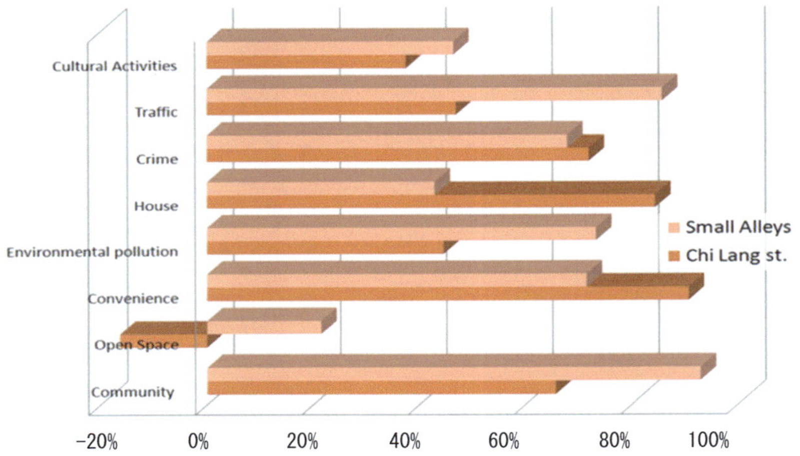

Fig. 5.11 Satisfaction with living environment

the inhabitants perceive that escalating traffic volume, swift urbanization, and ineffective policies are influencing their living conditions detrimentally.

Significantly, a larger proportion of residents within the small alleyways, as opposed to those on the main street, believe that rapid urbanization is the culprit behind major environmental issues. The area along these smaller alleyways is less urbanized compared to the main street locale, hence these residents have experienced firsthand the negative implications of urbanization emanating from the main street areas.

5.2.5.3 Transformation of Traditional Community Linkage

The area flanking the main street exhibits more signs of urbanization and is witnessing the gradual dissolution of traditional community ties. Table 5.1 presents the frequency at which residents engage in conversation with their neighbors. Contrasted with the majority of residents within the small alleyways, who interact with their neighbors on a daily basis, or at least monthly, some residents along Chi Lang Street do not engage in any communication with their neighbors.

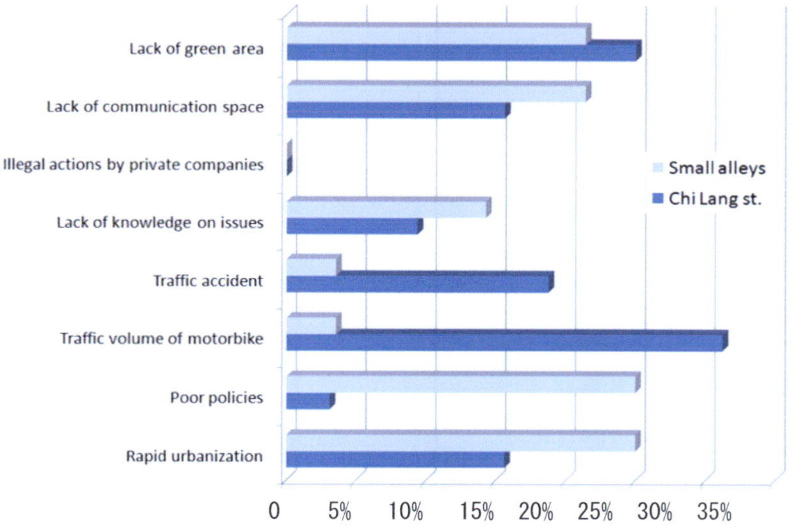

Fig. 5.12 Factors that lead to serious issues in living environment

Table 5.1 Frequency of communication of the respondents with neighbors on Chi Lang street and small alleys

	Chi Lang street (%)	Small alleys (%)
More than three times a day	20	42
A few times a day	33	27
A few times a week	27	19
A few times a month	0	12
A few times a year	10	0
Zero	10	0

One possible explanation for this could be the shortage of communal spaces. As suggested by Fig. 5.11, there is a notable dearth of areas conducive to neighborly communication along the main street. Additionally, high traffic volumes might be having a detrimental impact on interpersonal communication among residents.

As previously discussed, Cung Xom stands as a crucial traditional community activity, pivotal for fostering community ties. However, survey data[4] suggests that a significant 88% of main street residents no longer participate in Cung Xom. This could be attributed to several factors: (1) a lack of interest in traditional community activities from individuals who have migrated from other areas for business purposes, (2) an absence of designated community spaces for hosting Cung Xom, and (3) challenges in coordinating such activities due to the prevalent architectural style of tube houses along the main street. Consequently, individuals conduct worship activities independently within their homes. This leads to the inference that traditional community activities along Chi Lang street appear to be on the verge of extinction.

Field research in 2014 was conducted during a major seasonal festival, the Mid Autumn Festival, to examine the transformation of Xom, which organizes traditional community activities. This festival in Vietnam was originally organized mainly for parents to spend time with their children in lieu of the time lost during harvest season, as well as to celebrate the harvest. Parents offer fruits and sweets, locally called "*Mam co*," and children perform a lion dance and lantern parade for Xom residents, as shown in Photos 3–5. However, such traditional festivals have gradually transformed in recent times.

Figure 5.13 shows four Xoms in one administrative block on in west side of the main street

[4]The survey about Cung Xom in Chi Lang street was conducted with 24 additional households in August 2010.

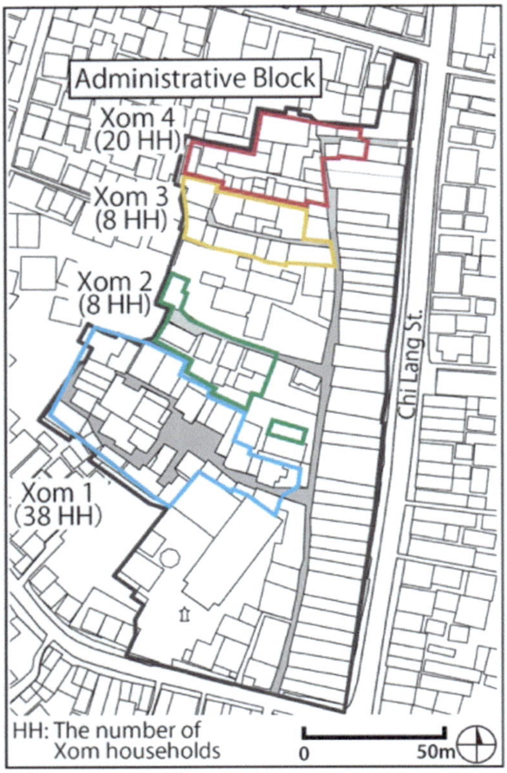

Fig. 5.13 Xoms along a small alley

of Chi Lang. The administrative leader living in the house facing Chi Lang street did not even know how the residents in small alleys organized Xoms. All four Xoms are basically composed of households along the same small alley, but the residents in each Xom are less actively cooperating than they used to and the number of Xom members is decreasing. With the fading of Xom's role, the traditional activities become more simplified and transformed. Much of the Mid Autumn Festival's activities in Hue city are not organized in these Xoms anymore. Upon examining 10 Xoms in the Gia Hoi area, it was found that only 2 still organize the festival, three others do it together, and five do nothing. Instead, many children in the area enjoy the festival in local temples, public and private schools, or individual big houses (Fig. 5.14).

These substitutes of the festival signify that much effort needs to be made to maintain the city culture in the transforming Xom communities, or else the Xoms may disappear in the near future, even though they serve a crucial role in the preservation of the city's historical culture.

5.2.5.4 Function of Small Alleys for Community Activities

Figure 5.15 shows Xom 4 located on the east side of Chi Lang street. There are 15 households that live in this community. In Xom 4, the width of the alley is small enough (less than 2 m) to create a convenient space for community activities among residents. They share information, communicate with neighbors, and cooperate with each other. The small alley among those households serves multiple functions. It serves as a place for children's playground, recreation and exercise of residents, socialization, and commercial market activities (Fig. 5.16). The residents often sit outside in the small alleys and it acts almost like a big living room of the community.

Moreover, if a community member is sick, any information regarding the person is quickly shared among the community, so that they can gather and visit the sick person. Such a level of socialization has improved the level of relationship that inhabitants share.

5.2.6 Conclusions

Gia Hoi, the historical ancient quarters of Hue, has undergone significant transformation, emerging as a bustling economic and commercial district. The establishment of Vietnam's independence and the termination of warfare have indeed left lasting imprints on this region. Furthermore, the advent of the Doi Moi policy has triggered the Vietnamese populace to explore expanded economic possibilities, thereby inevitably influencing their traditional habitats.

Such urbanization disrupts traditional living environments, evidenced by alterations of architectural preferences from wood-based to concrete structures, and the repurposing of community spaces. This progress has also

Organizing Model	Activity Location	Organizer	Participants	Activities
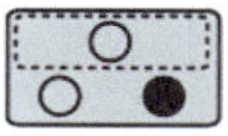	Xom	Xom's children	Xom's residents	Lion dance Lantern parade Offerings
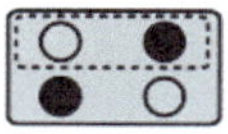	Temple	Members (Age: 6 -18)	Members (Age: 6 - 70)	Lion dance Lantern parade Offerings
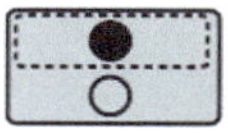	Public school Private school	Teachers	School children	Lion dance Games Providing sweets, etc.
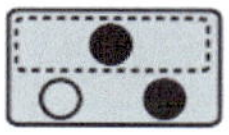	Private house	House owner	Neighboring children	Party Providing sweets

: Organizer : Participants ○ : Children (Under 15 years old) ● : Adult (Over 16 years old)

Fig. 5.14 Transformation in organization of mid autumn festival

Fig. 5.15 Floor plan and Facade of Xom 4 in 2010

Fig. 5.16 Xom 4 community's use of the small alley for their community activities

precipitated economic disparities and social vulnerabilities, such as households being bereft of the "Red Book" land ownership certificate and displacement of residents due to government-imposed relocation policies. Specifically, it has had profound effects on traditional community linkages, with a noticeable shift in interest from community-centric to individual economic activities. This change in priorities is leading to the gradual extinction of traditional community activities and linkages.

One contributing factor to this decline is the scarcity of communal spaces. The escalation in transportation demand has led to congested streets, as inhabitants strive to erect as many shops and residences as possible. Consequently, the availability of open areas designed for social interaction and community-based activities is dwindling. Historically, local festivities were organized collectively with neighbors; however, these are increasingly becoming private affairs, conducted in individual households.

Conversely, residents residing in small alleyways continue to orchestrate community events, given that these pathways double as communal gathering spots. A case in point is "Xom 4," where a Xom comprises not just familial and kinship bonds, but also includes outsiders, and the Xom leader is entrusted with managing its activities. These narrow lanes serve as crucial spaces for community events, functioning as children's play areas and as locales for neighborly socializing.

Community linkages enable collaboration, assist socially disadvantaged groups, and mitigate flood and crime damages through communal efforts. Such networks, thus, hold considerable significance for sustainable development. Nevertheless, the survey findings indicate that the burgeoning transportation needs are encroaching upon the spaces designated for communal activities. It can be inferred that the swift pace of urbanization and relentless pursuit of economic advancement have adversely affected traditional community ties in Vietnam. In order to preserve this social cohesion, it is imperative to safeguard community spaces, inclusive of small alleyways.

5.3 Traditional Living Environment of Garden Houses in Hue City, Vietnam

5.3.1 Introduction and Methodology

Hue, the historical capital of the Nguyen Dynasty (1802–1945) in Vietnam, is known as "the garden city" because of its green landscape, garden tombs, royal palaces, and especially its traditional garden houses. Houses in Hue, especially the traditional houses, are always closely connected with gardens, which are metaphorically utilized as spaces for linking humans with heaven and nature. Gardens can be found everywhere from urban spaces to residential houses and they contribute to the gracefulness and attractiveness of Hue city. Therefore, Hue Traditional Garden Houses (HTGHs) are integral elements of Hue residents' life and spirit.

HTGHs have been constructed since the time of the Nguyen Lords (1558–1774). Whereas the Nguyen Lords lived in their palaces and the Nguyen Kings (1802–1945) lived in the Purple Forbidden City, other members of the royal family and high-ranking mandarins lived in HTGHs constructed around the palaces of the Nguyen Lords and the citadel of the Nguyen Kings. As a result, HTGHs are widely distributed in the citadel area and in the surrounding ancient villages, such as Kim Long, Vy Da, Phu Cat, and Nguyet Bieu (Fig. 5.17).

Many different factors such as lifestyle, the building codes imposed by the Nguyen Dynasty, as well as Hue's climate and geography, have all had a great impact on the formation, spatial organization, built form, and living environment of HTGHs. This has resulted in HTGHs displaying unique characteristics that cannot be observed in traditional houses in other regions of Vietnam.

This work focuses on the traditional living environment in HTGHs based on the findings of a series of surveys carried out in and around Hue citadel and Kim Long area from 2010 to 2018. It aims to explore the historical value and unique

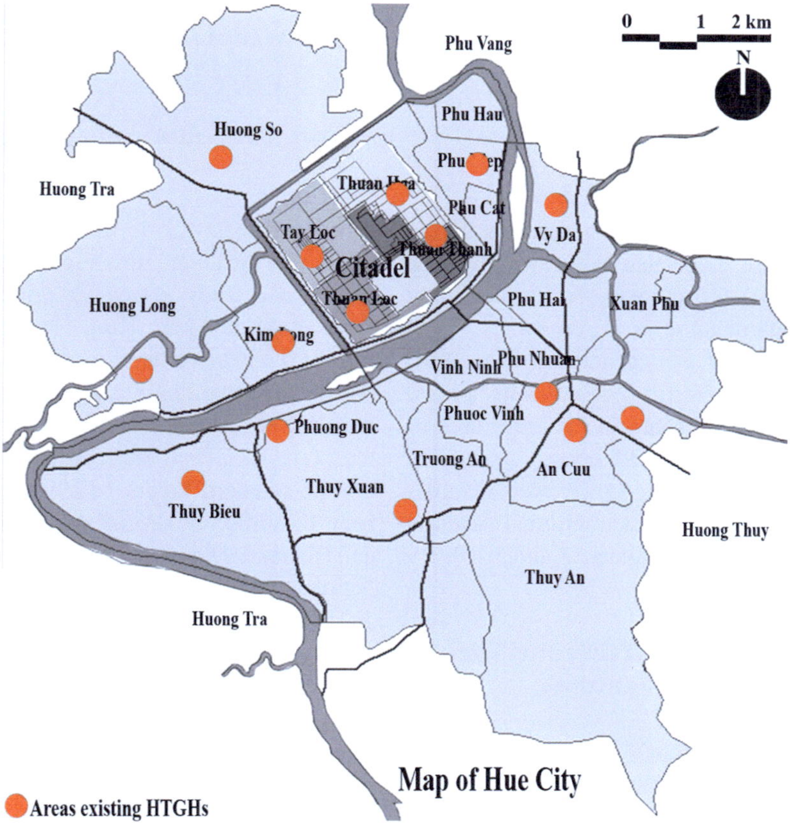

Fig. 5.17 Distribution of HTGHs in Hue city

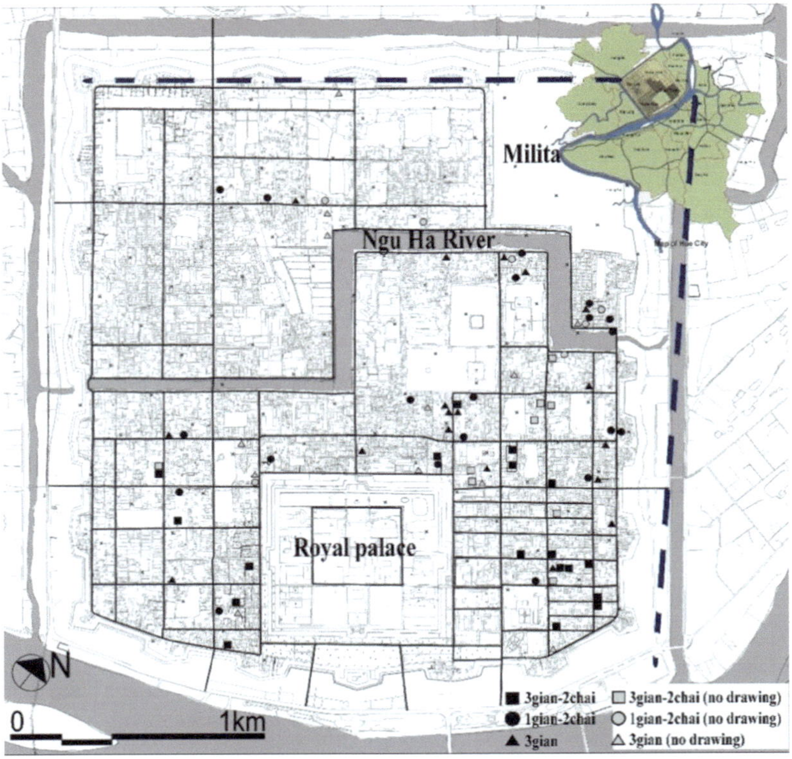

Fig. 5.18 Distribution of the 56 HTGHs surveyed in the citadel. *Notes Gian*—chamber; *Chai*—lean-to

characteristics of the living environment found in HTGHs in the hope that this research work can serve as a useful reference for further study of traditional garden houses in general.

Based on the surveys carried out from 2010 to 2018, the owners and inhabitants of 90 HTGHs (56 houses located in the inner area of Hue citadel and another 34 houses located in the Kim Long area) were interviewed in order to collect information about traditional function and lifestyle in the past (Figs. 5.18 and 5.19).

5.3.2 Formation of HTGHs and Their Architectural Features

According to (Nguyễn, Đ. V. and Nguyễn, Đ. Q. 2008: 137), traditional garden houses in Hue appeared in the seventeenth century during the time of the Nguyen Lords (1558–1774). They then spread rapidly during the time of the Nguyen Dynasty. It is thought that HTGHs in the citadel area were possibly constructed as recreation and study places for princes during the reign of Minh Mang King (1820–1840).[5] After that, HTGHs were later constructed as dwelling houses for royal families and high-ranking mandarins. Later, even ordinary people could build their own HTGHs if they could afford them.

The standard layout of a typical HTGH has been described in several previous studies such as (Hoàng 1999; Nguyễn 2008; Nguyen 2007; Nguyen and Kobayashi 2014; Nguyen et al. 2015). The standard layout of a typical HTGH normally consists of a *Nha Chinh* (main house),

[5] Quote in National History of the Nguyen Dynasty (Quốc sử quán triều Nguyễn) (1994: 95) when Ming Mang King granted the right to build HTGH to his princes: "*…grant a garden house for reading and recreational purposes, so that you (princes) can concentrate on study…*" (translated by the author).

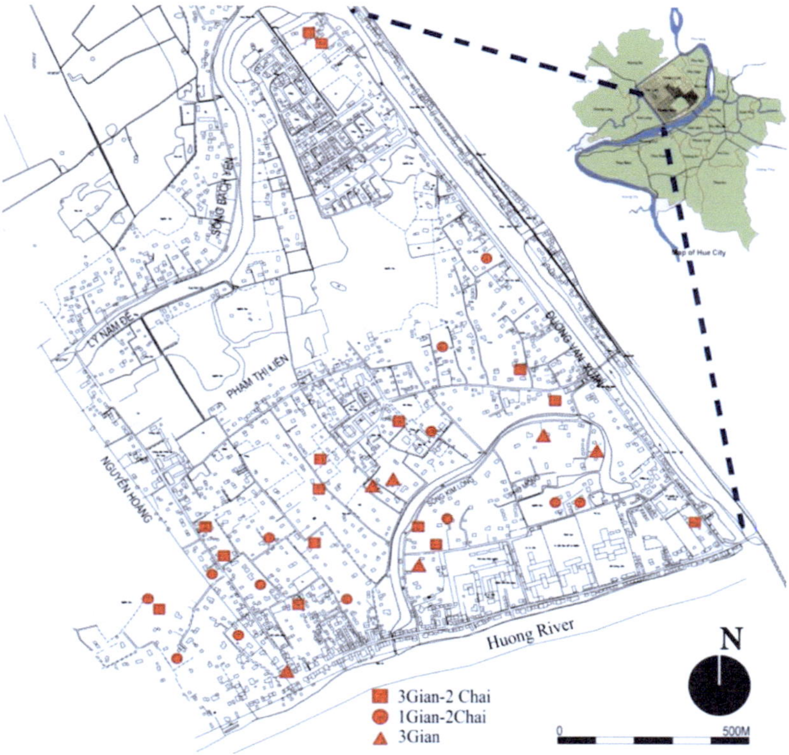

Fig. 5.19 Distribution of the 34 HTGHs surveyed in the kim long area

Nha Phu (annex house), *Binh Phong* (screen), *Be Can* (basin), symbols of the *Ta Thanh Long* (Left Azure Dragon) and *Huu Bach Ho* (Right White Tiger), garden, alley, hedge, and gate. Normally, the *Nha Chinh* serves as a place for worship, living room, and bedroom, while the kitchen, dining room, and storage area are in the *Nha Phu* (Fig. 5.20). The *Nha Chinh* can be classified into three types according to the number of *Gian* (chamber) and *Chai* (lean-to) in the *Nha Chinh*: 1 *Gian*—2 *Chai* (A type), 3 *Gian*—2 *Chai* (B type), and 3 *Gian* (C type), as illustrated in Fig. 5.21.

The *Nha Chinh* is normally made of timber called *Ruong*. The *Binh Phong* and *Be Can* are arranged in front of the *Nha Chinh*. In keeping with the principles of Feng Shui, the *Binh Phong* serves as a screen to prevent evil spirits (Dead *Qi*), while the *Be Can* represent the "Water Mouth," which can bring health and prosperity to the owner. The symbols of *Ta Thanh Long* and *Huu Bach Ho* serve as two guardians for the house. They normally take the form of two small trees or flowerpots located in front of two *Chai* of the *Nha Chinh* or on the two sides of *Binh Phong*. The *Nha Phu* is usually laid perpendicularly to the *Nha Chinh*.

Inside the *Nha Chinh*, the rear middle *Gian* is normally used as a place of worship (Fig. 5.22) The left *Chai* (commonly known as the *Chai Dong* or the *Chai Tren* in Vietnamese) is used as a place for men, while the right *Chai* (commonly known as the *Chai Tay* or the *Chai Duoi* in Vietnamese) abutting the *Nha Phu*, serves as storage space and a convenient bedroom for the women of the household when carrying out their daily activities. The space in front of the worship space is flexible. It is a multifunctional space, which can serve as a guest room, a living

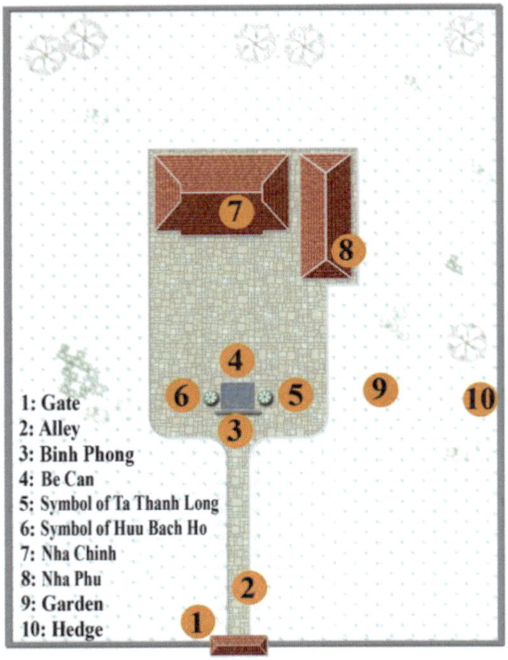

Fig. 5.20 Layout and composition of a typical HTGH

room, sometimes a dining room, or somewhere to take a nap during the middle of the day. The kitchen, dining room, and storage spaces are usually located in the *Nha Phu*, while the bathroom, toilet, and space for washing are normally all placed outside, at the rear of the building. Part of the *Nha Phu* was originally used as a bedroom for women as well.

5.3.3 Traditional Living Environment in HTGHs

It seems that the living environment in HTGHs was originally intended to express harmony among the buildings, nature, and humans. The *Nha Chinh*, *Nha Phu*, and other elements all harmonize with the surrounding garden to create a beautiful and natural picture. This is reflected in the layout arrangement, spatial organization, and lifestyle associated with HTGHs.

Currently, most gates and fences of the surveyed HTGHs are made of iron, brick, and concrete, with no roof. However, the gates would originally all have had a roof, along with green hedges of various plants such as Chinese tea, hibiscus, and brushwood. The roof would have provided travelers with a convenient place to shelter from sudden rain, while the green hedge would have created a softer, friendlier feature than that of concrete and iron.

Behind the gate, a small alley leads to the symbols of the *Binh Phong* and *Be Can* (Fig. 5.23). The roles of *Binh Phong* and *Be Can* are to delay the guest reaching the living room and allow the owner some time to get dressed, if necessary, so that they can welcome the guest in a more dignified manner in accordance with the traditional lifestyle of the Hue people.

The center of HTGHs is the *Nha Chinh*, which usually faces the southeast. According to the religious belief of Vietnamese people, the south

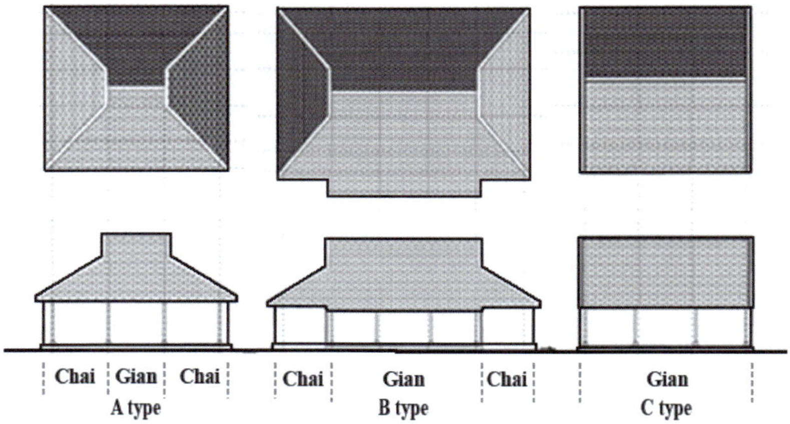

Fig. 5.21 Classification of HTGHs based on the number of *Gian* and *Chai* in *Nha Chinh*

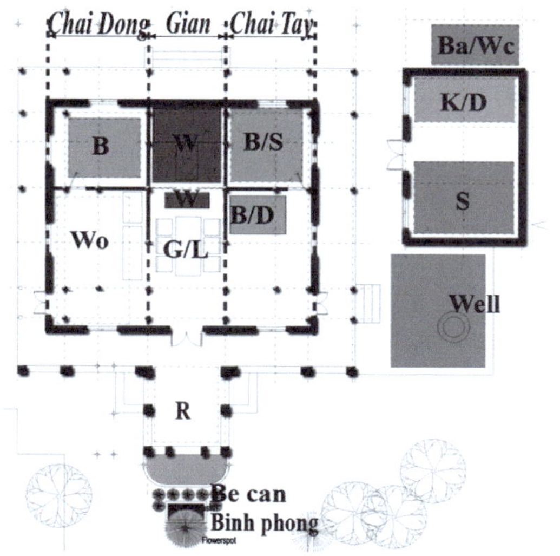

Fig. 5.22 Functional usage of space in an HTGH. *Notes* W-Worship; B-Bedroom; S-Storage; Wo-Working; L-Living room; G-Guest room; R-Recreation; K-Kitchen; D-Dining room; Ba-Bathroom; Wc-Water closet

and the southeast are considered as the best orientations for facing of a dwelling house because it can then receive good prosperity, health, and good natural ventilation (Nguyen et al. 2010: 487). Normally, the front wall of the *Nha Chinh* is a system of plywood doors called *Cua ban khoa* (Fig. 5.24). When the guest enters the *Nha Chinh*, he has to bend down and step over the *Cua ban khoa*. According to the traditional beliefs of the Hue people, this action is an expression of respect to the owner and implies that the guest should be careful to avoid hitting columns and furnishings inside the house. As part of the interview survey, some interviewees mentioned that the height of the *Nha Chinh* should be higher than that of the *Nha Phu*, or that the floor level of the *Nha Chinh* should be higher than that of the *Nha Phu*.

Figures 5.25, 5.26, and 5.27 illustrate the traditional functions of each space in *Nha Chinh* in type A, type B, and type C HTGHs, respectively. The interior of the *Nha Chinh* in these HTGHs can be divided and partitioned in various ways, depending on the separation between *Gian* and *Chai* and between the front and back spaces. Thus, the *Nha Chinh* of HTGHs in types A and C can be divided into six spaces, while 10 spaces are present in type B. A description of the functional usage, past and present, for each type of house is included below.

In type A house, all HTGHs originally placed the worship space in the rear *Gian* (space 2) and

Fig. 5.23 Symbols of *Binh Phong* and *Be Can*

Fig. 5.24 *Cua ban khoa* in an HTGH

located bedrooms in the rear *Chai* (spaces 1 and 3). However, space 3 in the nine houses was originally incorporated as storage space. This space was normally next to the *Nha Phu* and was a convenient area for women in the family to carry out their daily activities. Spaces 4 and 5 were formerly used as guest rooms, incorporating with the living room (10 and 13, respectively), while space 6 could be used as storage (1 case), a bedroom (2 cases), recreation (1 case), or a living room (1 case). The various functions and use of the *Nha Chinh* of HTGHs in type C houses in the past were quite similar to that in type A houses.

In type B houses, the space used for worship was always in the rear middle *Gian*, and this is still the case today. In the past, spaces 1 and 6 usually served as bedrooms for male members of the household, incorporating with working place, but nowadays these spaces also carry out other functions such as storage, recreation, and vacant space. Space 2 was formerly used for worship or for storage, such as a small book library for the owner (9 cases), or as a bedroom (11 cases). The functional usages of space 4 were mainly as bedroom (10) and worship space (7 cases), while space 5 usually served as a bedroom (22 cases). In 10 of the houses studied, space 5 was also incorporated with storage. Spaces 7, 8, and 9 mainly served as guestrooms and/or living rooms, but in some cases, they also served as worship spaces, dining rooms, bedrooms, recreation areas, and work spaces.

Based on the discussion of traditional usage outlined above, it can be surmised that the middle rear *Gian* in the *Nha Chinh* was only ever

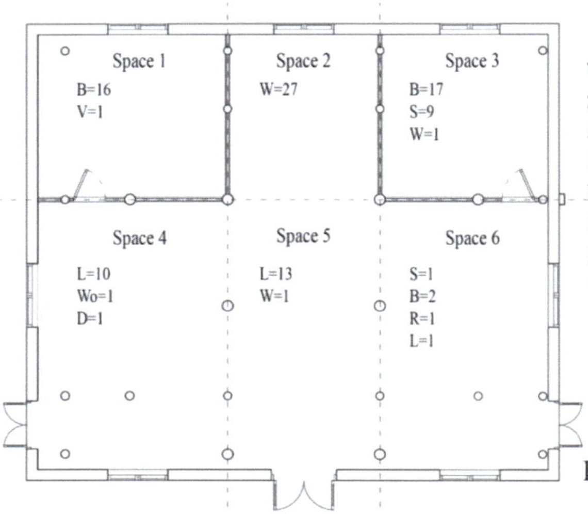

Notes:
W: Worship
S: Storage
B: Bedroom
L: Living room
Wo: Working space
R: Recreation
D: Dining room
V: Vacant space

HTGH 1 gian-2 chai

Fig. 5.25 Functional usage of space in a type A *Nha Chinh* (1 *Gian*-2 *Chai*)

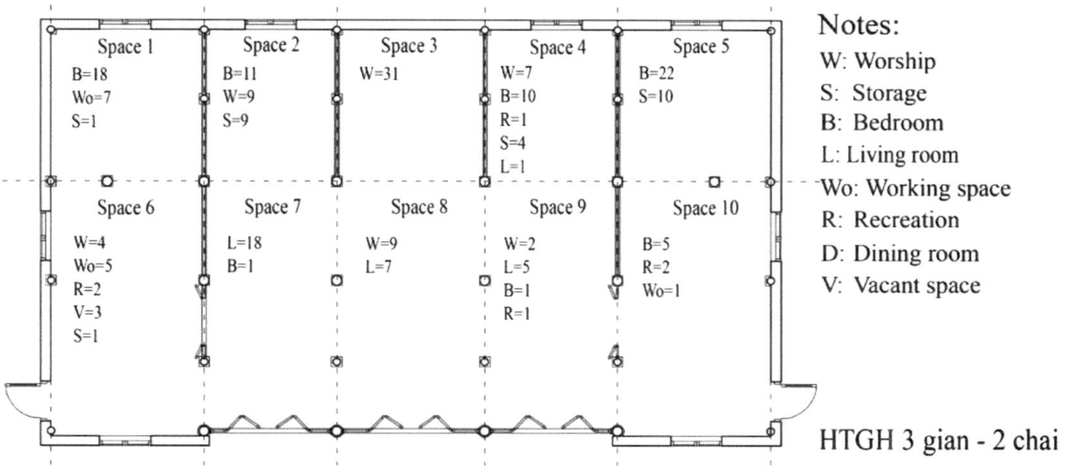

Fig. 5.26 Functional usage of space in a type B *Nha Chinh* (3 *Gian*-2 *Chai*)

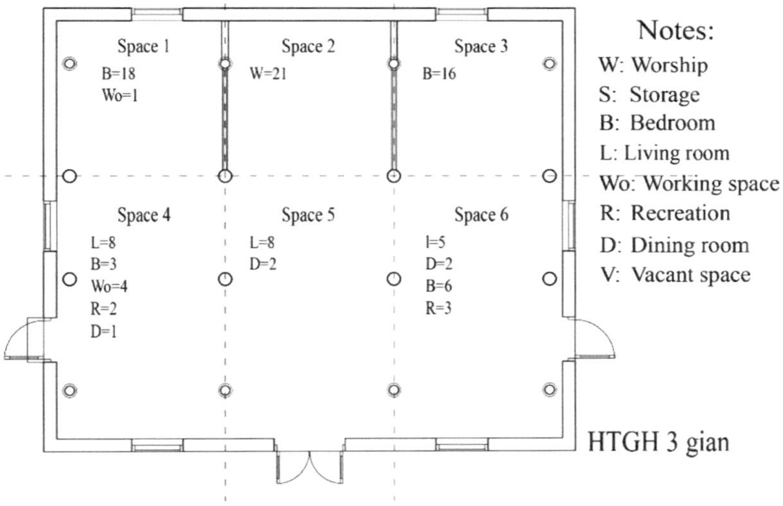

Fig. 5.27 Functional usage of space in a type C *Nha Chinh* (3 *Gian*)

used as a space for worship. Conversely, the spaces to either side of the middle rear *Gian* mainly served as bedrooms, although they could sometimes be incorporated with storage for the space next to the *Nha Phu*. The spaces in front of the worship space were used as a living room incorporating a guestroom, while the other spaces were multi-purpose areas that could be used as working spaces, recreation areas, dining rooms, etc. According to the survey results, the functional usage of most of these spaces in the *Nha Chinh* has not changed much over time, especially for spaces in the rear part. The space traditionally used for worship has kept its original function and it can sometimes be seen expanding into front space or into its sides when the house is used for clan worship. The remaining spaces in the front part of the *Nha Chinh* are becoming more multi-purpose in use-fulfilling a variety of functions in some cases.

According to the interview results, most owners considered that the most important space in

Fig. 5.28 Worship space in an HTGH

their house is still the worship space and that the middle rear *Gian* is, therefore, the most solemn. Vietnamese people have retained their customs and beliefs and they respect their ancestors and remember their origins. Thus, this is the space to connect the linkage between living family members and their ancestors (Nguyen et al. 2015: 78). This is the reason why the space for worship is always located in the middle rear *Gian*.

Inside the *Nha Chinh*, the owners usually arranged many old pictures of their ancestors on the worship tables (Fig. 5.28). Many valuable engraved boards and remembrances (wooden paintings and old appliances) can also be seen throughout the *Nha Chinh*. In some HTGHs, the owners, even though they do not know Chinese well, can read all the Chinese words engraved on the carved boards fluently. The simple reason for this is that these words are their father's advice for their descendants. In addition, these old pictures and remembrances that ornament the house imply that the ancestors are still present, even though they may have passed away long ago. This helps also ensure that the owners and the family members always remember their origins and heed the advice of their ancestors. This reflects some of the hidden charm and positive aspects of the traditional culture and lifestyle associated with HTGHs. In contrast, the *Nha Phu* is usually the space used for the kitchen, storage, etc. In the past, this space was mainly reserved for the females in the family, but it is now more commonly used for a variety of daily family activities.

5.3.4 Conclusion

Based on the survey results for HTGHs located in the Hue citadel and Kim Long areas, it is now possible to gain a better understanding of the traditional environment associated with HTGHs. The layout of a typical HTGH includes a *Nha Chinh*, *Nha Phu*, gate, fence, alley, *Binh Phong*, *Be Can*, symbols of *Ta Thanh Long* and *Huu Bach Ho*, and a garden. This arrangement signifies the owner's desire that the house's occupants will achieve wealth and prosperity, and avoid the attendance of evil spirits. It also expresses harmony among the house, the garden, and the people living there. In addition, it reflects the traditional lifestyle of people living in Hue. The green hedge and garden help create a welcoming environment in harmony with nature. The gate with a roof, the alley, and the location of the *Binh Phong* and *Be Can* imply a subtle appreciation of design on the part of the owner.

In terms of lifestyle, it can be seen that the functional use of space in HTGHs has been carefully thought out and arranged symmetrically. The space in the rear middle *Gian* of the *Nha Chinh* is always used for worship because it is the most solemn space in the house. Guestrooms incorporating the living room are usually placed in front of the worship space, and two *Chai* serve as bedroom and storage area. The rest of the space is available for other functions, and the *Nha Phu* is used for kitchen, storage, bedroom, and other purposes.

5.4 Urban Vulnerability to Climate Change in Danang City, Vietnam

5.4.1 Introduction

Climate change impact and responsibility have grown to become a primary concern in many countries all over the world, especially in those with coastal areas, which are currently facing a high risk of climate change and its consequences. Located in the coastal Southeast Asia region, Vietnam is reported to be one of the most significantly impacted nations in the world by climate change as measured by the global climate risk index 2019 (Eckstein et al. 2020). Although there have been numerous studies on the assessment of climate change and its stressors, it is important to update the climate change situation and comprehensively analyze its impacts based on the local characteristics of the study area. Studies on climate change vulnerability assessment have become a virtual need in any region since they help define an important baseline for planning solutions, strengthening the adaptive capacity, and mitigating the impacts of climate change. This section aims to build a practical framework for climate change vulnerability assessment and its effect on urban life in Danang city, which is a leading socio-economic region on the Central Coast of Vietnam.

Danang city is a coastal city located in Central Vietnam that is considered vulnerable to climate change and its consequences. A statistical report from the Danang People's Committee about disaster history in Danang from 1999 to 2020 has shown that floods and typhoons occurred annually from 1998 to 2020 with at least two floods per year. The economic and human loss caused by disasters in Danang city during this period is remarkable. In total, 237 people were injured, 205 people went missing and were killed, 15,633 households collapsed, 130,861 households were destroyed, and 9,700 billion VND was lost due to the natural catastrophes in the period of 1998–2020 (Danang People's Committee 2020). Meanwhile, Danang has been considered Vietnam's new economic dragon for its steady growth and global integration. This city has also attracted a large population that is concentrated in the city center and negatively impacts its climate change vulnerability. The highest population density is in the urban districts within a small and narrow coastal valley. The urban districts make up 20% of the total land area of Danang but support 87% of the city's population. The major social, economic, and political infrastructure of Danang is also concentrated in the urban areas (Nguyen et al. 2012), which are mostly located near the coast. These characteristics make the urban districts even more vulnerable to climate change and its stressors. This work, therefore, tried to assess, quantify, and map climate change vulnerability in Danang city, Vietnam, by integrating Geographical Information System (GIS) method and multi-criteria analysis. The main objective of the project is to determine the vulnerability of climate change to each area in Danang city based on a comprehensive assessment of three factors: exposure to hazards, local sensitivity, and adaptive capacity (IPCC 2014). The results of the project will help identify areas with different levels of vulnerability to climate change from low, medium, and high. The study also proposes several adaptive solutions for areas highly vulnerable to climate change.

5.4.2 City Profile

5.4.2.1 Natural Conditions and Climate Characteristics

Danang is located in a typical tropical monsoon region with a natural land area of 1,285 km^2, comprising 980 km^2 of mainland and 305 km^2 of Hoang Sa islands (Danang GSO 2020) (Fig. 5.29). Its climate is characterized by two seasons in a year: a rainy season from August to December and a dry season from January to July, with rainfall mainly concentrated from September to December. On average, this area is directly or indirectly affected by 1–2 typhoons and 1–2 serious flooding spells each year (Do and Nagasawa 2014). The average humidity in Danang city is 83.4%, the average temperature is approximately 26 °C, and the average rainfall is 2,505 mm per year (Danang People's Committee 2020). The topography of this city has great variation from flat to mountainous regions, with elevation ranging from 0 to 1,664 m above mean sea level. Due to the variation in topography and geomorphology, the natural conditions as well as occurrences of natural disasters in Danang city are also spatially diversified.

5.4.2.2 Socio-Economic Conditions

Currently, Danang is the third biggest city in Vietnam and the center of socio-economy in Central Vietnam. Danang's territory is divided into eight districts, including six urban districts (Hai Chau, Thanh Khe, Son Tra, Ngu Hanh Son, Lien Chieu, and Cam Le), one suburban district (Hoa Vang), and one island district (Hoang Sa) (Fig. 5.29).

The city's economy has grown continuously with gross domestic product (GRDP) in the period 2010–2020 estimated to increase on average 7.89% per year. The city's economic structure is shifting toward "Service-industry-agriculture." In 2020, the service sector is estimated to reach 64.56%, the industry/construction

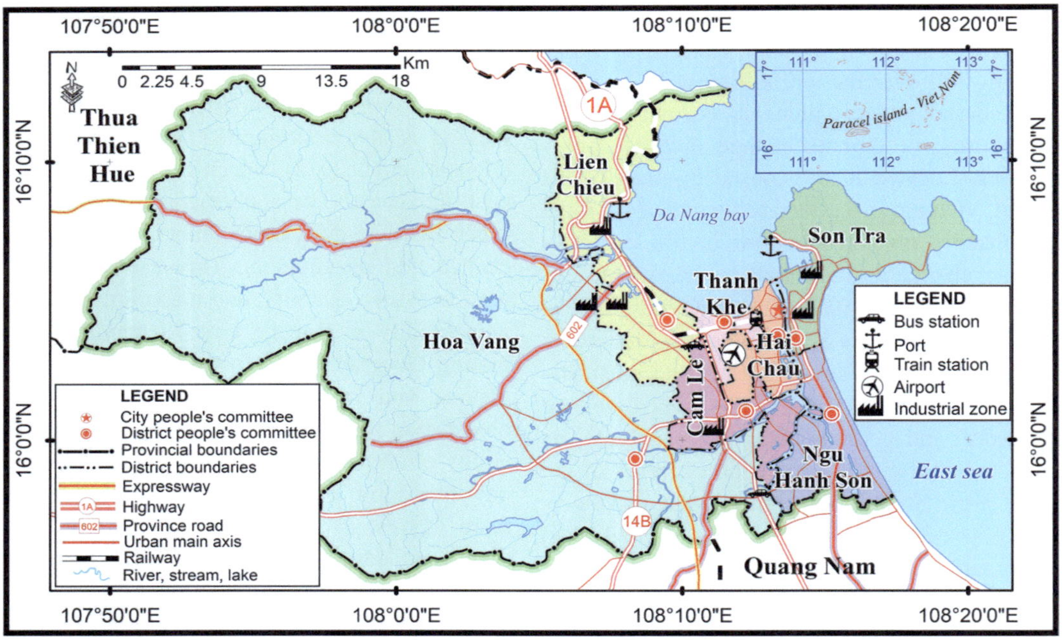

Fig. 5.29 Administrative map of Danang city

sector is estimated at 22.32%, and the agricultural sector is estimated at 1.72% of the City's GRDP. The service sector, including tourism, still accounts for the highest proportion of the economic structure of this coastal city. Danang attracted 8.7 million international and domestic visitors in 2019 (Danang GSO 2020), which makes this city a tourism hotspot in Central Vietnam.

5.4.2.3 Population and Urbanization

The average population of Danang in 2020 reached 1.16 million persons, an increase of 28,400 persons and nearly 2.5% increase compared to that in 2019, in which the urban population was 1,020,400 persons, accounting for 87.3% and the rural population was 149,100 persons, accounting for 12.7%. Danang city has been undergoing rapid urbanization during the past decades. Currently, it has the highest proportion of urban population among the provinces in Vietnam, with nearly 90% in 2020 while the average urban population rate of the whole country is only 37.3% (Danang GSO 2020).

5.4.2.4 Climate Change and Natural Hazards in Danang City

Due to the geographical location, the topographic and geomorphological characteristics of the study area, as well as the impact of global climate change, and the situation of natural disasters in Danang City are very complicated and tend to increase in quantity and severity. The common typical hazards that often occur in Danang are tropical cyclones, typhoons, floods, droughts, salinization, and landslides.

Typhoons and tropical cyclones (also named tropical depressions) in Danang usually occur during the period from May to December, mainly in October and November. These typhoons and tropical depressions are often accompanied by heavy rain. Therefore, in addition to strong winds, the mainland is also affected by floods and inundation.

Statistics from the Vietnam Disaster Management Authority show that from 1997 to 2020, 204 typhoons and tropical depressions appeared in the East Sea, of which 39 typhoons and 16 tropical cyclones affected Danang city; especially typhoon Xangsane in 2006, which is recorded as the biggest typhoon in the history of Danang, the typhoon Ketsana in 2009, and the typhoon Nari in 2013 directly affected Danang city, causing heavy loss of life and property (Fig. 5.30).

Flooding usually occurs after typhoons or heavy rain. Due to its geographical location in the downstream of Vu Gia–Thu Bon river basin, with the direct outlet of rivers flowing into the sea, Danang faces a high risk of flooding when there is heavy rain combined with sea-level rise in the context of climate change. In the past 20 years, Danang city has faced several historical floods, which have caused considerable damage to economic and urban life. Floods in 1998, with a peak of 10.4 m in the upstream, had caused 32 people deaths, 158 collapsed households, 19,029 inundated households, and the total loss was estimated at 182.3 billion VND. The 1999 floods are considered the biggest flood event in the history of Danang, where 37 people died, 61 people were injured, 4,579 households collapsed, 46,333 households were inundated, and a total loss of approximately 611 billion Vietnamese dong was experienced. The 2007 flood killed 3 people, injured 3 others, and 28,269 households were flooded, with approximately 9,500 tons of rice destroyed, approximately 760 hectares of vegetables ruined, apart from severe damage to traffic and drainage systems. The total loss in these floods was 1,524 billion VND (ISET and Danang City Government 2009). Most recently, the 2020 flood in Central Vietnam after continuous heavy rains in October was evaluated as the biggest flood in the past 20 years. In Danang city, this flood inundated 9 of the total 11 communes of Hoa Vang districts, with 3 missing people, 1 injured, 79.85 hectares of vegetables damaged, 24.73 hectares of fishing areas affected, and loss of 3 fishing ships, sinking of 1 ship, and damage to multiple tourist boats (Danang People's Committee 2020) (Table 5.2).

a. Typhoon Xangsane landed on Danang city in 2006

b. Life during floods in Danang city

c. Residents in Danang are boating on the street during flood in December 2018

Fig. 5.30 Urban life during disasters in Danang city. Photos by authors and referred from Danang City Government

Table 5.2 Impact of climate change and its stressors in Danang city by economic loss (Danang People's Committee 2020)

Damage category	Unit	2013	2015	2017	2018
Injured people	Person	19	1	5	0
House damage	House	7,048	70	1,109	925
Agricultural damage	Hectare	417	103	135.2	3,326
Total loss	Million VND	1,037,840	1,110	55,560	10,995

5.4.3 Climate Change Vulnerability Assessment for Danang City

5.4.3.1 Understanding the Climate Change Vulnerability

Assessing vulnerability to climate change is important for defining the risks posed by climate change and provides information for identifying measures to adapt to climate change impact (Satapathy et al. 2014). The definition of vulnerability has been mentioned in various studies in the literature (Balica 2012; Feyissal et al. 2018; Hinkel 2011; IPCC 2014; McCarthy et al. 2001; Satapathy et al. 2014). The Intergovernmental

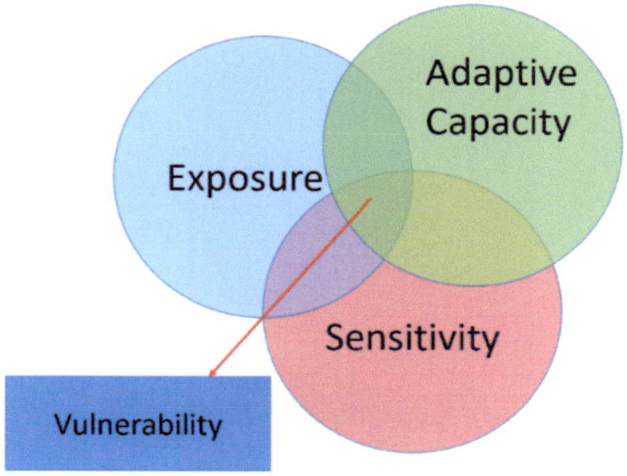

Fig. 5.31 Research approach for flood vulnerability assessment. Adapted from IPCC (2014)

Panel on Climate Change (IPCC) has defined vulnerability as "the propensity or predisposition to be adversely affected. Vulnerability encompasses a variety of concepts and elements including sensitivity or susceptibility to harm and lack of capacity to cope and adapt" (IPCC 2014). Although there have been many different definitions of vulnerability due to the differences in research fields and scientific viewpoints, people have agreed that vulnerability can be determined by three components: exposure, susceptibility, and adaptive capacity (Fig. 5.31). Exposure refers to "the nature and degree to which a system is exposed to significant climatic variations." Sensitivity refers "to the degree to which a system is affected, either adversely or beneficially by climate-related stimuli." Adaptive capacity refers to "the ability of a system to adjust to climate change—including climate variability and extremes—to moderate potential damages, to take advantage of opportunities, or to cope with the consequences" (McCarthy et al. 2001).

Vulnerability assessments are commonly distinguished as either following top-down or bottom-up approaches (Dessai and Hulme 2004). Top-down approaches are usually preferred at global, national, and regional levels, while bottom-up approaches start their analysis at the local level (e.g., households, villages, and communities). There is no one-size-fits-all solution. Assessment of the vulnerability should be conducted on a case-by-case basis and cannot generally taken by a single ready-made method. Thus, these methods are not always systematically related (Hinkel 2011). This study applied the vulnerability approach that was proposed by IPCC (2014) and can be simulated through the following formula:

$$V = f(E, S, AC) \quad (5.1)$$

Here, V is the climate change vulnerability, E is exposure to natural hazards, S represents the system sensitivity, and AC is the adaptive capacity. It is evident from Fig. 5.31 that vulnerability is an overlapping of three components, including E, S, and AC.

5.4.3.2 GIS-Based Multi-criteria for Climate Change Vulnerability Assessment

The primary methodology employed in this research is the combination of remote sensing, which involves the use of satellite image data, and geographic information system (GIS), along with field survey data. The integration of remote sensing and GIS techniques is considered an effective tool for evaluating the impact of natural disasters such as floods, droughts, and salinization. This approach has the advantage of providing fast and precise updates that reflect changes in the temporal and spatial dimensions of the disasters. In fact, remote sensing and GIS methods have been extensively utilized in studies related to climate change (Feyissal et al. 2018; Nguyen et al. 2016; Ouma and Tateishi 2014; Tran et al. 2017, 2021a, b). As climate change vulnerability assessment is generally conducted using an indices-based approach, it is highly recommended to employ GIS methods for generating a comprehensive set of indicators that includes both spatial and attribute information. Remote sensing data, which consists of a series of satellite images, allows for the computation of various indices that cannot be obtained through field measurements or statistical data. In this study, Landsat 8 images were utilized to generate the Normalized Difference Vegetation Index (NDVI) and the Land Surface Temperature (LST), which contribute to the

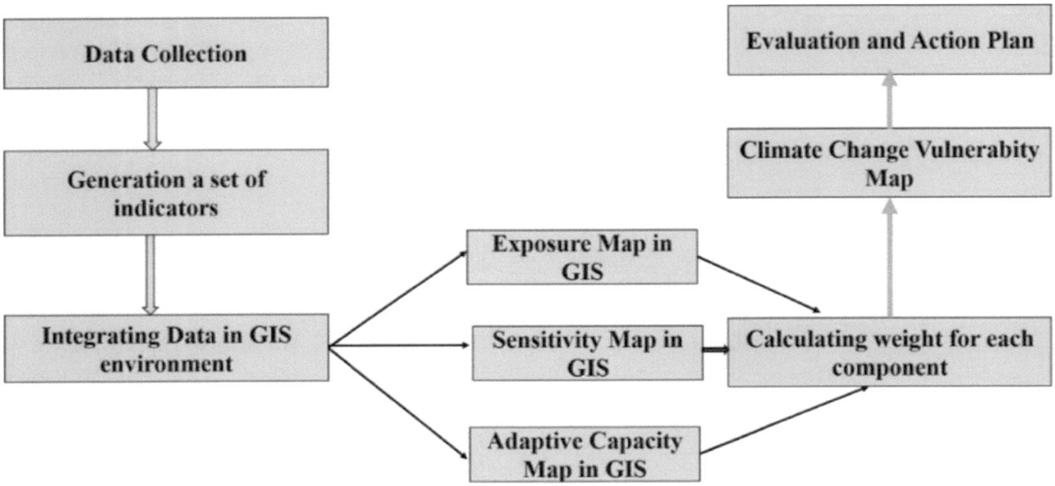

Fig. 5.32 Flowchart of data processing

exposure components. Other indicators were calculated and generated within a GIS environment, using statistical data and spatial data analysis techniques with the support of QGIS and GRASS GIS software.

Figure 5.32 represents the GIS-based approach utilized in assessing climate change vulnerability in Danang city. Initially, a set of indicators was developed to evaluate the components of climate change vulnerability, including variables that reflect the level of exposure (E), sensitivity (S), and adaptive capacity (AC). These indicators were then standardized and normalized into a unique scale ranging from 0 to 1 within a GIS environment. Finally, the research determined the climate change vulnerability index (V) of each district using a multi-criteria analysis, specifically the Analytical Hierarchy Process (AHP) method introduced by Saaty in 2008. The calculation of the V index has taken into consideration the weights of the driving factors.

5.4.3.3 Generation of a Set of Indicators

The development of a set of indicators for climate change vulnerability assessment in Danang city is based on field survey data integrated with statistical data on natural hazards and socio-economic conditions. A total of 18 indices were developed, in which six indicators including typhoon intensity, flooding, temperature, NDVI, wind speed, and elevation variation were selected for exposure component (E). The typhoon intensity index was calculated based on the distance to coastline and river channels, while the flooding index was measured by average flood depth by district using data from the Vietnam Institute of Meteorology, Hydrology and Climate Change (IMHEN). The temperature index was calculated using Landsat 8 data during the summer season to identify hotspots, whereas the NDVI index was generated using Landsat 8 data and the formula mentioned in Weier and Herring (2000). Wind speed data was collected from Windy.com, and elevation variation was generated from the SRTM digital elevation model available at (https://earthexplorer.usgs.gov).

Regarding the indicators for the sensitivity and adaptive capacity components used for climate change vulnerability assessment in Danang city, we have selected six indicators including population density, poverty, population structure, economic scale, land use, and temporary housing for the combination of the Sensitivity (S). These indicators represent the severity of the area's vulnerability to hazards. The adaptive capacity index

Table 5.3 Indicators used for climate change vulnerability assessment for Danang city

Component	Indicator	Detail	Weight	Sub-weight
Exposure	Typhoon	Distance to coastline and river channels (DIST)	0.16	
	Flooding	Average flood depth by district	0.28	
	Wind speed	Average wind speed during typhoon	0.28	
	Elevation variation	Difference in elevation per unit area	0.09	
	Temperature	Surface temperature during hot season by district	0.09	
	NDVI	Normalized Difference Vegetation Index	0.09	
Sensitivity	Population density	Population density by district	0.27	
	Poverty index	Percentage of poor household/ total household	0.15	
	Population structure	Percentage of women in the total population	0.27	0.20
		Percentage of people of working age		0.12
		Percentage of people over working age		0.12
		Percentage of children under 5 years old		0.20
		Percentage of disabled persons in the total population		0.36
	Economic scale	Total number of enterprises by district	0.15	0.5
		Total revenue by district		0.5
	Land use	Percentage of built-up area	0.08	0.5
		Percentage of agriculture area		0.5
	Temporary house	Percentage of temporary houses by district	0.08	
Adaptive capacity	Income index	Average income of each household	0.27	
	Education level	Percentage of people having high-education levels	0.15	
	Health care	Percentage of medical staff in the total population	0.15	
	Road density	Average road length per unit area	0.08	
	Drainage culvert	Average length of water pipe per unit area	0.08	
	Living condition	Rate of households having car	0.27	0.2
		Rate of households having boat		0.1
		Rate of households having Television		0.1
		Rate of households having mobile phone		0.1
		Rate of households having laptop		0.1
		Rate of households using clean water		0.2
		Rate of households having sanitation facilitates		0.2

is formed by six indicators: average income, education level, healthcare ability, living conditions, road density, and drainage culvert. These indicators represent the area's ability to adapt to hazards. The data used for calculating these indicators were obtained from the statistical yearbook data and the Department of Natural Resource and Environment of Danang city. QGIS and GRASS GIS were used for data processing. All indicators were calculated at the district level and are explained in detail in Table 5.3.

5.4.3.4 Climate Change Vulnerability Index

In the first stage of this study, all the indices were normalized into the same scale from 0 to 1 using the method mentioned by Tran et al. (2021b). These indices then were integrated with

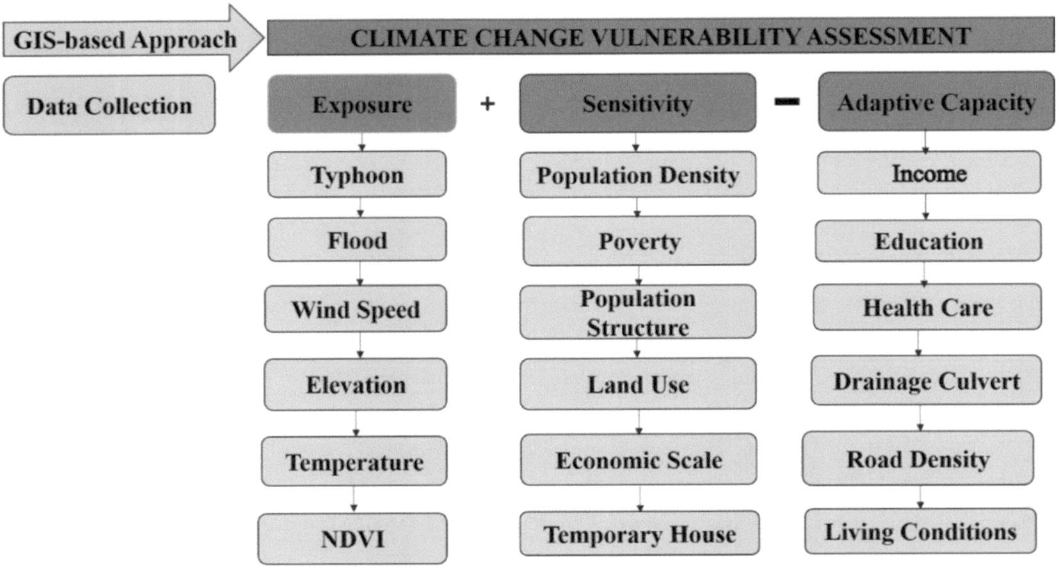

Fig. 5.33 GIS-based climate change vulnerability assessment approach

the AHP model (Saaty 2008) for weighted calculation and generating a combined index for climate change vulnerability. The calculation approach is shown in Fig. 5.33. Weighting by AHP is widely used in many studies. Details of applying the AHP method in flood hazard assessment are presented in Tran et al. (2017). This study developed a criteria system for climate change vulnerability assessment and the results of weight calculation for each element using the AHP method are described in Table 5.3. The last step involves checking the consistency of the subjective evaluations using the consistency ratio (CR). The comparison matrix is consistent if the resulting CR is less than 0.1 or 10% (Saaty 2008). All the weight schemes showed CR less than 0.1, which is reasonable in the AHP method (Saaty 2008). The results of weight calculation for parameters in climate change vulnerability assessment are shown in Table 5.3 (Fig. 5.34).

5.4.3.5 Identification of Areas Vulnerable to Climate Change in Danang City

The climate change vulnerability index (V) is calculated by combining the exposure, sensitivity, and adaptive capacity using Eq. 5.1, in which the weighting scheme has been taken into account. As a result, the climate change vulnerability index for Danang city was found to range from 0.18 to 0.6. Subsequently, V values were reclassified into four levels, from low, medium, high, and very high, with equal intervals. The division schemes using equal ranges are as follows: 0.18–0.28 (low), 0.28–0.38 (medium), 0.38–0.48 (high), and 0.48–0.6 (very high). In this study, we did not consider the very low level since the study area is considerably affected by climate change, as discussed previously. Moreover, compared to the vulnerability category in previous research (Balica 2012), the minimum value of V in Danang city is 0.18, which should be categorized as low level.

Figure 5.35 describes the climate change vulnerability levels in different districts of Danang city. The districts of Ngu Hanh Son and Son Tra are identified as very-high vulnerability areas due to their coastal topography, proximity to the coastline and river channel, high population density, and susceptibility to tropical cyclones with strong wind speeds. These areas are also tourism hotspots in Danang.

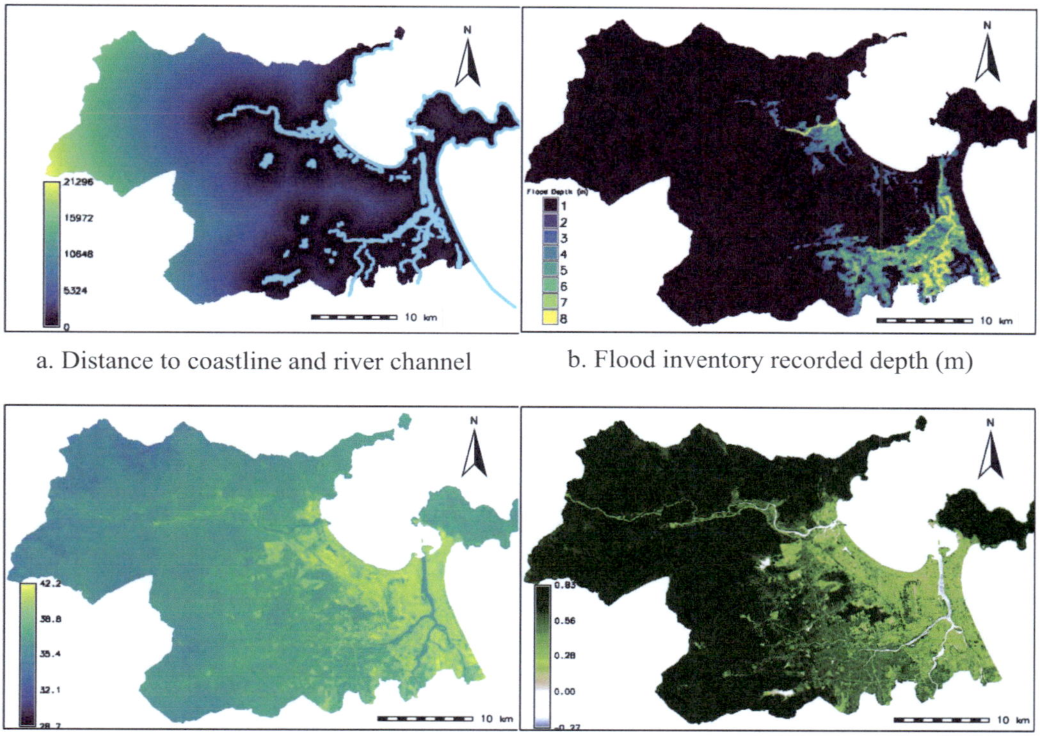

a. Distance to coastline and river channel

b. Flood inventory recorded depth (m)

c. Temperature stressor

d. Normalized difference vegetation index

Fig. 5.34 Parameters used to generate the climate change vulnerability index

On the other hand, the center city located in Hai Chau district has a low vulnerability level despite having a relatively high value of sensitivity. This is because Hai Chau district has a high rate of economic development and comparable adaptive capacity, as shown in Fig. 5.35c. It could be concluded that enhancing adaptive capacity is a key factor in responding to climate change vulnerability.

Hoa Vang district is a rural district of Danang city with high elevation in the west and lowland topography in the east. Although this district is categorized as being "low level" in the exposure map (Fig. 5.35), Hoa Vang has a very low value of adaptive capacity, which is negatively affected by the climate change vulnerability index. Therefore, Hoa Vang, along with Cam Le and Thanh Khe districts, is categorized in the high vulnerability level areas.

5.4.4 Urban Responsibility to Climate Change in Danang City

5.4.4.1 Enhancing Adaptive Capacity

It is obvious that enhancing the adaptive capacity is crucial for coping with climate risk. The Danang city government is already aware of this and has been investing in space, infrastructure, and urban architecture development since 2005. The guiding principle of this development has been "extending the coastline, stretching the river," which involves the construction of bridges over the Han river, investment in riverside and coastal routes, and making good use of land area to promote the value of urban lands. These initiatives are aimed at improving the city's adaptive capacity, which is essential for reducing vulnerability to climate change (ACCCRN 2016).

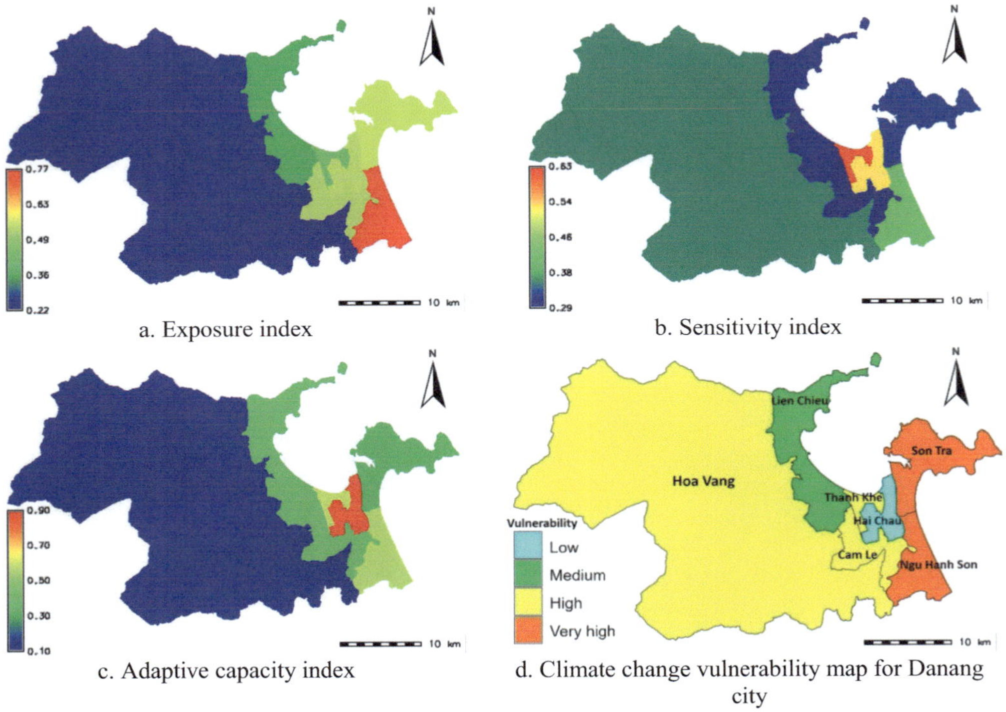

Fig. 5.35 Climate change vulnerability map and its components for Danang city

In addition to transportation infrastructure, the city has also promoted investment in the medical sector by encouraging the construction of hospitals such as Danang's Women's Hospital, Maternity and Pediatric Hospital, and Danang Oncology Hospital. Culture and tourism-focused projects have also been undertaken, including the construction of the International Exhibition Fair Center, Danang Historical Museum, Tien Son Sports Palace, and other infrastructure in districts throughout the city. These initiatives are aimed at promoting economic growth and improving the city's adaptive capacity to cope with climate risk.

Danang city government's efforts to focus on social welfare work in addition to construction investments since 2000. These efforts include finishing the construction and maintenance of 950 houses for social policy families and completing a plan for repairing the houses of veterans in 2014. The city has also provided housing assistance to veteran and ethnic minority families, with a total capital of 24 billion VND (1.08 million USD) in 2015 (ACCCRN 2016). Nearly 100% of relocated households received assistance to build a new house by the end of 2014. Furthermore, from 2011 to 2015, the Program for Storm-resistant Housing, funded by the Rockefeller Foundation and implemented by the Danang Women's Union, built or renovated 420 houses for typhoon resilience. These initiatives are aimed at improving social welfare and reducing vulnerability to climate change impacts such as tropical storms and typhoons.

Education is an important part of the adaptive capacity of Danang city. Various education programs and activities have been conducted on education for sustainable development (ESD) at different levels, from primary schools to universities (Kieu et al. 2016; Hoang and Kato 2016; Vo et al. 2013). Danang city has also made a significant investment in ESD projects in the local communities (Danang People's Committee 2019) with the goal of enhancing the resident's adaptive capacity to the risk of climate change.

In 2013, the city was selected by the Rockefeller Foundation as part of the first group of 33 cities to join the 100 Asian Cities Climate Change Resilience Network (ACCCRN). Danang is the first city in Vietnam to develop and release a resilience strategy; however, implementing city resilience requires more than just a document. The proposed strategy actions are expected to inspire greater collaboration between the Danang city government and the public in order to jointly enhance city resilience.

Danang city has expanded its partnership with 100 Resilient Cities to prepare its first resilience strategy. The strategy is based on Danang's vision for "a peaceful, dynamic, prepared, and connected city in the changing world," which identifies four goals to address resilience challenges. The first goal is to create a peaceful city that removes fear and anxiety from places where residents live, work, and play. The second goal is to create a dynamic city with a dynamic economy in partnership with people and businesses to meet needs in a changing world. The third goal is to create a prepared city with infrastructure systems that can recover and be well-prepared for challenges in the development process. The fourth goal is to create a connected city where knowledge is driven by meaningful information sharing. These goals are aimed at improving the city's resilience to various challenges and enhancing the quality of life for its residents (ACCCRN 2016).

5.4.4.2 Building Smart-Service for the City Development

Danang city focuses on its smart city project as a key driver of socio-economic development in the future, alongside a transition to technology-based city management. The city aims to be among the top three performers in digital transformation, information security, and e-commerce by 2030, according to a resolution issued by the city's Municipal Party Committee. To achieve this goal, the city is developing e-governance solutions to provide all online public administrative services at level 4, have 60% of level-3 and 4 non-business services provided on multiple digital platforms and supported on mobile devices, and have all governmental agencies provide open data. Currently, Danang is using 1022 public service portals that help people rapidly access all city information, including disaster and risk warnings for citizens. These initiatives are aimed at promoting digital transformation, enhancing information security, and improving public services to support the city's socio-economic development.

Danang has recently invested in a series of big-scale tourism infrastructure projects. A new terminal for the Danang International Airport was constructed in 2015 and operated in 2017, allowing the airport to serve an expected 6–8 million passengers, and 400,000 to 1,000,000 metric tons of cargo per year, according to the airport-designed planning. The airport has served up to 80 percent capacity in the year from 2017 to 2021 excluding the 2021 due to the Covid-19 situation. Until the year 2030, the airport is expected to see an estimated 10–12 million visitors per year (Danang International Airport 2021). Along with infrastructure development to attract more investment and tourism, Danang also desires to promote the industrial sector, especially high technology and clean industries (ACCCRN 2016).

In 2014, Danang city has decided to build a model of a smart and creative city (Danang People's Committee 2019). In 2020, Danang received the Vietnam Smart City Award 2020. Currently, the city is also running different projects focusing on smart city development covering different aspects, such as smart traffic, smart tourism, smart communities, etc. The common goal of these projects is to help the city become more resilient and safe under the impact of climate change and natural disasters and be more confident about its future sustainable development.

5.4.5 Conclusions and Recommendations

Danang city, which is located in the coastal and tropical monsoon regions and characterized by complicated geomorphological conditions, has become vulnerable to climate change and climate-related disasters. Urban development is one of the major socio-economic factors of the city, but it also makes the city become more sensitivity to climate change. The present study has generated a framework of GIS application integrated with multi-criteria analysis for climate change vulnerability assessment for Danang city, which addresses a recommendation made by the IPCC in 2014. The study used a multi-parametric AHP model to generate a set of indicators that integrated the natural and socio-economic conditions of the city and demonstrated the effectiveness of this approach for climate change vulnerability assessment. The climate change vulnerability assessment approach that developed from this study could be applied to other study areas that have similar geographical and social conditions. In future work, the authors plan to focus on developing an indicator system for climate change vulnerability assessment, integrating field surveys and interviews to improve the database on climate change impact and responsibility. Additionally, the authors plan to consider future scenarios for climate change in the comprehensive vulnerability assessment. These future efforts aim to improve the accuracy and usefulness of the climate change vulnerability assessment and inform strategies for climate change adaptation and mitigation.

5.5 Air Pollution in Vietnam: Recent Changes and Current State

As a result of remarkable economic growth, Vietnam is currently facing severe air pollution as well as other environmental problems. The National State of Environment 2016 (MONRE 2016) stated that air pollution is increasing in almost all big cities in the country. Air pollution is caused by the primary (or direct) emission of air pollutants such as particulate matter (PM), carbon monoxide (CO), nitrogen oxides (NOx), sulfur dioxide (SO_2), and volatile organic compounds (VOCs), and by photochemical reactions producing secondary organic/inorganic aerosols and oxidants such as ozone (O_3) and peroxyacetyl nitrate (PAN).

One of the main sources of air pollutants in urban areas is traffic. In Vietnam, motorcycles are commonly used as a means of transportation for people and the number of motorcycles has increased rapidly in recent years (Fig. 5.36). There were an estimated 43 million motorcycles in Vietnam in 2014 (VietNamNet Bridge 2015). In the capital city, Hanoi, there were 5 million motorcycles in 2016 and it was predicted that there will be 25 million motorcycles by 2025 with an annual growth rate of 10% (Sakamoto et al. 2018). Exhaust emitted from automobiles contributes to air pollution as well. Industrial processes and coal power generation are also considered to be large sources of air pollution. In addition to these, burning of agricultural waste such as rice straw, the so-called "biomass burning," is often carried out in rural areas near big cities (Fig. 5.37). Furthermore, other sources such as residential activities, construction, long-range transportation, etc., can also contribute to increased levels of air pollution. In order to work out a strategy to mitigate air pollution in big cities, a deeper understanding of factors affecting air quality is required.

There are many published reports in the literature on air pollutants measured through a network of continuous air monitoring, although the information is scattered in terms of time and space. This work therefore reviews previous reports on particulate matter and gaseous pollution, with an aim to contribute to understanding and dealing with current and future air pollution in Vietnam.

5.5.1 Particulate Matter Pollution

Particulate matter is the most concerning air pollutant as it can adversely affect human health, change rainfall patterns, negatively or positively contribute to climate change, affect agriculture, and reduce visibility. PM_{10} and $PM_{2.5}$

Fig. 5.36 Motorcycles on a main street (non-rush hour) (Photo by authors)

Fig. 5.37 Rice straw burning near Hanoi city. Photo by authors

are especially of more concern because of their effects on human health. The WHO guideline values for particulate matter are presented in Table 5.4. $PM_{2.5}$ poses a serious air pollution problem in several cities in Southeast and South Asia (Balasubramanian et al. 2017; Cohen et al. 2009; Dey et al. 2012; Hopke et al. 2008; Kim Oanh et al. 2006; Shen et al. 2017). Recently, besides $PM_{2.5}$ and PM_{10}, concerns regarding $PM_{0.1}$ (or nanoparticle) have also been raised mainly because, unlike other PMs, $PM_{0.1}$ can even pass the cell membrane.

The Ministry of Natural Resources and Environment (MONRE) of the Socialist Republic of Vietnam has reported severe air pollution in all big cities of Vietnam (MONRE 2010, 2016). Sampling data showed that 85–90% of total suspended particle (TSP) values in 2012–2016 in 37 main cities were higher than the national standard of 200 μg/m^3 for a 24 h average. All annual TSP levels for road sites in Hanoi, Ho Chi Minh City, Can Tho, Danang, and Hai Phong over 2012–2016 were higher than the national standard. They were the highest in Hanoi, and varied from approximately meeting standard to 3–4 times higher than the standard. The annual PM_{10} values from the continuous monitoring station in Nguyen Van Cu station in Hanoi over 2012–2016 were higher than the national standard of 50 μg/m^3. Similarly, the levels of Viet Tri were also higher than standard in all data available for 2013–2016. Other cities such as Danang, Hue, and Halong had almost all annual PM_{10} levels meeting standards. Annual $PM_{2.5}$ levels of six continuous monitoring stations were higher than the national standard of 25 μg/m^3 in Hanoi, Viet Tri (2–3 times), Ha Long, Hue, Nha Trang, and Danang. It is noticed that the $PM_{2.5}$ levels for 2016 were lower than those in previous years and the $PM_{2.5}$ levels in Danang and Nha Trang met the national standard after 2016.

Since the PM pollution in Hanoi, the capital city, has been severe, this review concentrates on the PM pollution there. Trends of TSP, PM_{10}, and $PM_{2.5}$ in Hanoi are shown in Fig. 5.38. Research published in the last two decades reported $PM_{2.5}$ levels in Hanoi to be

Table 5.4 WHO guideline values for particulate matter

Specie	Guideline value	Averaging time
$PM_{2.5}$	10 μg m^{-3} 25 μg m^{-3}	Annual 24 h
PM_{10}	20 μg m^{-3} 50 μg m^{-3}	Annual 24 h

some multiples higher than the WHO recommendation (Cohen et al. 2010; Hai and Kim Oanh 2013; Hien et al. 2011; Hopke et al. 2008; Kim Oanh et al. 2006). The mean of $PM_{1.0}$ was reported only one time at a residential area of 30.1 ± 13.9 μg m^{-3} between November 2015 and June 2016, encompassing three seasons to be lower than several cities in China and India (Hien et al. 2021). $PM_{0.1}$ level in Hanoi was reported to be 5 μg m^{-3} at an urban site and 6–10 μg m^{-3} at a traffic site, higher than in other countries (Thuy et al. 2016, 2017). PM as well as other air pollutants in Hanoi has a monthly variation following a typical cycle that is high in the dry season (winter) and low in the rainy season (summer). A periodic daily increase in $PM_{2.5}$ (episode) was also observed. Hien et al. (2002) found that $PM_{2.5}$ and PM_{10} were closely related to meteorological conditions, as air parcel trajectory from the north and northeast brings unfavorable conditions for air pollutants dispersion, e.g., temperature inversion. They also reported that regression models showed that meteorology could explain more than 50% of the variances in 24-h PM_{10} and $PM_{2.5}$ concentrations. Air pollutants (PM_{10}, SO_2, NO_2, and CO) were also reported to increase after cold surges, also known as monsoon surges in Hanoi. Sharp increases in concentrations of air pollutants (hazy period) due to unfavoured meteorology conditions were also observed in neighboring regions (Chen et al. 2017; Chuang et al. 2017; Zhang et al. 2013).

For controlling air pollution, source apportionment is important. Source apportionment of $PM_{2.5}$ in Hanoi was carried out by receptor modeling by some researchers. Cohen et al. (2010) noted that the main sources of $PM_{2.5}$ were: automobiles (40 ± 10)%, soil (3.4 ± 2)%, secondary sulfates (7.8 ± 10)%, smoke (13 ± 6)%, industry (19 ± 8)%, and coal (17 ± 7)%. Meanwhile, Hai and Kim Oanh (2013) noted that the main sources were: secondary mixed PM (40%), diesel traffic (10%), residential/commercial cooking (16%), secondary sulfate rich (16%), aged sea salt mixed (11%), industry/incinerator (6%), and construction/soil (1%). Moreover, Hien et al. (2021) reported that the dominant sources of $PM_{2.5}$ were: road dust (26.9%), long-range transport (24.1%), biomass burning (9.1%), vehicle (8.0%), sea salt (7.0%), construction (4.0%), and coal (3.0%), whereas the main sources of $PM_{1.0}$ were: long-range transport (32.3%), coal (15.4%), vehicle (10.7%), road (5.5%), sea salt (5.1%), road dust (5.5%), biomass burning (4.3%), and construction (1.0%). In addition, source apportionment of $PM_{0.1}$ was conducted for the first time in Hanoi by Nghiem et al. (2020). They noted that there are five main sources, as follows: traffic gasoline and diesel emission (46.28%), secondary (31.18%), resident/commerce (12.23%), industry (6.05%), and road/construction (2.92%). These findings suggest that traffic activities should be focused upon in order to control PM levels, especially $PM_{0.1}$.

Source apportionment at different air parcel trajectory directions was also carried out, suggesting the influence of meteorological conditions. Meteorological characteristics of $PM_{2.5}$ in Hanoi were investigated in Hanoi over July 2016–June 2017 using advanced highly sensitive sensors (Ly et al. 2018). The study reported that the monthly variation in the $PM_{2.5}$ concentration in Hanoi reflected the effects of regional meteorological conditions. In addition, the effect of direct emissions from transportation on $PM_{2.5}$ at the observation site is not dominant. During the winter monsoon, the effect of long-range transport from outside of the region was observed in Hanoi by using sensors and a machine learning approach (Ly et al. 2021). The results revealed

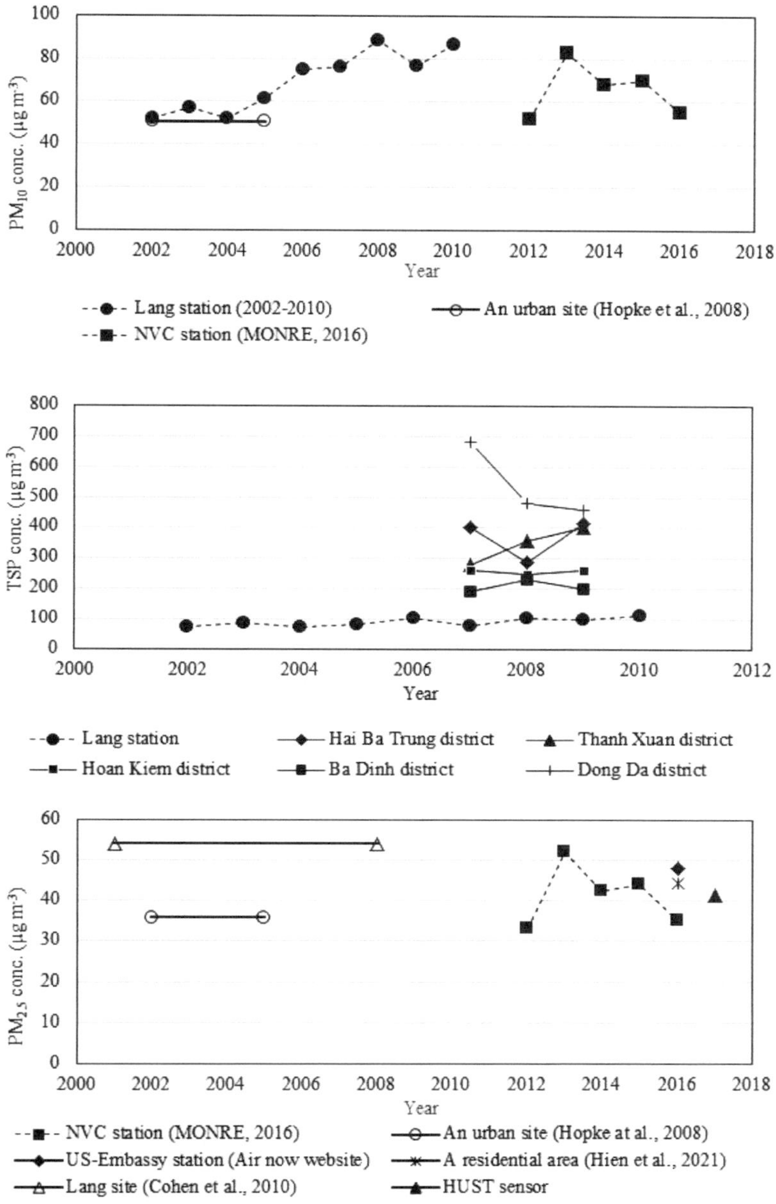

Fig. 5.38 Annual TSP (top), PM_{10} (middle), and $PM_{2.5}$ (bottom) levels (µg m^{-3}) at monitoring stations in Hanoi (Cohen et al. 2010; Hien et al. 2021; Hopke et al. 2008; MONRE 2016). The solid marks represent continuous monitoring data. The hollow marks represent filter-loaded samples averaged over the whole sampling period

that the effect of meteorological conditions and long-range transport from the outside on local $PM_{2.5}$ concentrations is significant. Current research by Ngoc et al. also demonstrated the effect of meteorological factors on haze events in spring in Hanoi (Ngoc et al. 2021).

5.5.2 Gaseous Pollution

Typical gaseous pollutants include NO_2, CO, SO_2, and VOCs as primary pollutants, and O_3 as secondary pollutant (or photochemical oxidant) formed by photochemical reactions of primary pollutants. Since gaseous pollutants pose a threat to human health as well as cause PM pollution, WHO has set guideline values for them in order to offer guidance in reducing the health impact of air pollution (WHO 2006, 2010, 2017); these guideline values are summarized in Table 5.5. MONRE estimated the contribution of each emission source of primary air pollution on a national scale (MONRE 2010). They report that the major source of CO and VOCs is traffic, especially motorcycles, contributing approximately 90% of their CO and VOC emissions, and buses and trucks contributing 30% of NO_2 emissions. Industrial activities contribute 70% of NO_2 emissions and almost all of SO_2 emissions. In urban areas, particularly in major cities such as Hanoi, the capital city, and Ho Chi Minh City (HCMC), the air quality is degrading, while new urban areas and suburban areas have not yet suffered from air pollution on a wide scale.

Table 5.5 WHO guideline values for selected gaseous air pollution

Specie	Guideline value	Averaging time
Primary pollutants		
CO	100 mg m^{-3} (87 ppm) 60 mg m^{-3} (52 ppm) 30 mg m^{-3} (26 ppm) 10 mg m^{-3} (8.7 ppm)	15 min 30 min 1 h 8 h
NO_2	40 μg m^{-3} (21 ppb) 200 μg m^{-3} (106 ppb)	Annual 1 h
SO_2	20 μg m^{-3} (7.6 ppb) 500 μg m^{-3} (191 ppb)	24 h 10 min
Toluene	8 mg m^{-3} (2.1 ppm)*	24 h
Secondary pollutants		
O_3	100 μg m^{-3} (51 ppb)	8 h

Unfortunately, reports on gaseous pollution in cities other than Hanoi are rarely available. Therefore, this text mainly discusses the gaseous air pollution in Hanoi. Although previous concentrations of SO_2, NO_2, CO, and VOCs in HCMC showed similar or slightly higher values than in Hanoi (Hien et al. 2011; Imamura et al. 2007), the recent air quality index suggests that the severity of air pollution in Hanoi is expected to be worst nowadays. Figure 5.39 shows annual levels of gaseous pollutants. The data was summarized from Lang station (Nhung and Dung 2012), NVC station (MONRE 2016), and our observation at HUST. In general, the annual levels of SO_2 and CO met the standard with a few exceptions during 2012–2016, as in other big urban areas (MONRE 2016). The annual levels of SO_2 increased over 2002–2006, reached the peak in 2006 at 18 ppb, and decreased in 2010 to 6 ppb. Within the next 5-year period (2012–2017), these levels rose gradually from 3 ppb in 2012, reaching their highest point in 2015 (11 ppb) before declining to 8 ppb in 2017. Similarly, the annual levels of CO declined between 2002 and 2007, varying in the range of 0.36–1.14 ppb, while the statistics for 2016 and 2018 were 1.3 and 1.7 ppb, respectively. Meanwhile, the annual level of NO_2 was larger than the annual standard, and the level of O_3 over 2012–2016 sometimes violated standards in the NVC station in Hanoi. NO_2 levels fluctuated within a range of 10–18 ppb in the first period (2002–2008), while they rose dramatically in the second period (2012–2016), varying between 15 and 28 ppb. O_3 concentration declined from the highest point in 2002 (21 ppb) to 11 ppb in 2003 and oscillated slightly within 7–11 ppb in the span of the next 7 years. Additionally, the O_3 concentration values observed at HUST were 18.8 and 8.6 ppb in 2016 and 2017, respectively.

Other groups also reported gaseous pollution in Hanoi. Hien et al. (2011) reported that annual mean concentrations of SO_2 and NO_2 at the National Hydro-Meteorological Center in 2003 showed similar values to those in HCMC,

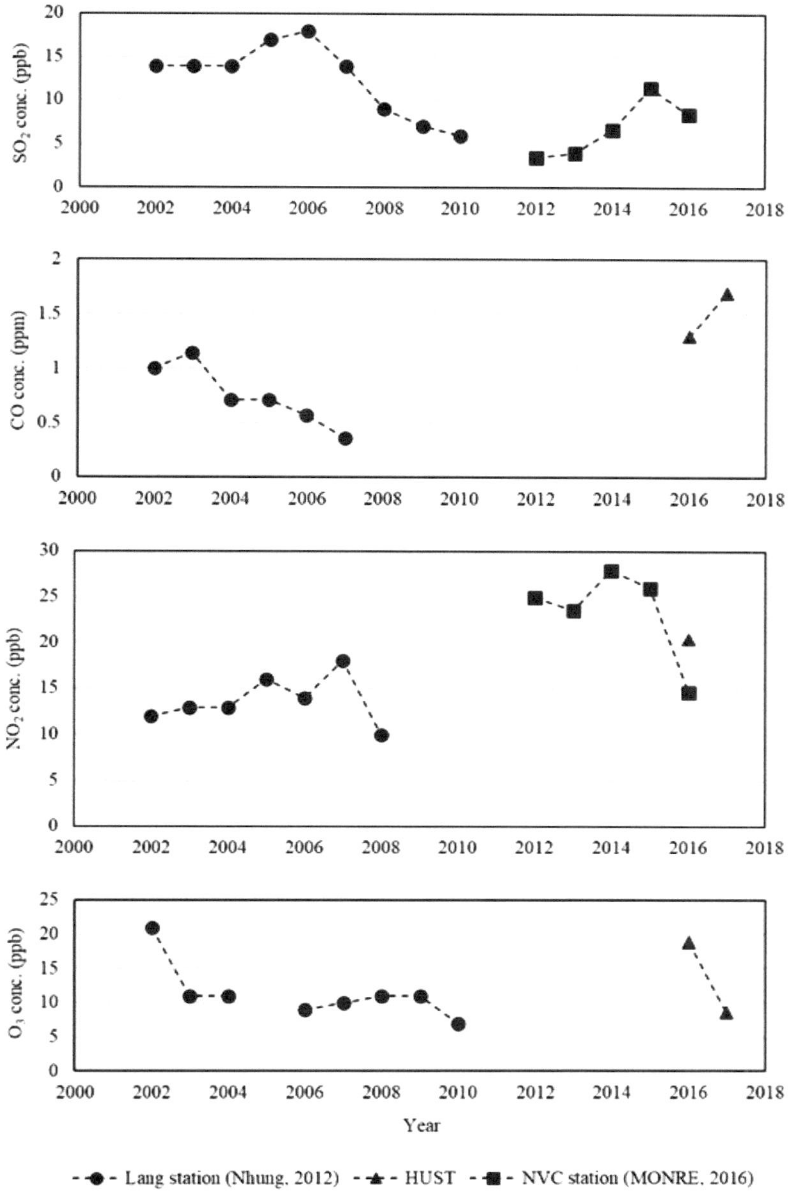

Fig. 5.39 Annual SO_2 (top), CO (second), NO_2 (third), and O_3 (bottom) levels in Hanoi

which are approximately 15 ppb, while the annual mean concentration of CO in Hanoi was 1.3 ppm, which is smaller than that in HCMC, 4.2 ppm. They also reported that annual mean values of NO_2 and SO_2 in 2007 ranged from 9 to 33 ppb and from 4 to 18 ppb, respectively, depending on the observation point in several inner districts (Hien et al. 2014). From October 2015 to December 2016, NO_2, CO, and O_3 were monitored at the same place, where the overall average values in the campaign period for these species were 21.4 ± 11.8 ppb, 1.3 ± 0.8 ppm, and 19.3 ± 15.3 ppb, respectively.2) In summary, NO_2 and O_3 concentrations in the inner districts in Hanoi sometimes exceeded the WHO guideline values. The pollutants have higher

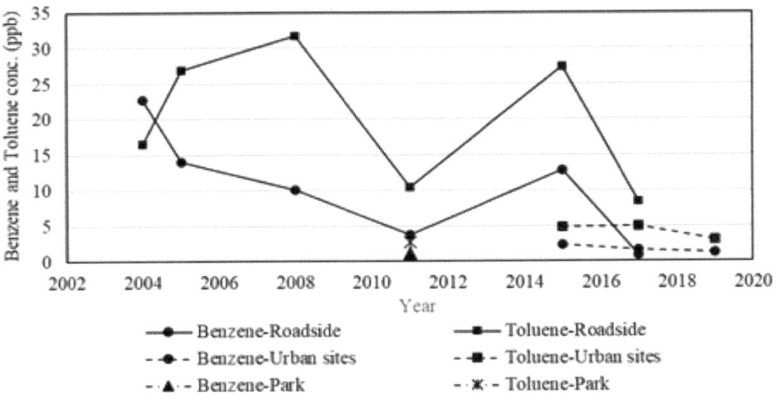

Fig. 5.40 Benzene and toluene concentrations in Hanoi

concentrations near pollution source areas such as the roadside, traffic hot spots, and industrial areas (Hien et al. 2014; Truc and Kim Oanh 2007), which would cause more severe health effects. Therefore, it is clear that their impact on health is non-negligible, and thus measures should be taken in order to reduce these pollutants.

The concentration values of benzene and toluene (two important traffic-related compounds) at the roadside, urban, and park sites in Hanoi from 2004 to 2019 are represented in Fig. 5.40. The data have been summarized from some previous studies (Do et al. 2013, 2015; Hien et al. 2021; Imamura et al. 2007; Ly et al. 2020; Ngoc et al. 2018; Sakamoto et al. 2018; Truc and Kim Oanh 2007). It must be noted that the concentrations are collected from different sites, so the concentration trends are "relative". It should also be noted that the VOC observation was conducted in the winter season (October–March) of 2004, 2015, and 2019 (Hien et al. 2021; Ly et al. 2020; Truc and Kim Oanh 2007) while it was carried out in summer (June–August) for other years (Do et al. 2013, 2015; Imamura et al. 2007; Ngoc et al. 2018; Sakamoto et al. 2018). The concentrations of the two pollutants are highest at the roadside, then urban, and the lowest at the park, which reflects well the fact that VOCs in Hanoi are mainly driven by transportation. Except in 2004, the concentration of toluene was higher than that of benzene for all investigated sites. At roadsides, concentrations of toluene and benzene have reduced two to threefold from 2004 to 2017.

In Hanoi, traffic has been considered the main contributor to primary pollutants. The average diurnal trends of CO and NO_2 in 2003 had both morning and evening traffic peaks, which are coincident with the rush hours for traffic (Hien et al. 2011). SO_2 peaked approximately an hour later than NO_2 and CO, which indicates that there are other sources of SO_2 other than the rush hour traffic. Roadside concentrations of BTEX (benzene, toluene, ethylbenzene, *m,p*-Xylene, and *o*-Xylene) were monitored along with the traffic volume during the dry season of 2004 (Truc and Kim Oanh 2007). Concentrations of BTEX, except for benzene, depended on the distance from the road and decreased with distance. For example, the roadside toluene concentration was three times higher than that at 50 m away from the road. This is because traffic is a significant source of BTEX, other than benzene. On the contrary, the concentration of benzene decreased as far as 10 m but then increased again at further distances, which may indicate that additional sources of benzene exist. Evaporation from parked cars/motorcycles may contribute to this behavior. In this study, the relationship between the hourly number of vehicles in each group (motorcycle, van, car, taxi, bus, and truck) and pollutant levels (BTEX, CO, NO_2, and SO_2) was investigated via principal component analysis. The analysis indicates that BTEX and CO are likely associated with motorcycle emissions, that emissions from buses, vans, taxis, cars, motorcycles, and trucks are possibly linked to NO, and that truck emissions are associated with SO_2. The concentrations of NO_2, SO_2, and benzene were measured at 96 sites across Hanoi city in 2007 (Hien et al. 2014). NO_2 and benzene had the highest concentrations at traffic hot spots and showed a positive relationship with population density in inner-city areas, suggesting that traffic is their main contributor. It was also found that spatial distributions of NO_2 and benzene were different, which suggests that benzene has an additional source other than traffic, i.e., evaporation from motorcycles at rest. In industrial areas, slightly higher NO_2 concentrations were observed, indicating that industrial activities also contribute to NO_2 emission, although their contribution was smaller than traffic and they were away from the inner-city area. Unlike NO_2 and benzene, SO_2 had the highest value in industrial areas, while it also had higher values near traffic areas, indicating that the main contributor of SO_2 is industrial activities such as coal and oil burning, with traffic being the main contributor in the inner-city area. Nghiem et al. (2021) suggested that direct emission from traffic was not the main source of SO_2 at a traffic site: Nguyen Van Cu station and heavy-duty vehicles for construction may cause an increase in SO_2. In 2015, a fuel use analysis was carried out based on observed VOC distribution (Sakamoto et al. 2018). The analysis showed that gasoline is the biggest contributor among the main three fuels, CNG, LPG, and gasoline, and that the VOC distribution follows the pattern of motorcycle traffic, which supports the current recognition that motorcycle is the main contributor of VOC in the inner city. In addition, a recent study by Le et al. (2021) showed that the concentration of $PM_{2.5}$ and CO during the social distancing period in 2020 reduced by 7–10% and 5–11% compared to those before this period. These results suggest that the apparent improvement of air quality in Hanoi during the social distancing period was in line with reduced traffic emissions while emissions from coal-fired power plants remained relatively stable.

Meteorological events may sometimes influence gaseous pollution levels as well. For example, in Hanoi, there are two monsoon seasons, the northeast monsoon in winter and the southeast monsoon in summer, and the winter monsoon is associated with multi-day pollution events (Hien et al. 2011). Hien et al. monitored SO_2, CO, and NO_2 from October 2003 to February 2004 and found that their concentrations rose and sometimes reached 4–5 times higher than the other days after winter monsoon surges arrived (Hien et al. 2011). In the days of the highest polluted event, hourly averaged concentrations of SO_2, NO_2, and CO increased to up to ~130 ppb, ~150 ppb, and ~12 ppm. This pollution event occurred in association with nighttime radiation temperature inversion (NRTI) and subsidence temperature inversion (STI), which are caused by the winter monsoon. The highest pollution event occurred in the NRTI period. A one-year observation campaign of gaseous pollution from October 2015 to December 2016 showed that ozone concentration increased to 100 ppb when the winter monsoon arrived (Sakamoto et al. 2018), which is larger than the WHO guideline value, as shown in Fig. 5.41. In the same period, CO concentrations also increased by up to approximately 6 ppm, which were the highest values in the observation campaign. Since the winter monsoon comes from the northern or northeastern direction bringing

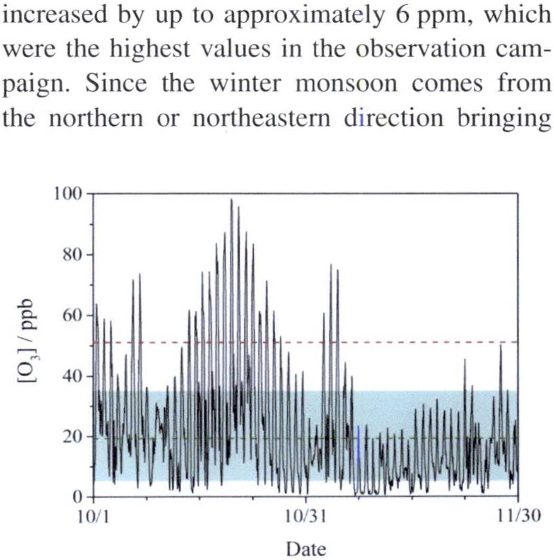

Fig. 5.41 Ozone profile in Hanoi from October 1 to November 31, 2015

dry and cold air during the period, the high concentration of CO can be explained by the pollution accumulation owing to NRTI or STI (Hien et al. 2011). Although there has not been clear evidence of the relationship between the winter monsoon and secondary pollutants such as O_3, there is the possibility that the accumulation of pollutants by NRTI or STI might enhance photochemical reactions, leading to high levels of O_3. This aspect of pollution studies merits further investigation to understand and control the air quality in the city.

5.6 Water Environment in Vietnam

5.6.1 Introduction

Vietnam has abundant water resources, with 16 major river basins and nearly 3,500 rivers in a dense and complex river network. The estimated total surface water discharge is 830–840 billion m^3/year, with 37% of those generated on Vietnamese territory and the rest coming from outside. The average annual rainfall over the entire territory is approximately 1,940 mm/year; however, owing to the influence of mountainous terrain, the rainfall is unevenly distributed across the country and varies significantly over time. Consequently, water flow varies greatly according to the season and region (MONRE 2018).

Despite its abundant endowment, Vietnam's water resources are already under stress because of rapid economic development, in addition to a shortage of physical infrastructure and financial capacity. This section provides an overview of the aquatic environment in Vietnam. Insights into water pollution in urban areas, including pollution sources, are also discussed.

5.6.2 Water Quality in Water Supply Sources, and Their Challenges

5.6.2.1 Rivers

Vietnam, which stretches vertically along its coastline, has numerous river basins, including the Red River Delta in the north and the Mekong Delta in the south. Based on their characteristics, these basins can be divided into eight regions. These characteristics are summarized in Table 5.6 and their locations are shown in Fig. 5.42. Upstream areas of most regions have good water quality due to low population density with the primary economic activities of forestry and agriculture, whereas densely populated and urbanized downstream areas of the Red River Delta and Mekong Delta have poor water quality, mainly due to the pollution of organic matter, nutrients, and microorganisms. Salinity intrusion has been recorded in estuaries in the northeastern region, Red River Delta, and most intensely in the Mekong Delta, where it is difficult to secure freshwater supplies.

Figure 5.42 shows the water quality in major river basins during the 2014–2018 period according to Water Quality Index (WQI) values. The WQI ranges from 0 (worst value) to 100 (best value) and has been developed by the Canadian Council of Ministers of the Environment. It is obtained by determining the percentage of water quality nonconformities in several water quality indicators and statistical processing (Newfoundland Labrador n.d). The Red-Thai Binh, Cau, Vu Gia-Thu Bon, Dong Nai, and Mekong River Basins have relatively good water quality, and many river segments are suitable for domestic water supply purposes. The Nhue-Day River Basin, located in the Red River Delta region upstream of Hanoi, is heavily polluted by organic matter, nutrients, and microorganisms. In the entire country, a moderate

Table 5.6 Water quality by region (2030 Water Resources Group 2017)

Region	Quality (Upstream/ Downstream)	16 main river basins (* other river basin)	Famous cities	Issues
Northwest Region	+++++/ ++++	Red-Thai Binh, Da*	Hoa Binh	-
Northeast Region	+++++/ ++	Bang Giang - Ky Cung, Red-Thai Binh, Cau*	Thai Nguyen	Urban pollution, saline intrusion, marine transport pollution risk
Red River Delta	++++/ ++	Red - Thai Binh, Nhue – Day*	Hanoi	Urban and industrial pollution, saline intrusion, agrochemical pollution, transport pollution risk
North Central Coast	++++/ +++	Ma, Ca, Gianh, Huong	Hue	Urban pollution, saline intrusion
South Central Coast	+++++/ ++	Thu Bon - Vu Gia, Ba, Tra Khuc, Kone, SERC	Danang	Urban pollution, saline intrusion
Central Highlands	+++++/ ++++	Se San, Sre Pok	Dalat	-
Northeast of Mekong	++++/ +	Dong Nai	HCMC	Urban and industrial pollution, saline intrusion
Mekong Delta	++++/ ++	Mekong	Can Tho	Saline intrusion, low pH in river (acid soils) agrochemical pollution, transport pollution risk

"+ + + + +" indicates excellent water quality, wherea "+" indicates unacceptable quality

value of the index, which indicates quality suitable for irrigation or other equivalent purposes, accounted for the highest percentage at 50%. Suspended Solids (SS) concentrations in the Ma, Red-Thai Binh, and Dong Nai River Basins are high, especially in the rainy season, resulting in higher percentages of medium and poor water quality there.

Most river basins in Vietnam have high values of SS and turbidity, exceeding the national technical regulations on surface water quality A2 grade (used for domestic water supply with suitable treatment, 30 mg/L) in the surface water quality standards (See Table 3.8 in Chap. 3), and many areas exceed B1 grade (used for irrigation, 50 mg/L), especially during the flood season. Pollution from organic matter is most common in downstream rivers. Water quality is of serious concern, mainly in river segments flowing through key economic-social centers, such as the downstream region of Cau River passing Thai Nguyen, Nhue-Day River passing Hanoi, and Dong Nai River passing Ho Chi Minh City (HCMC).

However, there are several other local pollution problems. Heavy metal pollution is mostly concentrated in the tributaries near mining and industrial production facilities, whereas oil pollution mainly occurs in reaches with navigational activities and those receiving industrial wastewater from production facilities and ports. Recently, contamination with emerging compounds, such as pharmaceuticals and personal care products, has been reported in areas with rapid population growth and intensive agriculture. Heavily polluted areas include river segments that flow through urban and industrial areas and craft villages.

5.6.2.2 Lakes and Reservoirs

Vietnam has about 3600 lakes and reservoirs of various sizes, of which less than 15% are large or medium (capacity of over 1,000,000 m^3 or a height of more than 10 m). Most of them in

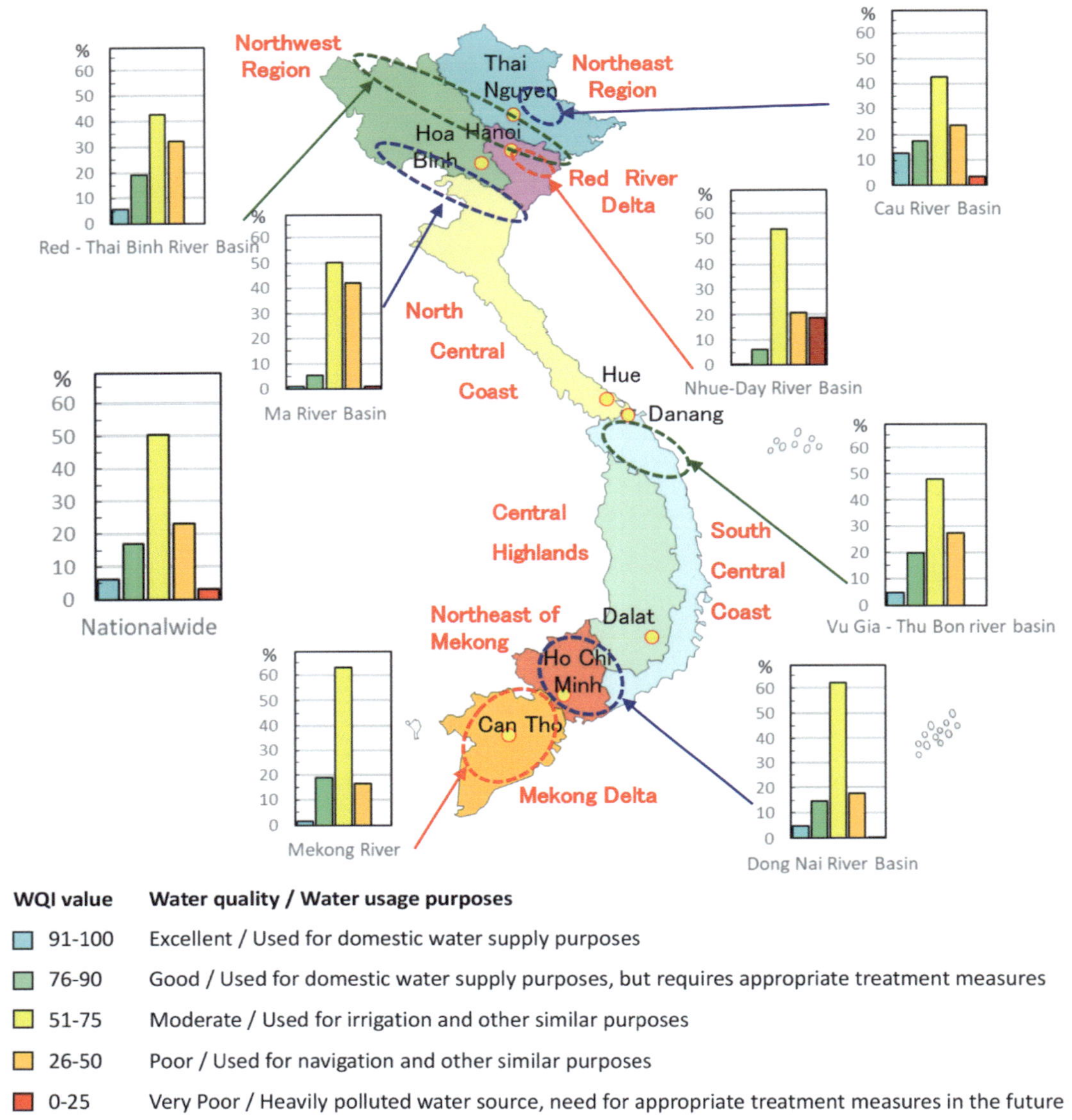

Fig. 5.42 Water quality conditions at major river basins in the 2014–2018 period according to Water Quality Index (WQI) (MONRE 2018)

urban areas are polluted with organic and nutrient compounds. Hanoi is surrounded by 120 lakes and ponds, of which 17 are situated in the inner city. These lakes are severely polluted and strongly eutrophic. Hoang et al. (2017) reported that five out of 15 investigated urban lakes were classified as hypertrophic. The lakes in the suburbs are mainly mesotrophic.

5.6.2.3 Groundwater

Groundwater is abundant in the Mekong Delta but is limited to the North Central Coast Region. The total groundwater potential in Vietnam is abundant with about 63 billion m^3/year, but only 7% of the groundwater is sustainably exploitable (2030 Water Resources Group 2017). In the northeastern region, scattered and diverse groundwater makes exploitation difficult. In the Red River and Mekong Deltas, groundwater is exploited beyond the recharge capacity around Hanoi and HCMC, resulting in declining water tables and further causing land subsidence and salinity intrusion, especially in the Mekong Delta.

Groundwater is an important alternative water supply because of surface water pollution. In general, the quality of groundwater remains good, but there are some pockets of contamination in parts of Hanoi, HCMC, and the Mekong Delta, where septic tanks, garbage dumping, and industrial effluents are poorly maintained and overexploitation occurs.

Tran et al. (2021a) conducted a comprehensive assessment of groundwater quality in the Mekong Delta using 282 samples from the dry and rainy seasons in 2017 and 2018, respectively. They found a saline water issue and significantly high nitrate contamination, exceeding the WHO drinking water standards. They rated the groundwater as excellent (41.5%), good (40.4%), poor (17%), very poor (1.1%), and unusable (0%) water for drinking purposes, by using Groundwater Quality Index (GWQI), which is estimated as the weighted average of the ratios of observed values to ideal values calculated for several water quality indicators. They also discussed the use of groundwater for irrigation by Wilcox diagram (Wilcox 1948) and found that 17.5% of total groundwater samples exceeded the permissible levels for salt-tolerant crops and solidity hazards. Furthermore, high concentrations of nitrate (>50 mg/L) and chloride (>1000 mg/L) were detected in shallow, intermediate, and deep aquifers.

Arsenic (As) contamination of groundwater is a serious problem in the Red River Delta. Winkel et al. (2011) indicated that 27% of the wells exceeded the WHO drinking water standard of 10 μg/L although the concentrations were found to vary greatly from less than 0.1 to 810 μg/L. Nga (2008) reported that theconcentrations in Hanoi were low (<2 μg/L) in the northern area, moderate (<30 μg/L) along the Red River, and high (43–112 μg/L) in the south. The long-term use of groundwater containing elevated concentrations of arsenic should be avoided by utilizing other sources of drinking water.

5.6.3 Water Pollution in Urban Areas

5.6.3.1 Current Conditions in Urban Rivers

In Vietnam, most urban areas are concentrated along main rivers such as Hanoi on the Red River, Danang on the Han River, and HCMC on the Dong Nai River. Population growth and urbanization in recent years have put pressure on the use of water resources and the environment of river basins because technical and social infrastructure systems are not synchronized. Most urban areas lack separate wastewater collection systems for domestic and industrial wastewaters. Wastewater is only partially treated and discharged directly into rivers and lakes, causing serious pollution to the surface water environment. Within cities, lakes and canals increasingly serve as sinks for domestic sewage, and municipal and industrial waste. Most lakes in Hanoi are seriously polluted with high biochemical oxygen demand (BOD) levels. Similarly, small rivers in Hanoi and HCMC have dissolved oxygen (DO) levels as low as 0–2 mg/l, and BOD and NH_4^+ levels as high as 50–100 mg/l and 3–25 mg/L, respectively (MONRE 2016).

In general, in rivers with large water flows, the water quality has decreased locally in the sections flowing through the urban area, but the water quality is still relatively stable, such as in the Red, Dong Nai, and Mekong Rivers.

Emerging pollutants, such as antibiotics, have also recently been found in lakes and canals in Hanoi, ranging from below the quantification limit to almost 50,000 ng/L. The median concentrations of amoxicillin, erythromycin, and sulfamethoxazole were more than 1000 ng/L in urban canals (Tran et al. 2019). In terms of the main rivers flowing through urban areas, the water quality is distinctly degraded.

5.6.3.2 Pollution Loading Sources

The surface water environment in Vietnam is affected by four main sources: domestic, medical, industrial, and agricultural wastewater. Based on the discharge permissions issued by MONRE in 2017, the total volume of wastewater discharged nationwide was estimated to be approximately 100 million m^3/day (MONRE 2018). Domestic wastewater accounts for a large proportion, but only 60% of urban households have connections to drainage systems, and only 12.5% of municipal wastewater is treated (MONRE 2018).

A detailed assessment of the industrial sector in 2008 indicates that the top three polluting industries in Vietnam were paper and wood production, chemical production and processing, and metal processing (2030 Water Resources Group 2017). The assessment also estimated that the food, drink, and textile industries would grow rapidly and become a large portion of the industrial pollution load. In addition, craft villages (See the column "Craft Village" in Chap. 3) contribute to pollution because these usually discharge wastewater directly into water bodies without any treatment. The effects are more serious in the Red-Thai Binh Basin, where 65% of craft villages are located.

By the end of 2015, 35 centralized municipal wastewater treatment plants commenced operation in Hanoi, HCMC, Danang, and other large cities, and their total capacity was approximately 850,000 m^3/day, which was equivalent to 12–13% of Vietnam's municipal wastewater treatment requirements (Bui 2015). In addition, it was estimated that several thousands of decentralized wastewater treatment plants had been constructed and installed for residential areas, hospitals, hotels, and office buildings (Nguyen 2010). However, wastewater treatment systems were only available in 50% of the hospitals and 7% of the 23,500 livestock farms as of 2014 (Bui 2015).

In the Hanoi metropolitan area, up to 750,000 m^3 of wastewater is produced each day, but only 10% is treated according to Vietnamese standards. In addition, industrial and service zones produce 300,000 m^3 of wastewater, and hospitals produce 6,000 m^3. The wastewater is conveyed with combined sewer systems to the southern part of the city and finally discharged to the Red River after being partially treated through pond systems (Nguyen and Tran 2020).

5.6.3.3 Countermeasures

Currently, water resource protection and management are carried out according to the provisions of three specialized laws: the Law on Environmental Protection 2020, the Law on Water Resources 2012, and the Law on Irrigation 2017, along with a series of documents under applicable laws and regulations. These laws form a full legal framework, which is the basis for the environmental protection and management of river basins in Vietnam. However, the water governance has not yet met the standards required to implement sustainable and integrated water resource management. At a meeting of the central government in April 2017, several solutions were proposed by MONRE to address environmental issues, such as promoting environmental planning, promulgating criteria for screening, and publicizing polluting and disgruntled industries. In particular, the urgent issues in enforcement were: (1) strengthening the activities of water pollution control, inspection, examination, and enforcement of compliance with the law on water environment protection; (2) applying economic tools, and applying scientific and technological solutions in water environment protection; (3) minimizing contamination of the water environment through prevention and control of pollution; (4) raising awareness and enhancing community participation and responsibility in water environmental management and protection, and (5) promoting international cooperation activities in water environmental management and protection, especially transboundary issues.

The principles of Integrated Water Resources Management suggest that the key solutions in river basin management are the assessment of discharge into river basins and consideration of pollution control from agricultural, industrial, and municipal sources. The Prime Minister approved three Environmental Protection Projects in the Cau River Basin, the Nhue-Day River Basin, and the Dong Nai River system to address environmental pollution by focusing on the planning of environmental protection, exploitation, use, and protection of the water environment of the localities for development and initial implementation. Environmental protection projects have been developed, approved, and implemented in all 22 provinces and cities across the three river basins.

In Hanoi, the city urban drainage master plan for the 2030–2050 period was approved in 2013, aiming at tailoring the drainage planning of Hanoi within the master plan of construction, step by step eliminating urban flooding, and increasing the proportion of sewage and wastewater treatment to 90% by 2030 and 100% by 2050.[6]

5.7 Domestic Water Use in Vietnam

5.7.1 Characteristics of Water Use in Danang

5.7.1.1 Danang City

Danang is the biggest city in Central Vietnam and is known for its fishery and tourism. In recent years, it has become more popular as a resort location. The population in Danang is rapidly increasing and topped one million in 2014. In recent years, Danang city has developed rapidly as an east base of Indochina East–West economic corridors. There is an economic audience rating of 5.4% and a high standard as of 2013, and as for the GDP, approximately 25 times in 1993 increase as much, and economic growth is remarkable. Measures such as sewage treatment have not kept up with the requirements of economic progress while water demand has surged with sudden urbanization, leading to water pollution. While sewer coverage is 47% in urban areas, it is almost non-existent in villages, where septic tanks are predominantly used.

The main water source of Danang city is the Han river flowing through its eastern part, wherein the Cau Do water intake ground acts as the water intake point. However, because its altitude is low, when the tide level goes down (and salt in water goes up), the water intake point is moved to the An Trach water intake ground located up the river. The upper streams of the An Trach water intake point are located in the Quang Nam province, which makes management of the water source by Danang city difficult. The tap water supply of Danang city was 42,164 m^3/day in 2002, which increased rapidly to 118,024 m^3/day in 2011. Water is supplied by Danang Sole Member Limited Liability Water Supplying Company (DAWACO). The water supply capacity as of 2012 was 155,000 m^3/day.

One of the factors to have a big influence on water demand is population; specifically, in the residential areas, the amount of water available per day per person is more important. This figure has been reported by Katmandu as being 36.9 L/person/day in 2012, which is approximately one-eighth that of Japan. In comparison, the average amount of water supply per day per person for 2003 has been reported to be 120–200 L/person/day in the Kingdom of Thailand, 77 L/person/day in Kanpur (India), and 116 L/person/day in Kolkata (India). Thus, the amount of water supply per day per person differs greatly among cities.

Fecal pollution of the water environment is also becoming a serious issue because of urbanization, increasing amounts of untreated wastewater, and inadequate infrastructure. The demand for water has been increasing alongside economic development and population growth, and the water supply capacity is close to its limit. Water shortages thus seem inevitable. To supply enough water for a long period, it is necessary to understand the precise nature of future water demand. However, for this, it is important

[6]Prime Minister, 2013. Decision 75/QD-TTg on the Approval of Master plan on Ha Noi Capital's drainage through 2030 with a vision toward 2050.

Fig. 5.43 Questionnaire survey in progress

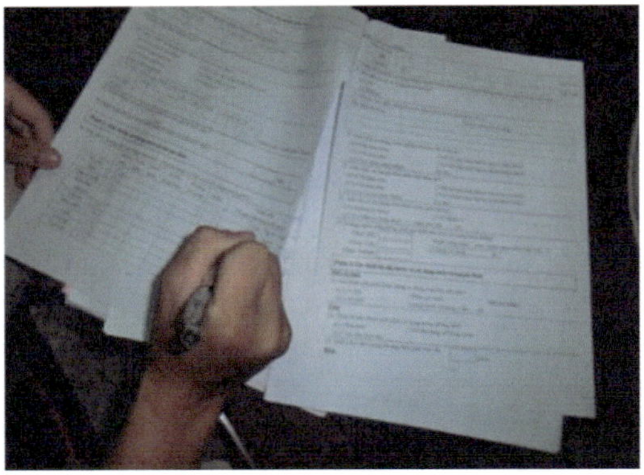

Fig. 5.44 Survey questionnaire

to know the current situation of water use first. The main objective of this work is to introduce the characteristics of water use in residential areas, which account for 70% of water use, and to analyze some of the factors that affect water demand.

5.7.1.2 Investigations

Two questionnaire-based surveys, on the characteristics of water use and on volume, were carried out in the rainy season of 2011 and the dry season of 2012 in Danang. Figure 5.43 shows the survey in progress.

308 households were selected for the survey, which was carried out in Vietnamese to collect information related to water use, frequency of use, and equipment and home appliances needing water and their use. Figure 5.44 shows the questionnaire itself. For the water-use application, 11 items (bathing, washing clothes, cooking, dishwashing, drinking, washing face, gardening, watering, car/motorcycle washing, and floor cleaning) were selected.

Water-use quantitative surveys were then carried out in 97 households. Table 5.7 shows the number of households and population in 7 districts. Hai Chau, Thanh Khe, Son Tra, Lien Chieu, Ngu Hanh Son, and Cam Le are in the urban area, while Hoa Vang is in the rural area. From each district, 12–15 samples were obtained. The survey was divided into two parts: Water consumption survey over 3–6 days using a water meter and an investigation of water-use frequency and amount of water supply.

Figure 5.45 shows the objectives and survey items. For 10 items, except drinking, water consumption per one time of use was measured. There are three ways of bathing (bathtub, shower, and bowl) that people practice in Danang. In order to quantify the water consumption per bathtub user, the volume of bathtub and number of baths per day were surveyed. For estimating drinking water consumption, findings from the questionnaire were used. In order to quantify the water consumption by shower users, the flow rate of shower and flow time were measured. Some people use a bowl for bathing, so the volume of the bowl and the number of baths were measured. Regarding toilet water use, people in Danang employ two types: flush and pour. In order to quantify the water consumption by flush toilet users, the manufacturer of the toilet and product ID were investigated. By using information obtained by product ID, water volume per flushing was obtained. In order to quantify the water consumption by pour toilet users, the volume of the pouring bowl and the number of pours were obtained. Similarly, to quantify the water consumption by laundry machine users, the manufacturer name and product ID were investigated. For people who wash the clothes by hand, the volume of the bucket and number of washes/day were obtained. When people wash dishes, cook, wash face, water gardens, wash

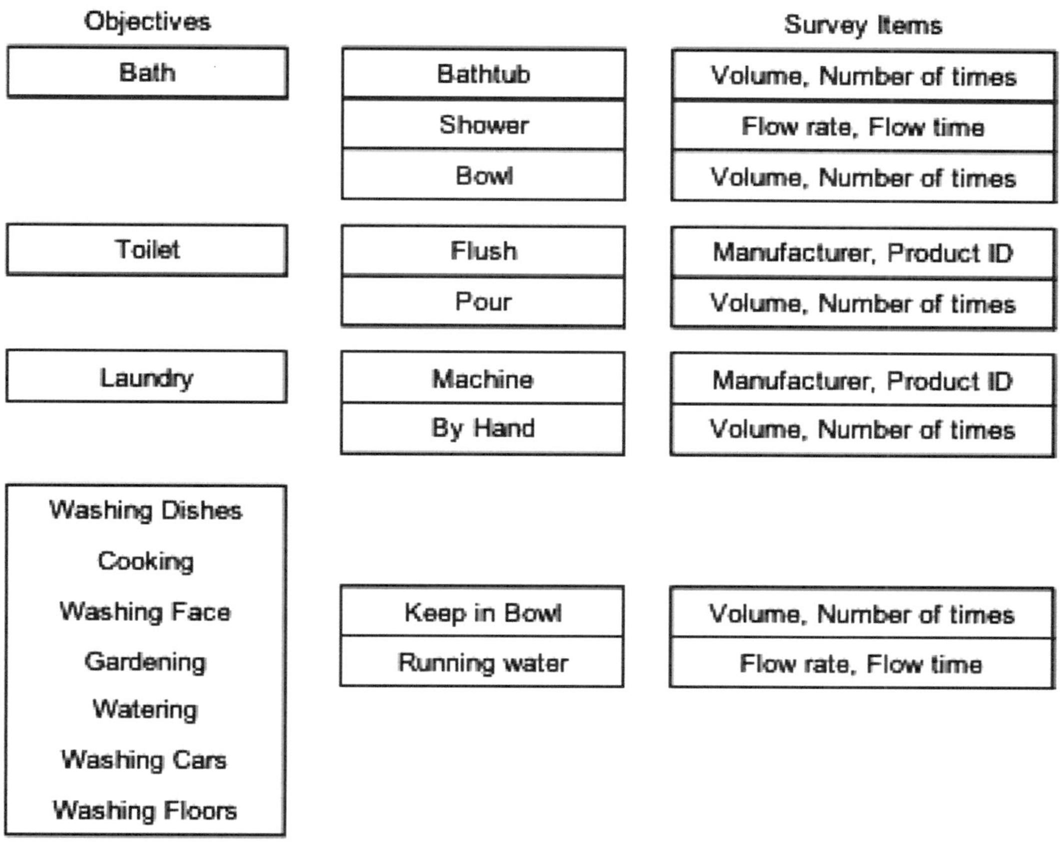

Fig. 5.45 Objectives and survey items in Danang

Table 5.7 Number of households and population in Danang (2010)

District	*n*	Population	
Hai Chau	13	196,098	Urban
Thanh Khe	15	178,447	
Son Tra	15	132,944	
Lien Chieu	13	68,270	
Ngu Hanh Son	12	136,737	
Cam Le	15	92,824	
Hoa Vang	14	120,698	Rural
Sum	97	926,018	

cars, or wash floors, they keep water in a bowl. Thus, the volume of the bowl was measured, and the number of times it was refilled was surveyed. In case of users with running water, flow rate and flow time were measured.

5.7.1.3 Tap Water Consumption

As can be seen in Fig. 5.46, the amount of daily water consumption in Danang was 101.8 L/cap/day. The median values of each application were 39.6 L (bathing), 13.2 L/cap/day (toilet), 10.8 L/cap/day (laundry), 10.6 L/cap/day (washing dishes), 10.5 L/cap/day (cooking), 5.7 L/cap/day (washing face), 3.5 L/cap/day (gardening), 3.3 L/cap/day (floor cleaning), 2.3 L/cap/day (watering), 1.5 L/cap/day (drinking), and 0.9 L/cap/day (washing cars and bikes).

Water Source of Bathing

Figure 5.47 shows the water source for bathing in Danang. Three kinds of water (tap water, groundwater, and rainwater) were used for bathing. 73.7% (227/308) citizens used only

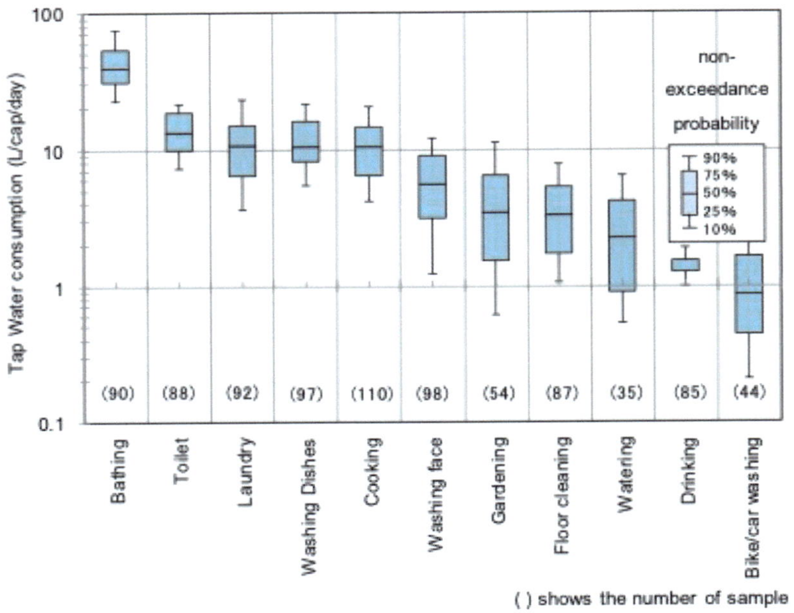

Fig. 5.46 Daily tap water consumption in Danang, Vietnam

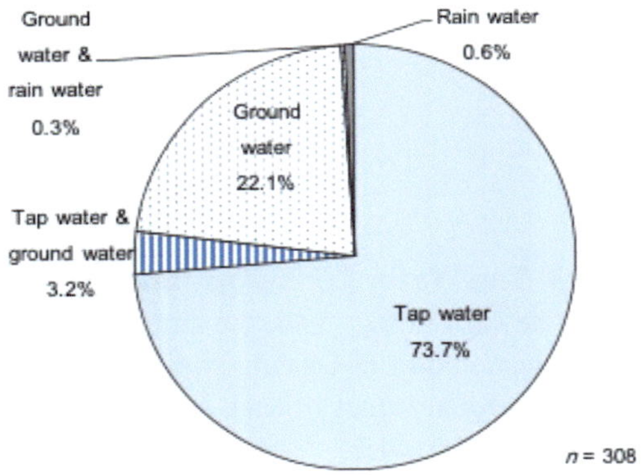

Fig. 5.47 Water source for bathing in Danang, Vietnam

tap water for bathing, while 22.1% (68/308) citizens used only groundwater, 0.6% (2/308) citizens used only rainwater, and 3.2% (10/308) citizens used both tap water and groundwater. In Danang city, many households have faucets for groundwater, which reduce the burden of water bills.

Shower User and Bowl User in Bath

Figure 5.48 shows the preferred bathing method in Danang: 53.2% of residents take a shower, 24.4% use a bowl, 20.8% use both a shower and bowl, and only 1.6% of citizens use a bathtub.

Figure 5.49 shows the non-exceedance probabilities of water use in baths. The median water flow from the shower was 4.4 L/min, and Danang residents took a shower for five minutes on average once per day. On the contrary, households used tubs of various sizes—from 1.1 to 76.0 L (median = 16.2 L). The number of times the tub was used and the height at which hot water was put together were multiplied by d) the capacity of the tub. Each person used 27.5 L of tap water at the median price per day. Together, e) shows the amount of tap water used for bathing. In the calculations for households using tap water only, the median value was 39.6 L/person/day. On the contrary, the median was calculated to account for households using groundwater, which was 34.6 L/person/day. The results

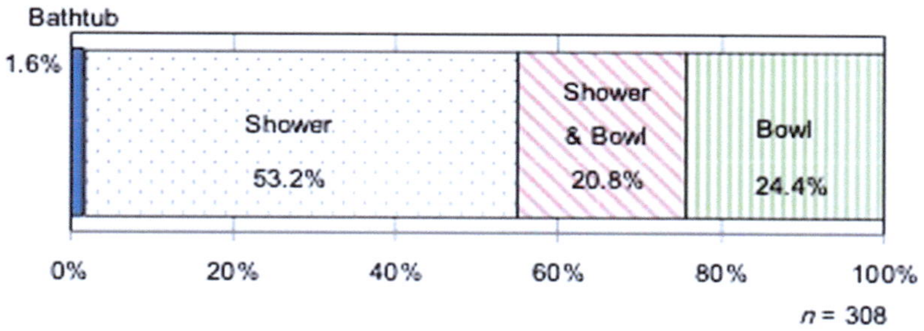

Fig. 5.48 Preference of bathing method in Danang, Vietnam

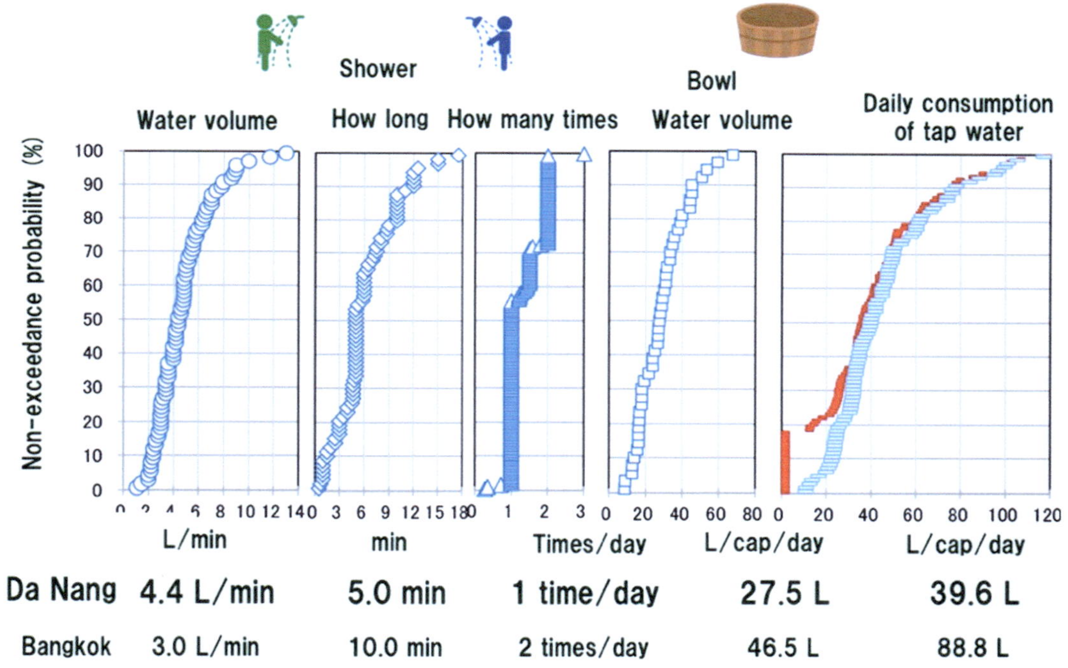

Fig. 5.49 Bathing statistics for Danang, Vietnam

indicate that the amount of tap water-used per person per day will increase in the future as groundwater users shift to tap water with economic development.

Figure 5.50 shows the daily water consumption of shower users and bowl users. Shower users consumed 42.0 L/cap/day, while bowl users consumed 32.1 L/cap/day. The annual income of shower users is significantly higher than those of bucket users ($p < 0 .01$). These results reveal that when household income increases, the water-use method for bathing tends to change from tub to shower.

As rapid economic growth continues in Danang, citizens are expected to prefer showers over bowls. This, daily water consumption per person will increase further for bathing. For reference, residents in Bangkok use 201.7 L/cap/day. The results of this work clearly suggest that residents of Danang would face water shortage issues in the future unless appropriate measures are taken.

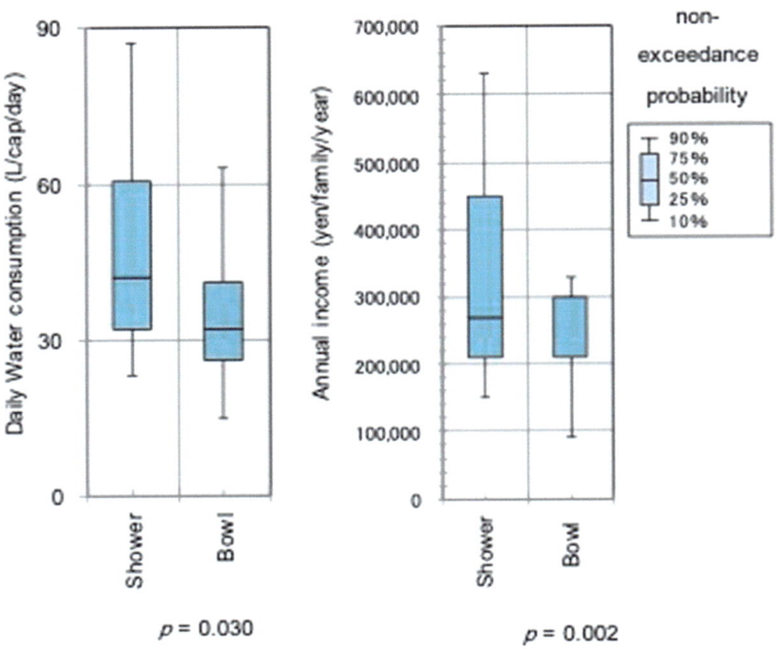

Fig. 5.50 Daily water consumption and annual income in Danang

5.7.2 Characteristics of Water Use in Hanoi

The speed of urbanization in Hanoi is relatively fast, creating a rapid shift in housing types and lifestyles. Hanoi historically had old single-unit houses and old apartment buildings but has been more modernized with the development of new single-unit houses and apartment buildings. Many of the old single-unit houses had kitchens and toilets outside of the living units and sometimes these shared facilities with other householders. The old apartment buildings were built during 1960–1990 and often included up to six floors without elevators. These old accommodations were sometimes equipped with a pour-flush toilet with no shower so that a bucket and water dipper were used. In contrast, the new housing types are often self-contained with modern water facilities, that are, cistern-flush toilets, showers, bathtubs, and kitchen sinks. This transition to modern life will have a definite effect on water use.

In many developed countries, water distribution consists of collection, treatment, distribution, and consumption. Water is first collected from its sources such as surface water or underground water and pumped into a water treatment plant where it is treated and then distributed to customers through a network of pipes under pressure (Derrible et al. 2021). In Hanoi, the first steps of the water distribution are similar to the conventional process. The difference is that until 2008, the water supply system was intermittent due to the lack of water; therefore, almost every house had a water tank in the basement (typically 3–5 m^3) to store water when the system was not running (Derrible et al. 2021). Moreover, because of the basement tanks, the water pressure of the system could be low; thus most houses in Hanoi had another smaller tank (typically 1–2 m^3) placed on roofs (Fig. 5.51). Water from the basement tanks is pumped automatically to the rooftop tanks and then supplied throughout the house. Even though Hanoi now enjoys a continuous supply of water, this practice remains a habit.

Fig. 5.51 Rooftop water tanks—a popular practice in Hanoi. Photo by Pham Nguyet Anh, 2021

5.7.2.1 Daily-Per-Capita Water Consumption

According to our questionnaire survey of 100 households situated in Ba Dinh, Cau Giay, Dong Da, Hai Ba Trung, Hoan Kiem, Hoang Mai, Long Bien, Tay Ho, and Thanh Xuan districts of Hanoi in December 2012, the daily water supply mode is composed of public suppliers, private drilled well water, and bottled water. Tap water users from public water supply and from privately drilled wells accounted for 95 and 9% of total surveyed households, respectively. Among tap water users from public water supply, 4% kept using privately drilled wells even after their houses had been connected to a public water supply network. There were 10% of total surveyed households buying 18–20 L water bottles for drinking. From water bills for three consecutive months and the number of people living in the households, the calculated water consumption amount for tap water users showed a large range, but 55% of the households consumed between 100–170 L/cap/day. The average ± standard deviation and median of consumption were 146 ± 58 L/cap/day and 133 L/cap/day, respectively. This number is quite similar to those for Central Asia (136 L/cap/day) (IDI-Japan 2004) and Southwest Asia (144 L/cap/day) (IDI-Japan 2004) and slightly lower than that for East Asia (151 L/cap/day) (IDI-Japan 2004). It is, however, much higher than that for Africa (63 L/cap/day) (Japan Ministry of Land, Infrastructure 2009) and much lower than those for Japan (314 L/cap/day) (Japan Ministry of Land, Infrastructure 2009) and the US (382 L/cap/day) (Tchobanoglous et al. 2003).

When investigating the average daily-per-capita water consumption of households using only tap water from public water supply by different lifestyles (Table 5.8), we found that residents with traditional lifestyles tended to show reduced daily-per-capita water consumption; for example, pour-flush toilet compared to cistern-flush toilet, laundry by hand compared to that by a washing machine, bathing by pouring water or by using a shower compared to that by a bathtub, and kitchen equipped with a stainless-steel sink compared to that using simple plastic tubs. These traditional lifestyles were dominantly observed in households where elderly people are living. Furthermore, as compared to households with and without the elderly on the same

Table 5.8 Daily-per-capita water consumption of households using only tap water from public water supply by different lifestyles ($n = 91$)

Category of water-related lifestyle	n	Daily-per-capita water consumption (L/cap/day)[a]	Significant difference
Elderly people			
Living with elderly people	36	121 ± 38	$p < 0.001$
Living without elderly people	55	163 ± 63	
Toilet			
Pour-flush toilet	4	104 ± 28	n.s
Normal cistern-flush toilet	51	167 ± 61	
Water-saving cistern-flush toilet	36	123 ± 43	
Laundry			
By hand	1	90	n.s
By washing machine	90	148 ± 58	
Bathing/showering			
Pour-bathing[b]	4	139 ± 42	n.s
Showering	85	145 ± 58	
Bathing using a bathtub[c]	2	225 ± 35	
Kitchen usage			
With plastic tub(s)[d]	7	113 ± 32	n.s
With a stainless-steel sink	84	150 ± 59	

[a]Average ± standard deviation;
[b]Feeding water from a tap into a bucket and pouring the water on the body;
[c]Bathing by water fed from a shower into a bathtub;
[d]Washing food and dishes in plastic tub(s);
n.s.: not significant

lifestyle conditions; for example, toilet use (normal cistern-flush toilet or water-saving cistern-flush toilet), laundry manner, bathing manner, and kitchen usage, the presence of the elderly tended to correlate with less water consumption than those without elderly. This shows that the presence of the elderly may induce different types of water-saving practices in a household because the elderly used to live in water-shortage conditions.

Sole-parent households are increasing in developed countries and the young generation is also likely to live alone (OECD 2011), which implies that multi-generational households are more commonly found in developing cities than developed ones. Traditional water-saving methods may conserve more water in multi-generational households, and the presence of an elderly in the house may indicate consciousness regarding water saving in such families in developing countries. At the same time, the number of multi-generational families is decreasing even in developing cities, which could mean that water wastage may increase gradually.

5.7.2.2 Hourly-Per-Capita Water Consumption

The hourly water consumption pattern reflects activities associated with lifestyles. The hourly water consumption pattern of residents in Hanoi showed the highest peak in the evening, the second highest in the morning, and almost no consumption at late night (Fig. 5.52). Typically, office and school hours in Hanoi are from 7–8 AM to 4–5 PM. The morning and evening peaks thus corresponded to before-and-after working/schooling hours. Even though people in Hanoi often have breakfast outside, the second-largest peaks were observed in the morning. Another small peak around 10 AM–2 PM was also observed, which can be explained by the fact that a considerable proportion of people working/school have lunch in their house, except primary school students who have school lunch. When we divide a day into four sessions, water consumption in the morning session (4–10 AM) accounted for 18.7% (27 L/cap) of total consumption within a day, in the daytime session (10 AM–5 PM) for 19.6% (28 L/cap), in the evening session (5–11 PM) for 57% (82 L/cap), and in midnight session (11 PM–2 AM) for 4.7% (7 L/cap). Indeed, household water was consumed mainly in the evening session in urban Hanoi.

Comparing with hourly water-consuming activities such as hygiene, toilet flush, bathing/showering, laundry, cooking, and dishwashing,

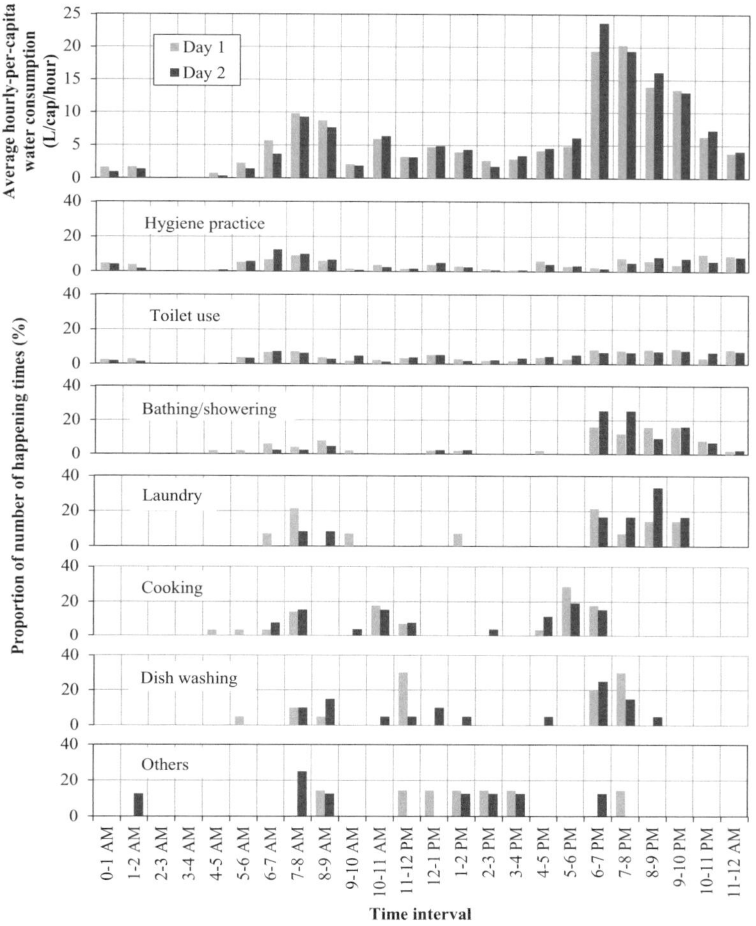

Fig. 5.52 Hourly-per-capita water consumption associated with water-use activities for two consecutive days. Other activities are bonsai watering and house cleaning

we found that the clear peaks of bathing/showering and laundry in the evening and morning occurred at almost the same time as the distinct peaks of hourly water consumption. The peak of dishwashing (8–9 AM, 11 AM–12 PM, and 7–8 PM) happened one or two hours after cooking (7–8 AM, 10–11 AM, and 5–6 PM), which is generally understood. Interestingly, the number of bathing/showering in the evening (5–11 PM) accounted for 78% of total number of bathing/showering within a day; the peak of bathing/showering (6–8 PM) happened one hour after cooking (5–6 PM), which reflects that urban residents in Hanoi mostly have a bath/shower after dinner. In the past, Hanoian often took a bath before dinner possibly due to lack of hot shower. However, as all houses in this study were equipped with a hot shower/bathing facility, residents likely have changed their lifestyle by having a bath/shower after dinner even at late night.

Laundry also occurred mainly in the evening (6–10 PM). In the past, residents used to wash clothes in the morning to dry outside but they have changed the habit to the evening because the washing machine allows the washed clothes to dry. The peak for laundry seemed to occur after cooking and dishwashing. It is interpreted that the residents do laundry after most of their daily chores in the evening. In fact, drying laundry on the roof-covered balcony was often observed in the evening in urban Hanoi.

Toilet use and hygiene are the two dominant activities at midnight, indicating residents' bed-time preparation. This trend is similar to that mentioned in another study investigating the significant contribution of toilets to total water flow at nighttime (Butler et al. 1995). In terms of water consumption by toilet flush, previous studies estimated toilets contributed to 20–40% of total water consumption (Butler et al. 1995; Almeida et al. 1999) and each person used the toilet 4.0 times per day (Friedler et al. 1996). In the case of Hanoi, modern apartments are mostly equipped with water-saving flush toilets (6 L per flush). If each flushing consumes 6 L of water, since the average frequency of one-day toilet flush was 5 times in this study, the water consumption is then estimated as 30 L, which is equivalent to about 23% of the total water consumption.

Our study investigated hourly water consumption and assessed its relationship with residential lifestyles in Hanoi, Vietnam. These details provide useful data for water supply planning as well as an explanation for the concentration of associated water discharge.

5.8 Wastewater and Fecal Sludge Management in Vietnam

5.8.1 Wastewater Management in Vietnam

Domestic wastewater management is a big problem in most urban areas in developing countries, including Vietnam. Although the country has put much effort into the investment, improvement, and construction of sewerage systems to increase the number of centralized wastewater treatment plants up to 63 in 2020 (with a total design treatment capacity of 1.34 mil. m^3/day) compared to 33 plants (with 500,000 m^3 capacity) in 2015, only 60% of urban population is connected to the public sewerage system and only 15% of total wastewater is treated (Thành 2021). In the absence of centralized wastewater treatment plants, septic systems play an important role as the major decentralized wastewater systems in many cities in developing countries; for example, Mandalay, Myanmar (90%) (Naing et al. 2020); Hanoi, Vietnam (84%) (Brandes et al. 2016); and the Citadel, Hue, Vietnam (100%) (Anh et al. 2016a).

Currently, most urban areas in Vietnam are using a combined sewerage system, which collects both rainwater and wastewater via pipeline collection networks and/or drainage canals. Besides the low coverage ratio, it is worth mentioning that many of these systems were established a long time ago (when Vietnam was occupied by French colonialists) and have now aged and deteriorated; moreover, they do not function properly due to poor maintenance. As a result, a high proportion of wastewater goes underground through the exfiltration process and only a part of the wastewater entering the sewerage system reaches the sewer outlet (Watanabe et al. 2021). Moreover, even the wastewater that is transported by the sewer pipes system is directly discharged into water bodies due to a lack of wastewater treatment plants, causing water pollution.

The movement path of wastewater from households in urban areas is illustrated in Fig. 5.53. Household waste streams include greywater (i.e., all wastewater streams generated in households except for the wastewater from toilets, including wastewater from sinks, showers, baths, washing machines, and dishwashers) and human excreta (including urine and feces). In most Vietnamese urban areas, the wastewater from the toilets goes to septic systems (septic tanks and cesspools), and their effluents are then discharged to the surrounding environment or sewer pipes. Similarly, greywater is discharged either directly to the nearby land, surface water bodies (e.g., ponds, lakes, rivers), or to the public sewer system. Notably, some households even discharge their greywater right in their gardens or to the surface ground in front of their houses, or at a distance from their houses via a simple channel, which sometimes

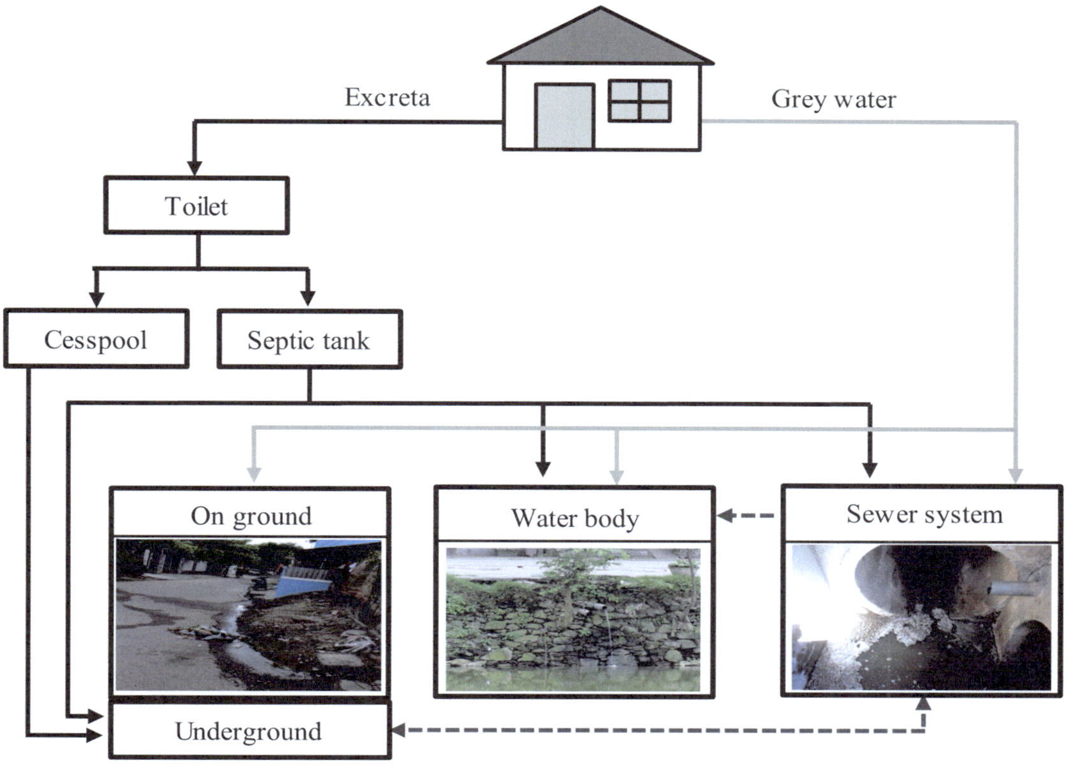

Fig. 5.53 Illustration of household wastewater discharge. Modified from Anh et al. (2016a). Photos by Tran Nguyen Quynh Anh, 2015

leads to conflicts between households. In Hue Citadel, Hue city, the percentage of households discharging their greywater to the sewer system, the surface ground, and water bodies are 52, 29, and 19%, respectively (Anh et al. 2016a).

Figure 5.54 illustrates the structure of septic tanks and cesspools, two of the most common septic systems in Vietnam. A septic tank is a watertight tank and is often located underground. One septic tank generally has 2–3 compartments working as preliminary treatment in which the wastewater is anaerobically fermented. The solids in the wastewater settle down to the tank's bottom, while oils and fats float to the top of the tank to form a scum layer. Under anaerobic conditions (i.e., in the absence of oxygen), bacteria digest the solids to produce sludge that accumulates in the bottom. Another product of the anaerobic treatment is biogas, which is mainly composed of methane (CH_4) and carbon dioxide (CO_2), two well-known greenhouse gases (GHGs), and other gases such as hydrogen sulfide (H_2S) and water vapor. To obtain good performance from the septic tanks, it is suggested that the septage (mixture of bottom sludge, liquid, and scum layers in the septic tanks) should be pumped out every 2–5 years. However, most septic tanks are not regularly desludged, causing high concentrations of contaminants in the septic tanks effluent as a result. The concentrated effluents are then discharged underground, directly to nearby water bodies, or indirectly to water bodies via canal or sewer pipes. Another common type of septic system includes cesspools. Cesspools are less popular than septic tanks and are often used in older houses. In Hue Citadel, Hue city, approximately 20% of households use cesspools as their septic system (Anh et al. 2016a). Unlike septic tanks, cesspools simply capture and store the waste but

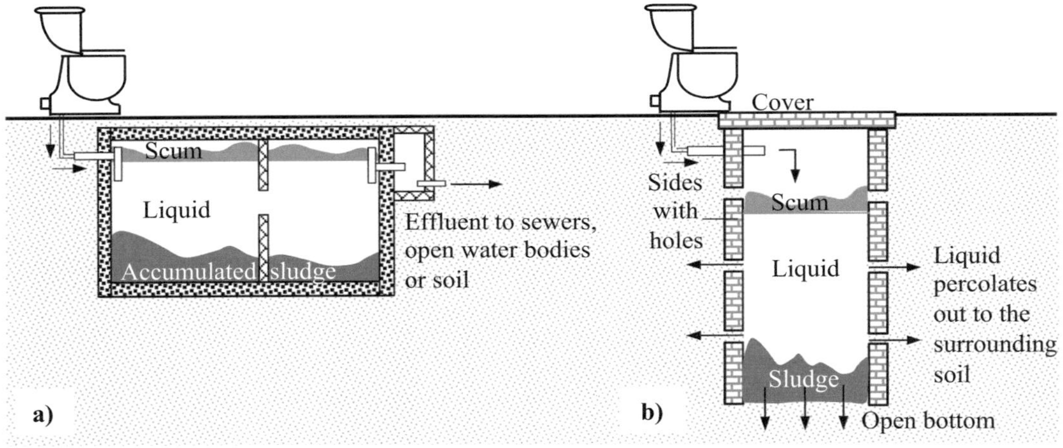

Fig. 5.54 Structure of a septic tank (**a**) and a cesspool (**b**)

do not treat it. Cesspools usually have a cylinder shape and are buried underground. Most cesspools are made of concrete with many holes on the wall-sides and an open bottom. Sanitary waste is dumped into the cesspool, wherein solids (sludge) settle down to the bottom while liquids percolate out into the surrounding soil through holes on the wall-sides and the open bottom, which may lead to soil contamination.

The following text presents and discusses the wastewater characteristics of two cities, Hanoi and Hue, based on the results of our 3-year study with the aim of understanding the wastewater and fecal sludge management scenario in Vietnam.

5.8.2 Domestic Wastewater Characteristics

5.8.2.1 Wastewater Characteristics and Pollution Load on the Water Body—A Hanoi Case Study

Most domestic and industrial wastewater in Hanoi is discharged into open water areas such as lakes and rivers. There are four rivers in Hanoi functioning as drainage canals for the city; one of them is To Lich river, which is 30–40 m wide and 3–4 m deep, and has the largest watershed area of 38.3 km^2 (Dao et al. 2010) with the largest population of 892,102 (Hanoi Statistical Office 2011). The river starts its flow from West lake water and ends its 14.8 km journey at Yen So lake, where the river water is drained into the Red river via pumps (Fig. 5.55). The inlet of To Lich river was considered at Hoang Quoc Viet street, where the river first exposes its open surface, and the outlet of the river was examined at Tuu Liet bridge.

Our wastewater characteristics survey in Hanoi investigated five single-unit houses alongside a canal, which is a branch of the Kim Nguu river that conveys wastewater. These households discharge greywater and septic tank effluent through two separate PVC pipes into a pit (approx. 30 L) and then the wastewater from the pit is discharged to a canal through one pipe (Fig. 5.56). The wastewater was collected continuously within 24 h. During collection, the wastewater volume was measured every hour. Wastewater samples were analyzed every 3 h via standard methods, and the parameters analyzed were: biochemical oxygen demand (BOD), chemical oxygen demand (COD), total Kjeldahl nitrogen (TKN), total phosphorus (TP), and suspended solids (SS). BOD is a measurement of the dissolved oxygen used by microorganisms in the biochemical oxidation of organic matter. The standard five-day BOD

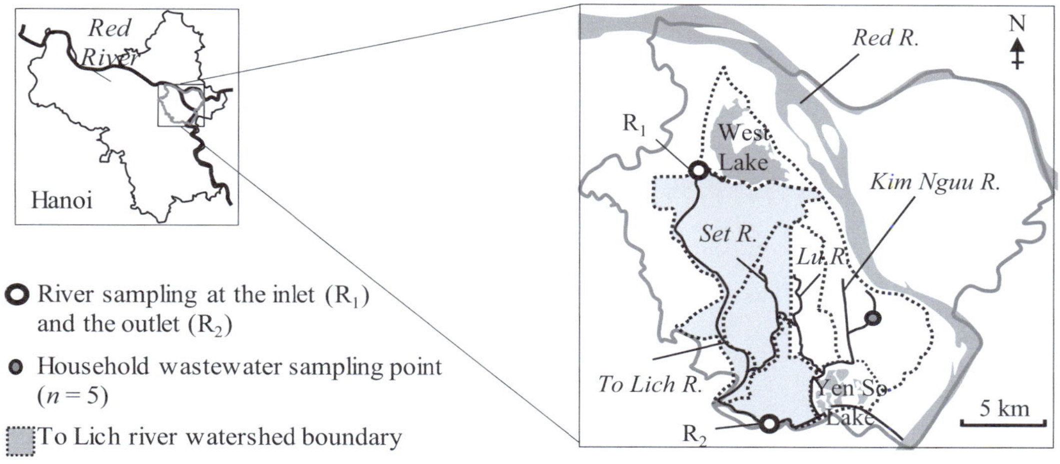

Fig. 5.55 To Lich river watershed and sampling points

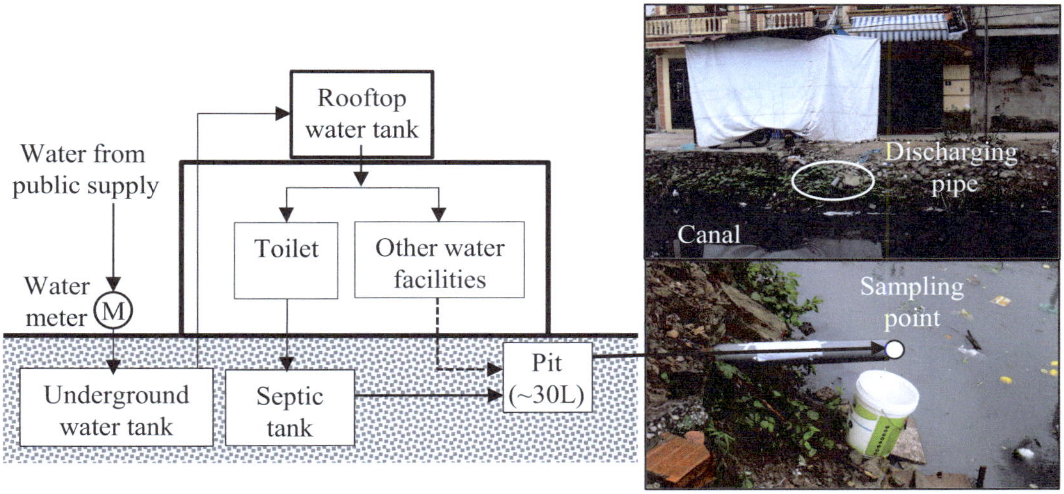

Fig. 5.56 Illustration of water flow and sampling method for the five households. Photos by Pham Nguyet Anh (2013)

test (BOD_5) is the most widely used parameter of organic pollution applied to wastewater. The COD test is used to measure the oxygen equivalent of the organic materials in wastewater that can be oxidized chemically. The chemical used in the oxidation of organic matter in the COD test can be dichromate (COD_{Cr}) in an acid solution or permanganate (COD_{Mn}). Total nitrogen (TN) is comprised of organic nitrogen, ammonia, nitrite, and nitrate. The Kjeldahl method determines the total of organic and ammonia nitrogen, and therefore, excludes nitrite and nitrate. The TP test measures the phosphorus in all its forms in the water sample, including orthophosphate, polyphosphate, and organic phosphate. Suspended solids (SS) refers to the portion of total solids retained on the paper filter (usually with a pore size of 1.58 μm) after drying at a specific temperature (usually 105 °C) (Tchobanoglous et al. 2003).

Similar to water consumption, the wastewater discharged by households also varies hourly

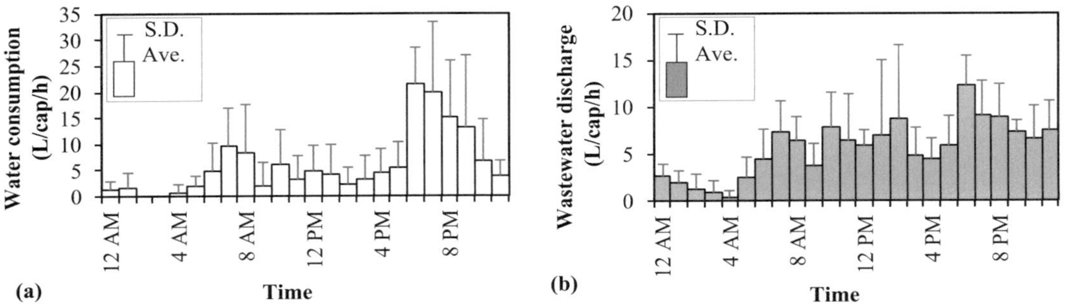

Fig. 5.57 Hourly water consumption of ten apartment households (**a**) and wastewater discharge pattern of five single-unit houses (**b**)

but, in general, it fluctuates less. Figure 5.57 describes the hourly discharge of five households with two peaks; the morning and the evening peak; which are considered to be the same as the consumption pattern mentioned in Sect. 5.7.2. The small fluctuation may be explained by the functions of the pit (~30L) that make the discharge amount more equalized. The average discharge amount was 134 L/cap/day, which is lower than that of the average consumption amount (146 L/cap/day). Although the water consumption survey in Sect. 5.7.2 conducted for apartment households and wastewater survey conducted for households in single-unit houses, these two are modern housing. Moreover, those living in new apartments often used to live in on-ground houses before moving to apartment buildings, and therefore, their water-use habits may not change much.

The wastewater quality of the five households (Fig. 5.58) shows that the highest concentrations were found at 12–1 PM when the wastewater was only slightly diluted due to low water consumption at this duration. The wastewater concentrations then sharply decreased at 6–7 PM, which can be understood as greywater from bathing and laundry becoming diluted by the rapid increase in water consumption. During nighttime (0–1 AM), the wastewater remaining in the plumbing system was still gradually discharged. The average concentrations of COD_{Cr}, BOD_5, TKN, and TP were 471, 225, 57, and 8 mg/L, respectively. The suspended solids concentration (80 mg/L) was low, indicating the role of septic tank in removing solid matter. COD_{Cr}, BOD_5, and TKN were found mainly in the soluble phase, whereas a large part of TP was found in the solid phase.

Since the household wastewater finally comes to To Lich river, we also conducted a survey of flow velocity measurement and river water quality analysis at the inlet and the outlet of the river to quantify the contribution of the households to river pollution loads. We estimated the river flow rate (m^3/day) at the inlet and the outlet of the river from the measured river flow velocity (m/s) and river cross-sectional area (m^2). The pollution loads in the river water were calculated based on the river flow rate (m^3/day) and pollutant concentration (g/m^3). The pollution loads from the households were the product of the wastewater concentration (g/m^3), wastewater discharge amount (L/cap/day), and total population (cap) living in To Lich river watershed.

The estimated flow rate of To Lich river near the outlet was 626,000 m^3/day, which is approximately 30 times higher than the flow rate at the inlet (21,000 m^3/day) (data not shown). Based on the average daily water consumption (146 L/cap/day) and the total population of the watershed, it was calculated that the household wastewater discharge amount was 130,000 m^3/day, which accounts for approximately 21% of the total discharge amount into the river (605,000 m^3/day). Figure 5.59 shows the serious water pollution of To Lich river. Organic contents,

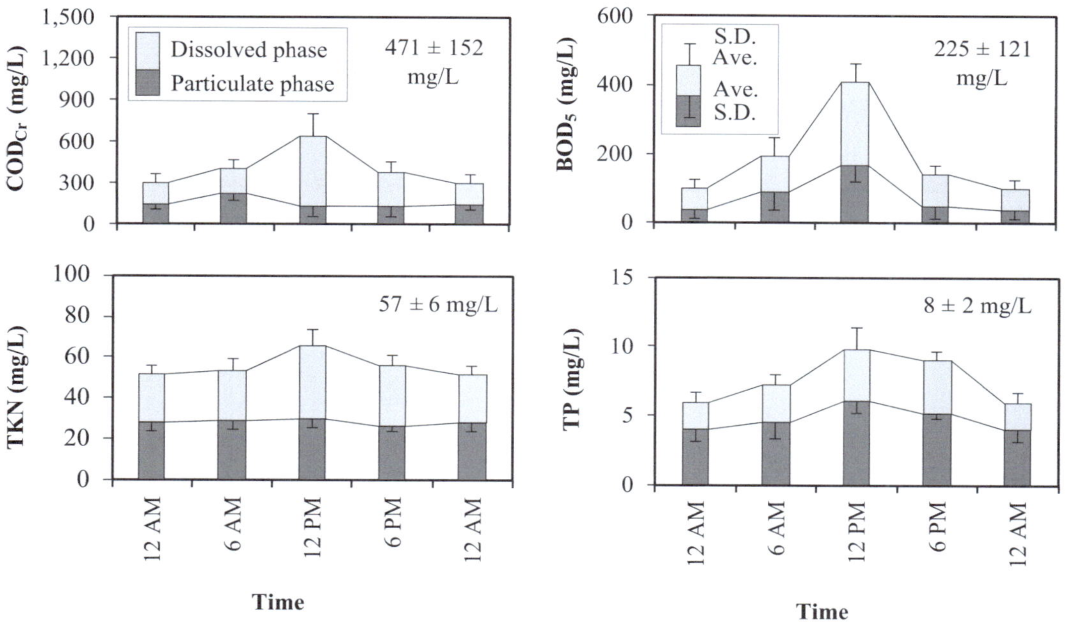

Fig. 5.58 Household wastewater concentration patterns for five households in dissolved and particulate phases. Figures in each panel show the average ± SD of the sum of dissolved and particulate phases

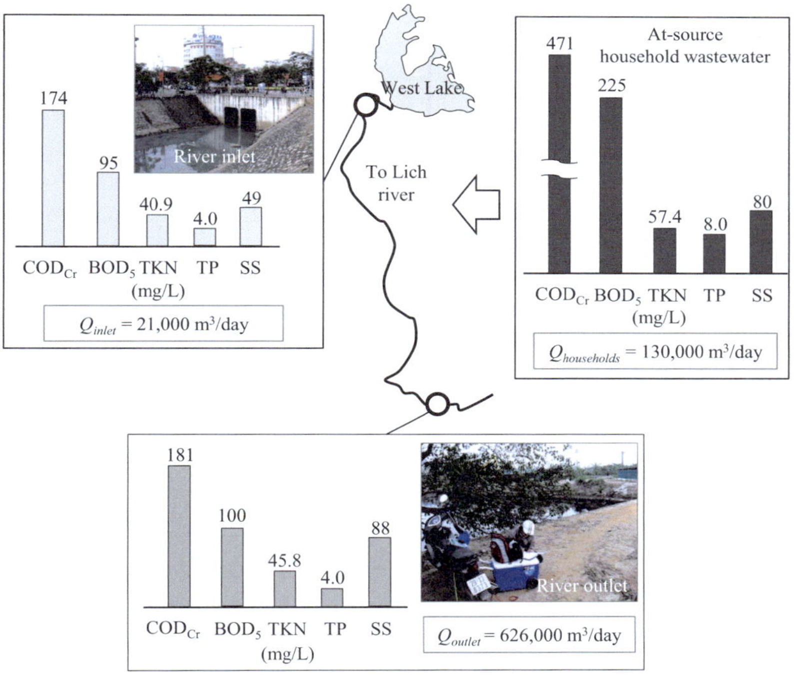

Fig. 5.59 Characteristics of household wastewater and To Lich river water. Photos by Pham Nguyet Anh (2013)

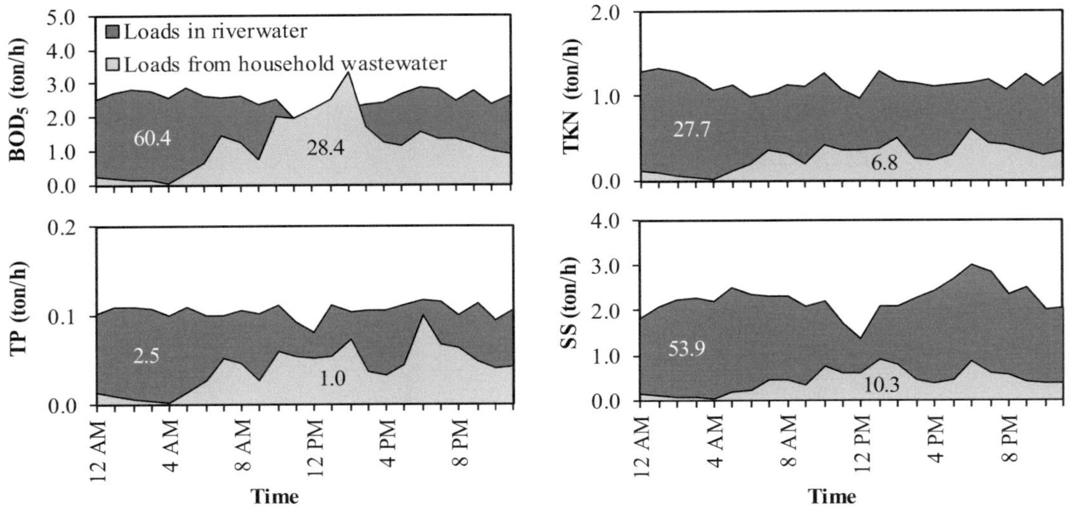

Fig. 5.60 Estimated pollution loads from household wastewater and river pollution loads. The numbers written in white color represent the total loads (tons/day)

in terms of COD_{Cr} and BOD_5, and nutrients in terms of TKN and TP exceeded thresholds regulated in National technical regulation on surface water quality—class B2, which is the maximum allowable values for surface water quality applied for navigation and other purposes with low quality requirements. SS concentration at the outlet (88 mg/L) was two times higher than the concentration at the inlet (49 mg/L), which can be explained by the accumulation of solid matter downstream during river flow. The quality of river water at the inlet (data not shown) and household wastewater indicate similar peaks at 12–1 PM, showing that the river was affected by household discharge.

Figure 5.60 shows the hourly fluctuation of household discharge and river water accumulation loads. The BOD_5, TKN, and TP loads were high in the evening when most water-consuming activities happened, as investigated in water consumption research (Sect. 5.7.2). The estimated BOD_5 load was 28.4 tons/day, which accounts for 47.0% of the BOD_5 load from the watershed. The BOD_5 load from households peaked in noon and afternoon; toilet flush and kitchen activities intensively happened at the same time. The BOD_5 in household wastewater (225 mg/L) was higher than that in river water (95 mg/L at the inlet and 100 mg/L at the outlet of the watershed). The TKN and TPloads from household wastewater, respectively, contributed 24.5% (6.8 tons/day) and 40.0% (1.0 tons/day) to the loads in the river. Although TP in household wastewater was mainly part of solid matter (Fig. 5.58), it still largely contributed to TP loading in the river. In contrast, TKN in household wastewater (57 mg/L) was mostly in the form of dissolved matter (Fig. 5.58) and TKN concentration was still high in the river water; 40.9 mg/L at the inlet and 45.8 mg/L at the outlet. This indicates that nitrogen removal might not have effectively occurred during the flowing of household wastewater to the river, and also TKN from other sources has largely entered the river. The SS concentration in household wastewater contributed 19.1% to the SS load of the river water. The SS load in the river was substantially high, about 54 tons/day, and could be explained by the sedimentation of solid matter in the wastewater.

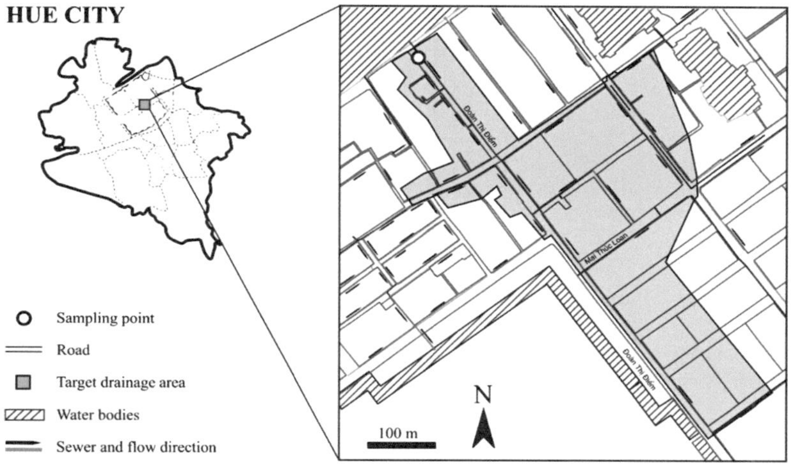

Fig. 5.61 Study site in Hue city. Modified from Anh et al. (2016b)

5.8.2.2 Wastewater Characteristics at the Sewer Outlet—A Hue Case Study

Hue city uses a combined sewerage system that covers approximately 40% of the city area on average (HEPCO 2013), mainly concentrating on the more developed wards in the center of the city. In the northern part of the city, domestic wastewater, after being conveyed through the sewer pipes, is finally discharged to water bodies. Therefore, it is important to understand the quantity and quality of wastewater discharged at the sewer outlet for better management. A residential drainage area in Thuan Thanh ward of Hue city was selected as the study area for surveys from 2014 to 2016 (Fig. 5.61). This area covers 11.2 ha, having a population of 1,452 people living in 363 households in 2015 (People's Committee of Thuan Thanh Ward 2015). Domestic wastewater in this area was collected by a sewer pipe system with different diameters (from 400–1000 mm) and finally discharged into a lake (Tinh Tam lake).

The average amount of wastewater discharge at the sewer outlet on no-rain days in the dry season was 2.3 ± 0.4 m^3/h (equivalent to 55 ± 11 m^3/day or 38 ± 7 L/cap/day). The discharge amount varied during the day with two peaks from 6 AM–4 PM and from 5 PM–11 PM and the lowest discharge values in the early morning (1 AM–6 AM) (Fig. 5.62b). This fluctuation in the amount of wastewater discharge was rather similar to the water consumption pattern of people living in the study site (Fig. 5.62a). Specifically, in the morning, a large amount of water was consumed for hygiene activities (e.g., bathing, toilet use) and breakfast preparation before people went to work. Then, the water continued to be used for lunch preparation since most Hue residents come back home to have lunch. In the late afternoon and evening, people come back home again and continue consuming a large amount of water for bathing, cooking, and washing. The general sleeping time is from 11 PM to 5–6 AM. Therefore, high water consumption in Hue was found in the morning and from afternoon to evening, while the lowest water consumption was from midnight to early morning. This pattern of wastewater discharge was observed to be rather similar for all days in 2015 and 2016, which indicated that the quantity of wastewater discharge might be different among days, but the trend of discharge did not change. The diurnal variation of domestic wastewater discharge amount in urban Hue was rather similar to that in Hanoi and Kobe, Japan (Japan Sewage Works Association 2013). Moreover, it should be noted that the wastewater amount discharged at the sewer outlet accounted for only 29% of the total wastewater introduced into the sewer system. The old and

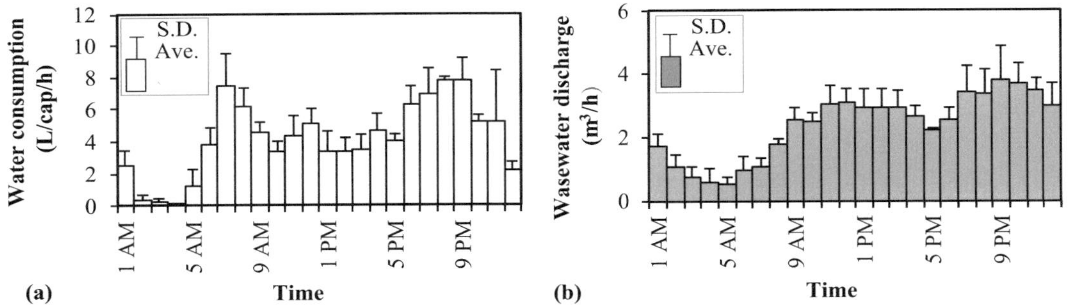

Fig. 5.62 Hourly water consumption of households (23 households on 4 days) (**a**) and hourly wastewater discharged at the sewer outlet (363 households on 4 days) (**b**) in the dry season (2015). Modified from Anh et al. (2016b)

incomplete sewer pipe system being used might be the reason for the loss of wastewater during transportation to the sewer outlet owing to the exfiltration process on no-rain days in the dry season. According to Watanabe et al. (2021), the exfiltration ratio was up to 65.6% for the dry season in the study area, leading to a large amount of wastewater seepage and subsequent contamination of soil and groundwater.

Sewer discharge concentration patterns on no-rain days in the dry season of 2015 are presented in Fig. 5.63. Unlike in Hanoi, the discharge concentrations did not show strong fluctuation over a day. Even in the nighttime and early morning (1 AM–5 AM) when little amount of water was consumed, pollutant concentrations were not too different from those in the daytime. One of the reasons for this trend might be the many complex phenomena (e.g., infiltration, exfiltration) occurring when the wastewater is being transported through the old sewer pipes, which affects the concentration of wastewater at the sewer outlet. The average concentrations of COD_{Cr}, BOD_5, TN, and TP at the outlet were 154 ± 16, 93 ± 16, 46 ± 5, and 3.5 ± 0.4 mg/L, respectively, which are much lower than those of at-source wastewater in Hanoi and also lower than the values specified in the National Technical Regulation on Domestic Wastewater (QCVN 14: 2008/ BTNMT) except for organic matter concentration. Besides the impact of exfiltration/

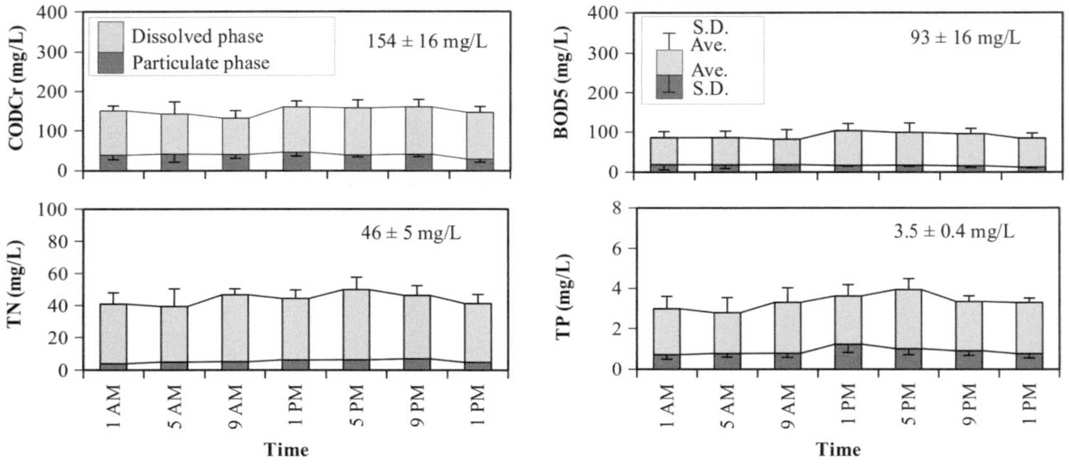

Fig. 5.63 Wastewater quality at the sewer outlet in 24 h on no-rain days (n = 4) in the dry season (2015). Figures in each panel show the average ± SD of the sum of dissolved and particulate phases. Modified from Anh et al. (2016b)

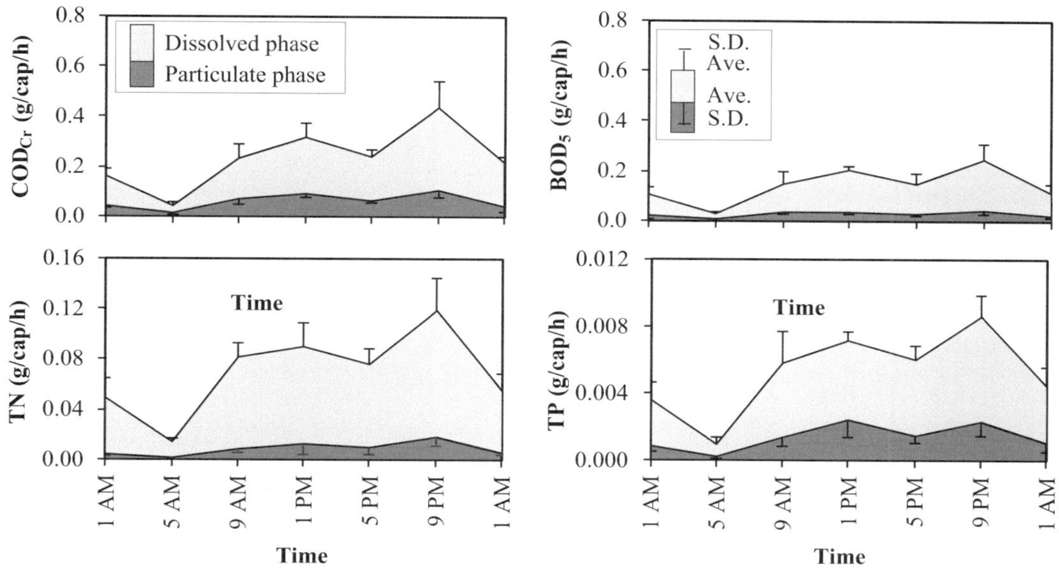

Fig. 5.64 Average pollution loads at the sewer outlet on no-rain days (n = 4) in dry season (2015). Modified from Anh et al. (2016b)

infiltration processes as mentioned above, another possible reason could be the role of the settling process in the sewer pipe system caused by the low velocity of wastewater; in-sewer wastewater velocity in the studied sewer system was only 0.009 ± 0.003 m/s, which is also reflected by the low concentration of suspended solids (SS: 38 ± 5 mg/L) and the low concentration of pollutants in particulate phase compared to the dissolved phase (Fig. 5.63).

Pollution loads at the sewer outlet showed the same pattern for all parameters with two peaks of discharge loads (5 AM–5 PM and 5 PM–1 AM) and the lowest loads were in the early morning (1 AM–5 AM) (Fig. 5.64), which is rather similar to the trend of pollution loads from households in Hanoi. A large amount of pollution load came from dissolved matter, which showed the dominance of dissolved matter in wastewater constituents as well as the role of the in-sewer settling process. Unit pollution loads of all parameters are shown in Table 5.9. Pollution loads from the sewer outlet in Hue were many times lower than those in Hanoi and Japan. This remarkable difference might be explained by the difference in living situations among the areas (e.g., living standards, water consumption habits, water consumption facilities, etc.), which might affect the wastewater characteristics of each area. Moreover, the management of the sewerage system (e.g., type and components of the sewerage system, maintenance of sewer pipes, etc.) is also a very important factor affecting the sewer discharge quality. In the case of Hue city, low pollutant loads at the sewer outlet can reduce the amount of pollution discharged into open waters, but create a high risk of causing serious pollution to soil and groundwater due to in-sewer processes such as sewer leakage.

Our studies' results showed that the wastewater in Vietnam has distinct characteristics in terms of both quantity and quality in comparison to other countries; for example, Japan. Moreover, the wastewater characteristics are also different between urban areas of Vietnam, such as between Hanoi and Hue. Therefore, we should carefully consider the issue of localization in developing the sewerage system, especially in designing wastewater treatment plants for Vietnam so that they can be more suitable for the specific characteristics of wastewater of the country and each area.

Table 5.9 Unit pollution loads from household and at the sewer outlet

Item	Unit pollution load (g/cap/day)			
	Hanoi (at households)	Hue (at sewer outlet)	Japan (at households) *	Japan (at the inlet of WWTPs)**
COD_{Cr}	65.6 ± 10.4	5.9 ± 0.4	23 [(b)]	n.a
BOD_5	31.9 ± 5.3	3.4 ± 0.3	45	40–45
TN	7.6 ± 0.9 [(a)]	1.8 ± 0.2	9.3	1–3
TP	1.1 ± 0.1	0.1 ± 0.0	1.1	0.15–0.4
SS	11.6 ± 5.1	1.4 ± 0.1	37	n.a

Note n.a. not available; [(a)]: Total Kjeldahl Nitrogen; [(b)]: COD_{Mn}
*Kumokawa et al. (2019)
**Tchobanoglous et al. (2003)

5.8.3 Septic Tank and Fecal Sludge Management in Vietnam

As mentioned in the beginning of this section, when the sewerage system coverage is limited, on-site treatment plays a crucial role in wastewater treatment and management. The septic tanks are important to pre-treat the wastewater, especially to remove the solids before the wastewater is discharged into open receiving waters. Although Vietnam has a technical guideline for septic tank construction[7] and a regulation for fecal sludge management,[8] compliance is low. Many septic tanks are usually constructed based on practical experience and desludging (pumping accumulated sludge out of the septic tanks) is only conducted when the septic tanks have problems such as blockage or release of bad odor. Moreover, most fecal sludge is not properly managed due to the lack of treatment plants, and therefore, it is often dumped illegally into the environment (Harada et al. 2016).

5.8.3.1 Septic Tank Structure and Management

All the septic tanks in our survey were in urban areas: 46 septic tanks ($n=46$) in Hanoi and 36 septic tanks ($n=36$) in Danang received only blackwater from toilets. The septic tanks in Hanoi were connected to cistern-flush toilets and were mostly rectangular in shape. Only one round septic tank was found at a household that utilized a biogas system previously. The average and standard deviation of septic tank size was 3.4 ± 1.2 m^3 with three compartments and the average number of users was five. The situation in Danang was quite different, as human excreta of 36 households were discharged into cistern-flush toilets (50%) and pour-flush toilets (50%), and then these toilets were connected to the septic tanks. Most septic tanks were rectangular (89%), with the remaining (11%) in cylindrical shape. The average septic tank volume was 3.4 m^3 with 2–3 compartments. It can be said that the structure of septic tanks in Hanoi and Danang was not bad as it satisfied the requirement of Vietnamese code in terms of size and number of compartments. The septic tank volume was quite big since most people often considered that a big tank means good performance. This size is even bigger than the US or European standards for a septic tank serving a single house of five members.

Despite the size and compartment number, the septic tanks were not managed in a proper manner. Most householders did not understand why the septic tanks should be desludged

[7]Ministry of Construction, 1999. Decision 47/1999/QD-BXD, Approving the regulation on the system and facilities of water supply and discharge in households, Hanoi, Vietnam.

[8]Vietnam Government, 2014. Decree 80/2014/ND-CP on Wastewater Drainage and Treatment, Vietnam.

frequently. The average septage storage period (the number of years since the last emptying or since septic tank construction if the tanks had never been emptied) of the septic tanks in Hanoi was 10.2 years, and that of the septic tanks in Danang was 9.5 years. One hundred percent of the septic tanks in Hanoi had never been desludged before, while 40% of the septic tanks in Danang had been already desludged. When the householders in Danang were asked about the reasons for desludging, 52% reported blockage and/or bad odor. Thirty-nine percent of the reasons were house construction or repair as the residents took this opportunity to desludge the septic tanks. There were several septic tanks desludged just because it was provided free of charge under the pilot project sponsored by the World Bank, accounting for approximately 7% of total desludging reasons. Only 2% of the householders thought that the septic tanks operated long enough and needed to be desludged.

Due to long operation without desludging, the sludge in the septic tanks was over-accumulated. It was found that in Hanoi, the methane emission rates from septic tanks accumulating septage for >5 years were significantly higher than those at 0–5 years (Huynh et al. 2021). The over-sludge accumulation shortens the hydraulic retention time of the wastewater in the septic tank, one of the factors affecting settling performance and anaerobic digestion efficiency. The performance of the septic tanks, therefore, is expected to be limited. Harada et al. suggested that the septic tanks should be emptied frequently to reduce the effluent concentration (Harada et al. 2008). In fact, septic tank effluents were highly polluted, which can be explained as the anaerobic digestion did not effectively decompose organic matter. It can also be understood that the settling process in the septic tanks was very limited and a considerable amount of solid matter was carried over in the effluent. The second compartment of the septic tanks is expected to contain excessive sludge accumulation as well. This reason, together with the scum layer, which could not be addressed in the measurement, can be the explanation for the low average sludge accumulation rate (0.04 L/cap/day) as compared to other studies (Pham et al. 2014). Since most septic tank effluents in Hanoi eventually flow into the water environment and many of the septic tank effluents in Danang soak underground, these effluents will pollute the receiving environment.

In Hanoi and Danang, desludging service is provided by state-owned as well as private companies. When the householders want to desludge septic tanks, they make a phone call to the services, which are advertised everywhere, such as on an electric pole or house wall on the street (Fig. 5.65a), or are introduced through word of mouth. If the negotiation succeeds, the service

Fig. 5.65 Desludging work: **a** Advertisement of desludging service on house's wall; **b** toilet floor broken for suction pipe access. Photos by Pham Nguyet Anh (2014)

provider will come to the house to negotiate the price again, check the site, and prepare for desludging. The sludge pumping is executed by a vacuum truck and the truck operators (often 2–3 people) perform the desludging. Since the householders do not have awareness of desludging, they usually do not prepare an access port to the tanks. All the septic tanks in Hanoi and 81% of the septic tanks in Danang had no access port. Only 11% of the septic tanks in Danang had an openable and closable port, which was provided by the pilot project mentioned earlier. The remaining (8%) had a cement-fixed port, which meant that the septic tanks had been already desludged before or the householders marked the tank access even if it was not desludged.

Without accessing the port, desludging activities become tough to work. The drivers often break the toilet floor manually or drill through the floor using a drilling machine under which the septic tanks are located (Fig. 5.65b). Once the tanks are accessed, suction pipes are used to pump out the sludge. Since the septic tanks are often operated over a long time; the scum layer is very thick and the septage is highly condensed. In that case, water is added to dilute the septage so that the sludge pumping is easier. At the end of the desludging, water is added again for cleaning. Of course, not all the sludge is pumped out because the suction pipe cannot access every part of the septic tanks. Moreover, this method can clean only the first compartment of the septic tanks. When the desludging work is completed, payment is made, and the vacuum truck transports the sludge to a discharging site. There were certain instances wherein the houseowners did not know the exact location of the septic tank and the desludging workers could not find the septic tank as well, and the desludging work could thus not be carried out. In another interesting case we encountered when following the vacuum truck driver, it was found that the septic tank was located under the bedroom or kitchen floor instead of the toilet floor. The desludging work is more difficult in highly populated parts of Hanoi and Danang where small lanes make the trucks unable to access. For this reason, the vacuum truck must park on the roadside and more pipes need to be connected to the septic tanks, creating a low pressure that is not sufficient for successful sludge pumping (Fig. 5.66).

5.8.3.2 Fecal Sludge Characteristics, Transportation, and Disposal

The sludge pumped out of the septic tanks varies by septic tank size and management, but in general, it contains high contents of organic matter,

Fig. 5.66 Desludging for a septic tank located in a small lane in Danang: **a** A vacuum truck parking at the main lane; **b** suction pipes connection to the septic tank via narrow and long lane. (Photos by Pham Nguyet Anh (2012)

Fig. 5.67 Fecal sludge treatment at Khanh Son landfill site, Danang: **a** discharging fecal sludge to receiving tank; **b** gravitational settling chambers with scum layer at the top. Photos by Pham Nguyet Anh (2012)

nutrients (nitrogen and phosphorus), and pathogenic bacteria and needs to be treated (Pham et al. 2012). If the household size and the septic tank volume were fixed, a longer desludging period could provide a higher septage concentration (Pham et al. 2012). When the desludging work was completed, payment was made, and the vacuum truck transported the fecal sludge to a discharging site. Only about 10% of the households in Hanoi could answer that the fecal sludge is transported to a treatment plant, while the majority did not know the discharging site. In Danang, this result was 28 and 72%.

In Hanoi, the fecal sludge treatment is just a component located at the Cau Dien composting plant under the management of the city's Urban Environmental Company (URENCO). Only the fecal sludge collected by Hanoi URENCO is transported to the treatment place, while that collected by private companies is usually illegally disposed of (Brandes et al. 2016). According to our investigation, the fecal sludge could be dumped in a flower garden or tree garden as a soil supplement. In this case, the vacuum truck's drivers are able to get money twice, one from the household and another from the garden's owner. Sometimes, the fecal sludge is illegally discharged to the city's sewers or canals or dumped at a vacant site far from the city center. In Danang, the fecal sludge is treated at the Khanh Son landfill site, where it is discharged into the gravitational settling chambers to separate the liquid that can be treated together with the landfill's leachate (Fig. 5.67). The solid is then disinfected with lime and dumped together with other solid waste.

Another possible way of fecal sludge treatment is to recycle the sludge into compost, as it is rich in organic matter and nutrients. However, when people were asked about their acceptance of using fecal sludge-derived-compost, more than 80% of the respondents accepted, but some said that they only accept if the fecal sludge is hygienically converted into compost, and a few disagreed because they worried about its safety. During our survey, we also observed that the workers conducting desludging often used bare hands and did not wear masks, which may expose them to potential health risks.

Based on the above, it can be said that the septic tanks in Vietnam were fit from a structural point of view, but the management was not proper. Moreover, desludging the tanks is tough work as the septic tanks are not accessible. Improper septic tank management and fecal sludge disposal may potentially deteriorate the receiving environment and affect public health.

5.8.4 Concluding Remarks

The current design of the wastewater treatment systems in Vietnam is mostly based on referenced data from other countries or based on wastewater quality at the inlet of existing treatment plants where the wastewater is much diluted. Our data, therefore, will prove to be essential in designing treatment plants and planning for wastewater management in Vietnam. Since the full development of a sewerage system requires a long time and Vietnam will surely achieve a higher percentage of sewerage system coverage in future, centralized wastewater treatment alone is not a solution for the whole country. It is suggested that wastewater management should combine centralized and decentralized approaches depending on the specific situations. On-site treatment, that is, septic tanks still play a crucial role in pretreating the wastewater, especially to remove the solids where household wastewater is not collected and transported to centralized wastewater treatment plants. However, since septic tanks can become pollution sources if they are not properly managed, more attention should be given to the public and government. The public should be aware of the desludging period to recover the septic tank performance, thus reducing the pollution load of the septic tanks on the environment. To ensure that desludging is regularly conducted, desludging activities and fecal sludge disposal need to be well managed and should receive more investment. Our study provided necessary information about at-source household wastewater quantity and quality, sewer outlet wastewater quantity and quality, pollution loads from wastewater, as well as the role of the septic tank and fecal sludge management in an attempt to contribute to the improvement of water environmental quality and enhancement of public awareness regarding environmental protection in Vietnam.

5.9 Solid Waste Management in Vietnam

In Vietnam, the amount of solid waste generation has been increasing in the last decades with population growth. Since further population growth is predicted in Vietnam over the next few decades, proper waste management systems need to be consolidated with the help of knowledge of global environmental studies. This chapter mainly discusses general information about waste generation and treatment in Vietnam as well as geoenvironmental challenges related to waste treatment and management as well as steps necessary for the protection of the geoenvironment. First, waste generation and management in Vietnam are introduced. Then, municipal solid waste (MSW) treatment technologies such as incinerators, composting, thermal disposal, as well as recycling construction and demolition (C&D) waste, will be explained. This is followed by a study type of landfilling and specification design for MSW landfill in Vietnam. Finally, the chapter ends with the proposed solutions and perspectives for the waste landfills in Vietnam in the future.

5.9.1 Waste Generation and Management in Vietnam

5.9.1.1 Waste Generation in Vietnam

Solid waste generated in Vietnam can be grouped broadly into four categories, which are municipal solid waste, industrial solid waste, agricultural solid waste, and medical solid waste. Municipal solid waste (MSW) comprises residential waste, commercial waste, market waste, street waste, etc. In 2003, the total MSW generation in Vietnam from both urban and rural areas was 12.8 million tons, which accounted for over 80% of total waste in Vietnam (World Bank 2004). Half of that (6.4 million) was urban MSW, which is equivalent to 17,534 tons/day. In 2016, the National Environment Report stated an estimated urban MSW generation of 38,000 tons/day in 2015, which is almost double the MSW amount in 2003 (MONRE 2016).

Figure 5.68 shows the generation of urban MSW over the years and forecasts for the future (MONRE 2011). Data regarding MSW in rural areas is limited due to the limited information on waste management in such areas. It can be seen from the figure that urban MSW generation has been increasing gradually over many years with

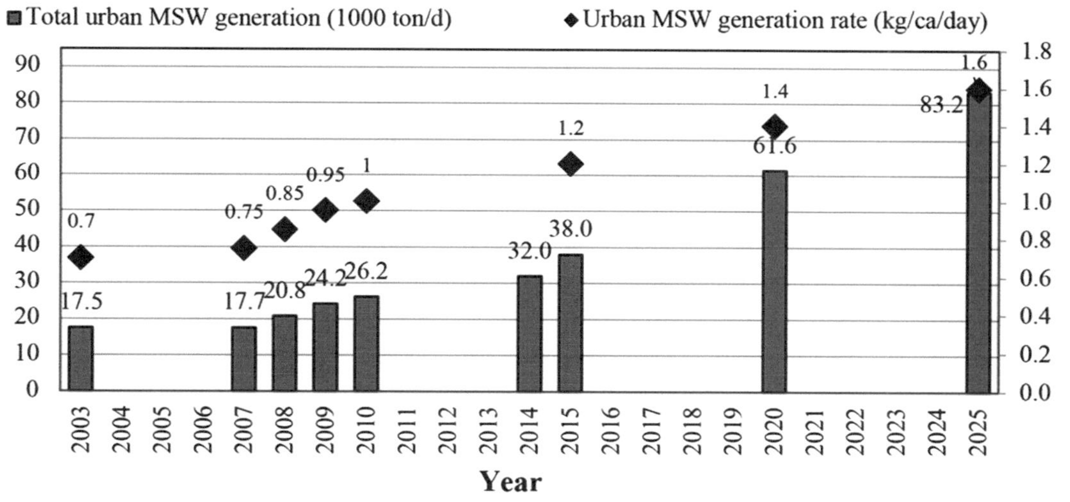

Fig. 5.68 Urban MSW generation: past and future data Data taken from World Bank (2004) (for 2003), MONRE (2011) (for 2007–2010), MONRE (2016) (for 2014 and 2015), MONRE (2011) (for 2020 and 2025)

an average increasing rate of 10% during the period of 2007–2010 (MONRE 2011) and 12% in the following period of 2011–2015 (MONRE 2015). However, it is estimated that waste will increase drastically in the next few years. Urban areas are major generators of MSW in Vietnam as they contain approximately 24% of the country's population but produce 50% of the country's MSW (World Bank et al. 2004). Urban MSW in Vietnam is largely generated from the two biggest cities, Hanoi city and HCM City, which accounted for 45.2% of the total urban MSW amount in Vietnam in 2007 (MONRE 2011).

The average urban MSW generation per capita was 0.7 kg/ca/day in 2003 (World Bank 2004), which increased to approximately 1.0 kg/ca/day in 2010 and is estimated to be 1.4 and 1.6 in 2020 and 2025, respectively (MONRE 2011). It is noted that MSW generation per capita varies among municipalities. Tables 5.10 and 5.11 show that richer urban areas have higher MSW generation per capita. In rural areas, MSW generation per capita is much lower; it was 0.3 kg/ca/day in 2003 (World Bank 2004) and it is estimated to be 0.45 kg/ca/day in 2019 (MONRE 2019).

5.9.1.2 Municipal Solid Waste Composition

MSW includes inorganic wastes, partly biodegradable organic wastes (such as glass, ceramic, metals, rubber, plastic, textile, electronic waste, toys, etc.), and easily biodegradable organic waste (yard waste, fruit and vegetable waste, food residues, animal corpse, etc.). Urban MSW in Vietnam was characterized by easily biodegradable content (kitchen waste) of approximately 55–77% in 2010 (Table 5.12). Currently, plastic waste is a big concern because of the increasing utilization of plastic bags in daily life and their poor degradability (MONRE 2015). Data for some selected cities (Table 5.12) shows

Table 5.10 Urban MSW generation in selected cities in 2010 (JICA 2011)

Item	Unit	Hanoi	Hai Phong	Hue	Danang	Ho Chi Minh City
MSW generation	Ton/Day	4,773	1,280	224	736	7,048
Population	Persons	6,451,909	1,837,173	337,169	887,437	7,162,864

Table 5.11 Average urban MSW generation in 2007 (MONRE 2011)

Types of Urban	Unit	Special class (Hanoi and Ho Chi Minh City)	Class 1 (incl. Hai Phong, Hue and Danang)	Class 2	Class 3	Class 4
MSW generation per capita	kg/Person/Day	0.96	0.84	0.72	0.73	0.65

Table 5.12 Waste composition of household waste in selected cities (JICA 2011) (Unit: %)

Types of waste	Hanoi	Hai Phong	Hue	Danang	Ho Chi Minh City
Kitchen waste	70.9	55.51	77.25	63.92	65.40
Paper	3.8	3.45	2.30	1.97	6.77
Textile	1.6	0.95	1.21	2.40	1.78
Wood	1.3	12.85	1.70	2.57	3.96
Plastic	9.0	6.10	13.99	13.82	16.07
Leather and Rubber	0.7	0.29	0.40	1.68	0.81
Metal	0.4	0.44	0.49	0.77	0.68
Glasses	1.3	0.29	0.48	1.84	0.51
Ceramic	–	–	0.25	2.15	0.18
Stone and sand	–	4.66	0.01	0.50	0.11
Briquette coal	6.8	–	0.00	2.46	0.69
Dangerous	0.5	–	0.01	0.50	0.11
Diaper	3.3	–	1.87	2.17	2.55
Others	0.3	15.46	0.05	0.58	0.14
Total	100	100	100	100	100

that plastic is the second major component of MSW, with Nylon bags contributing a large share of plastics from households.

There is little information regarding hazardous waste composition in MSW. The estimated amount is approximately 0.02–0.82% of the total MSW (MONRE 2016). Hazardous wastes generated in daily life are batteries, accumulators, tube lights, mercury thermometers, needles and syringes by injecting drug users, etc. At present, hazardous waste in household waste is not collected and processed separately and mixed with MSW in landfill sites.

Table 5.13 summarizes the main sources of solid waste in Vietnam in 2003 and 2008 (MONRE 2011). In 2003, these sources were MSW from urban and rural areas, contributing approximately 12,800,000 tons/year, followed by industrial solid waste (2,638,400 tons/year) and medical solid waste (21,500 tons/year). Agricultural waste was approximately 64,560,000 tons/year, which was the highest source of solid waste in 2003. The amount of MSW increased dramatically from 2003 to 2008: from 12,800,000 to 21,888,000 tons. Among the 26 million tons of waste generated in 2008, MSW from households, restaurants, markets, and offices accounted for over 80% share.

Figure 5.69 shows a typical solid waste management workflow in Vietnam. Generation and storage of waste are mainly by households, markets, industrial and commercial institutes, street/public facilities, and healthcare institutions. In

Table 5.13 Solid waste generation in Vietnam in 2003 and 2008 (MONRE 2011) (Unit: t)

Name	Source	Type	2003	2008
Municipal waste	Residence, Commerce, Markets	Household waste, Plastic, paper, Glass	12,800,000	21,880,000
Industrial waste	Industries	Metals, wood, Fuel oil, Sludge, organic chemicals	2,638,400	4,786,000
Medical waste	Hospitals	Tissue, Samples blood, Syringes	21,500	179,000
Agricultural waste	Cultivation, Livestock	Plants	64,560,000	(No data)

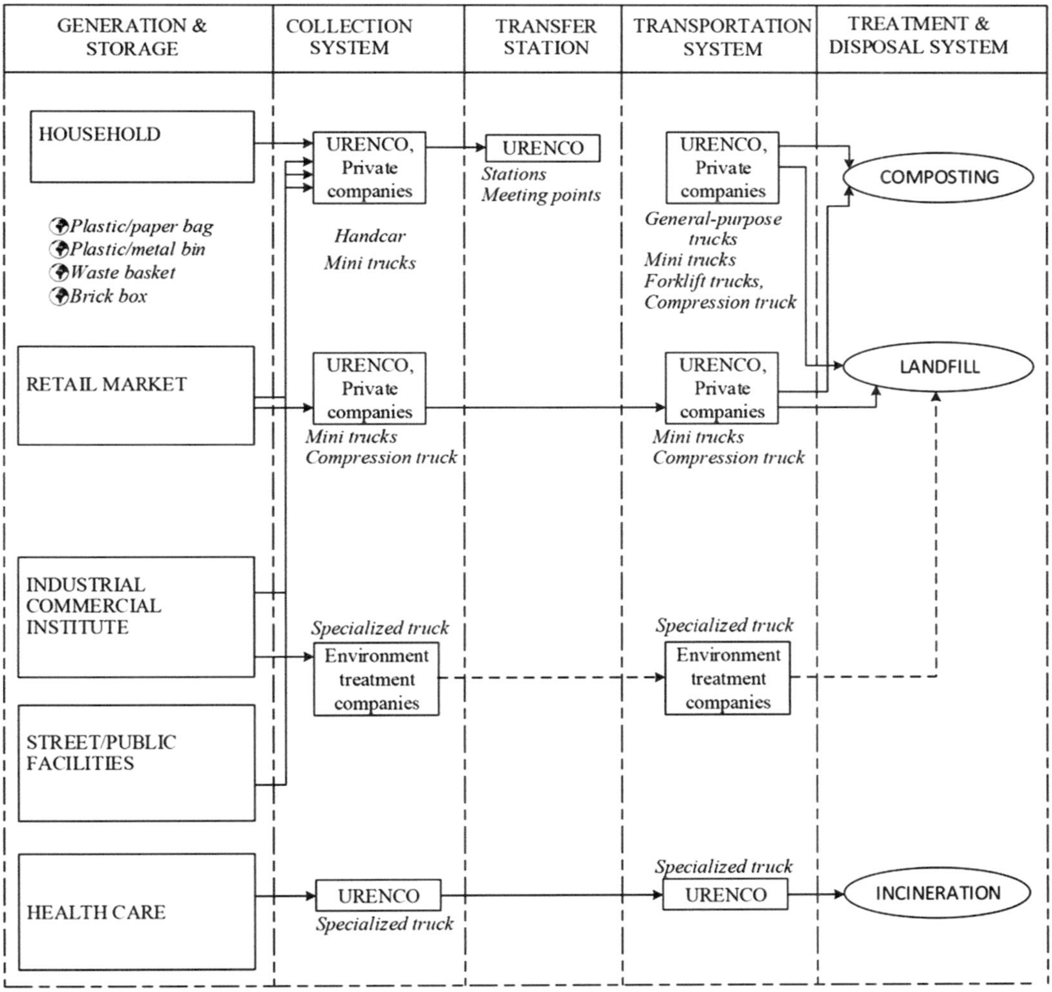

Fig. 5.69 Flow diagram of typical solid waste management in Vietnam. Modified from Thanh et al. (2011)

(a) (b)

Fig. 5.70 MSW collection point in Hanoi: **a** A local collection point in Dong Da district, Hanoi, **b** MSW collection by a garbage truck in Hanoi. Photos by Lan Chau Nguyen

the urban areas, the waste collection systems are well organized and functioning. However, in other regions, the collection is not well organized. In urban areas, the Urban Environment Company (URENCO) is responsible for the management of MSW. This company is responsible for the collection, transportation, and treatment of MSW from all sources (World Bank 2004). In addition, private recycling centers also participate in the management and treatment of MSW in big cities.

Fig. 5.71 Medical waste bin (yellow bin) in a hospital, Hanoi. Photo by Lan Chau Nguyen

5.9.1.3 Collection System

In the residential areas, MSW is collected from trash bins designed in different forms and with different sizes and colors. In general, these trash bins have volumes in the range of 5–100 L and are made of plastic, metal, or temporary containers (plastic bags, paper bags, or boxes). Waste in these containers is picked up by URENCO staff from each household or street (Fig. 5.70). Industrial waste is also collected by URENCO following a service contract with firms. Hazardous waste is collected and treated by waste treatment plants that are licensed by the local Department of National Resources and Environment (DONRE). Moreover, medical waste from hospitals is collected in special storage tanks (120 L) with a special seal and logo on top (Fig. 5.71).

5.9.1.4 Source Separation

In urban areas, the national average collection rate for MSW rose from 65% in 2000 to 85.3% in 2016, showing a significant improvement in waste collection in urban areas over the last decades (Table 5.14). However, there is still 15–16% of urban MSW that is not collected and is disposed of spontaneously into land, pond, lake, or openly burned. In suburban areas, the average MSW collection rate is approximately 60%. In rural areas, the collection rate is 40–55% on average, whereas it is only approximately 10% in remote areas (MONRE 2015). It must be noted that nearly 70% of Vietnam's

Table 5.14 Urban MSW collection rates of recent years (Unit: %)

Year	2000	2003	2004	2008	2010	2013–2014	2016
Urban MSW collection rate	65	71	72	80–82	83–85	84–85	85.3
Data source	World Bank (2004)		MONRE (2010)		MONRE (2011)	MONRE (2015)	MONRE (2016)

population lives in rural areas, and MSW in rural areas accounts for approximately 50% of national solid waste. Yet, the MSW collection rate in rural areas is remarkably low.

Generally, urban MSW collection rates are higher in larger and richer cities. By 2016, cities such as Ho Chi Minh City, Danang, and Hai Phong have achieved a 100% MSW collection rate; the rate is 98% in the urban area of Hanoi. MSW collection in urban class 2 and class 3 areas has significantly improved and is over 80–85%. However, there is little improvement in MSW collection in urban class 4 and 5 areas (MONRE 2016).

In both urban and rural areas, source separation of MSW is mostly not applied. However, the part of recyclable waste materials that is valuable or is easily sorted (such as paper, metal, and glass) is sold to junk buyers, who sell it to recycling facilities. The remaining waste, which includes putrescible waste, inert waste, remaining recyclable waste, and hazardous waste from households, is mixed together at the collection site. However, there is very little recyclable material left once the waste reaches the treatment and disposal sites, as scavengers (waste pickers) and garbage collectors already collect most recyclable material such as cans, PET bottles, scrap metal, plastic bags, wiring, and paper to sell to the recycling facilities. The flow of MSW and recyclable wastes is illustrated in Fig. 5.72.

While there have been projects launched for source separation of household waste in Hanoi (JICA funded, 2007–2010) and Ho Chi Minh City

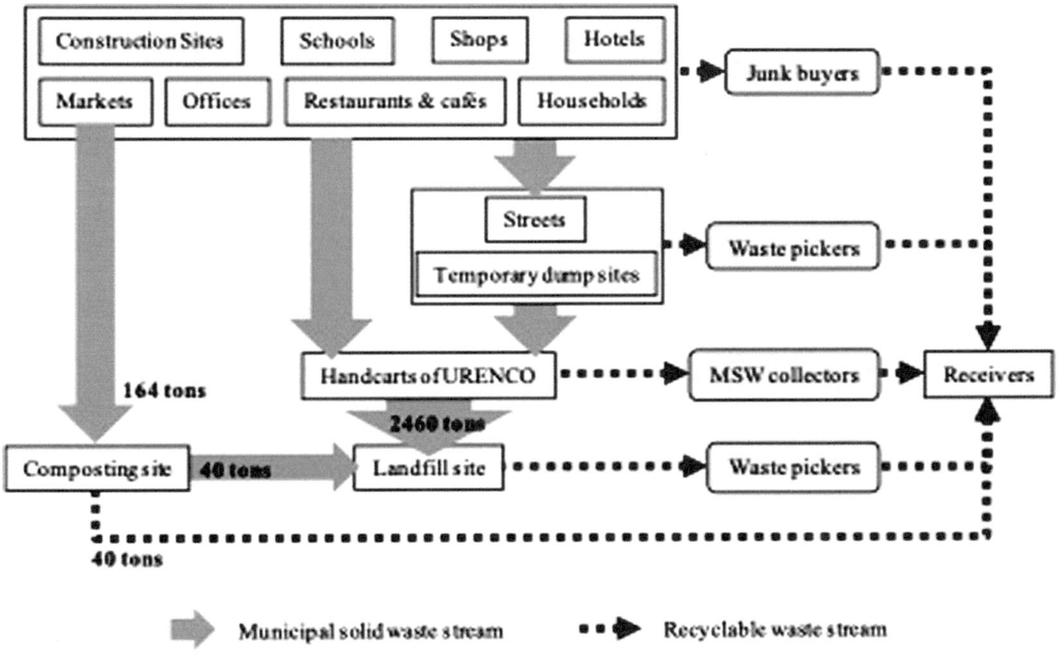

Fig. 5.72 MSW stream and recyclable waste stream in Hanoi

(EU funded, 1999–2007), they have achieved limited results. Since 2013, source separation trial projects for certain areas in HCM City brought remarkable results and, since 2017, the city started to expand the implementation of source separation in all 25 urban and suburban districts.

Hazardous waste from household waste is not separated at the source. It is thus not removed during the MSW stream and reaches final disposal. It thus causes problems in composting or further processing, leading to issues in the landfill site. In landfill sites, through several transformation processes, toxic elements can get dissolved in leachate and disperse into the air. In 2018, the revised National Strategy for Integrated Solid Waste Management has set a target of 85% for hazardous waste generated in households and individuals that must be collected, transported, and treated to meet the requirements of environmental protection.

5.9.2 Treatment of Waste

There are several methods of waste treatment in Vietnam, including landfilling, composting, and incineration. At the end of 2013, Vietnam had 458 landfills; however, only 26.5% were sanitary landfills and the others were of open dumping type. There were 26 solid waste treatment plants under operation in big cities such as Hanoi, Danang city, and Ho Chi Minh City. The types of waste treatment plants are shown in Table 5.15 (Nguyen and Vuong 2016).

Table 5.15 Methods applied in waste treatment plants in Vietnam (as of 2013) (Nguyen and Vuong 2016)

Treatment method	Number
Incineration	3
Incineration and composting	3
Composting and landfill	20
Total	26

5.9.2.1 Incineration

Incineration refers to the controlled and complete combustion process of MSW. The temperature inside incinerators typically ranges from 980 to 2,000 °C. One of the notable advantages of incineration is its ability to reduce the volume of combustible solid waste by 80–90%. However, incineration is not widely used for MSW treatment in Vietnam compared to other methods. Several incinerators have been set up in the North (e.g., Hanoi city, Thai Binh city, and Nam Dinh city) and the South (e.g., Binh Duong city) of Vietnam. In Hanoi, specific incinerator plants are operated by private companies such as Thanh Cong Cooperative, Thang Long Urban Environment JSC, and Thanh Quang Investment JSC, with a treatment capacity of less than 1,000 tons per day (Fig. 5.73).

Recently, an industrial waste incinerator with a capacity of 75 tons per day was constructed by URENCO within the Nam Son Complex. This project was funded by NEDO, a Japanese organization. As reported by URENCO, heat from combustion is recovered for generating

(a)

(b)

Fig. 5.73 Incinerator plant in Nam Son landfill: **a** Waste for incinerator plant, **b** incinerator plant. Photos by Lan Chau Nguyen

(a) (b) (c)

Fig. 5.74 Nam Son landfill site: **a** Nam Son landfill at active cell, **b** Nam Son landfill at closed cell with an HDPE cover on top, and **c** Soc Son waste power plant constructed in Nam Son landfill. Photos by Lan Chau Nguyen

electricity with a capacity of 1930 kW. Figure 5.74 shows the incinerator in the Nam Son landfill.

Currently, the MSW generation in Hanoi is approximately 6,400 tons/day. The MSW is treated mainly by landfilling in eight municipal centers and waste treatment sites at the district level. The largest landfilling site in Hanoi is the Nam Son landfill, which can receive and process approximately 3,800 to 4,200 tons daily. In this landfill, MSW is compacted and buried as shown in Fig. 5.74a, b. In large-scale facilities such as Nam Son landfill, Xuan Son landfill, and Kieu Ky center, sanitary landfill processing is applied for waste disposal, that is, leachate from these landfills is collected and treated. However, in other landfills, the leachates are not treated or simply handled, and dumped in a natural lagoon, which causes groundwater pollution.

The Soc Son Waste Power Plant project is located in the Nam Son Waste Treatment Complex and was approved by the Hanoi People's Committee in 2017. The investor is Thien Y, Energy Environment Joint Stock Company (Thien Y Company), and the contractor is MCC, China. It is expected that with 4,000 tons/day of waste incinerated, the amount of electricity generated would be 75 MW, in which the amount of electricity reused will be 15–20% and the remaining will be supplied to Vietnam Electricity Corporation. The plant has 5 incinerators with 3 generator sets (Fig. 5.74c). But, only incinerator No.3 with a capacity of handling 800 tons of MSW, has been operated since March 2022. Other incineration plants (No. 1, No. 2, No. 4, and No. 5) are expected to be completed in May 2022. Wet MSW will be incinerated into fly ash or bottom ash. The raw materials could be used as construction material.

5.9.2.2 Composting

Composting is potentially a very useful form of recycling organic waste to produce a clean soil conditioner and could help increase the recovery rate of recycled materials. MSW in Vietnam should have a high potential for composting because it contains a large proportion of food waste and other biodegradable waste. Increasing the proportion of MSW to be composted could contribute to the target of reducing direct landfill disposal of MSW to less than 30% by 2025, as per the Revised National Strategy on Integrated Solid Waste Management.

In fact, several centralized composting facilities have been operating in Vietnam over the last two decades. In 2011, MONRE reported 10 composting facilities with a capacity of approximately 200 tons/day, and 1 composting factory with a capacity of 600 tons/day (MONRE 2011). It was estimated that with the operation at full capacity of all existing composting plants, the composted quantity of MSW was still smaller than 2,500 tons/day which was less than 10% total quantity of MSW generated (MONRE 2011). In fact, these plants have not operated at enough designed capacity since the consumption of organic fertilizer has encountered many difficulties.

Thanks to the National Strategy and Policy on Integrated Solid Waste Management, the proportion of MSW composting is increasing. In 2019, there were 37 compost-producing plants in operation, concentrated mainly in urban areas (MONRE 2019). The total capacity of processing will be approximately 7,500 tons/day (the average capacity is 100–200 tons/day), and major processing technologies employed will be composting production (25 facilities) and combination with other technologies such as combustion or combination (JICA 2018).

Still, there are several drawbacks relating to composting activities in Vietnam. First, most MSW going to composting plants is not source separated; thus, the residue after the composting process to be disposed of in a landfill reaches approximately 35–40% of input waste. Moreover, the mixed MSW must be mechanically sorted in the facility, which makes the compost less pure and reduces its market. Composting operations in other Asian countries have failed when relying on mixed MSW as their main feedstock (Hoornweg et al. 2000). In addition to the poor quality of feedstock, inadequate attention to the biological process requirement also contributes to the poor quality of the fertilizer. On the contrary, the compost production market is not encouraged due to a lack of attention by the marketing organization to the distribution of the compost and consumer feedback. Without successful widespread composting, efforts to expand or sustain source separation will be less convincing, although it can still be targeted to recycle materials and promote general awareness. Domestic composting technologies, such as Seraphin, Ansinh-ASC (both approved by the Ministry of Construction), and Tam Sinh Nghia, can contribute to an increase in composting. However, these technologies are not sustainable when applied in practice, as they lead to secondary pollution. In order to increase the share of composting in overall MSW management, classification of MSW at source is mandatory; thus more attention to process control will ensure the quality of the fertilizer product and the development of a strong market.

5.9.2.3 Thermal Treatment and Energy Recovery of Municipal Solid Waste

Open Burning

Open burning of MSW is quite a common practice at unsanitary and uncontrolled landfill sites. It is estimated that approximately 50% of MSW in unsanitary landfill sites is burned at the sites (MONRE 2011). In such sites, after the garbage is filled, chemical spraying is periodically conducted for disinfection and deodorizing. The garbage is left there until is dry; it is then burned using oil. However, during the rainy season, garbage cannot be burned or burned thoroughly. Evidently, it also causes air pollution.

Small-Scale Incinerator

The National Environment Report 2015 stated there were approximately 50 MSW incinerators in 2015. Most of them are small scale (less than 500 kg/h) and two thirds of them are domestic installations. Burning of waste in a small-scale incinerator is an increasingly popular solution in many rural areas used to quickly solve the problem of MSW. However, there are concerns regarding gas emission control as, depending on technology and operation, it can lead to the emission of harmful gases into the environment, especially dioxins and furans.

Recovery of Waste as Energy Source

Since the National Strategy of Integrated Solid Waste Management was approved in 2009, shifting treatment/disposal methods from landfilling to other technology such as waste-to-energy (W2E)/electricity from waste (EfW) has been encouraged. The potential for energy recovery from MSW is large but most of the projects for MSW incineration with energy recovery have not started yet. Some of the projects include the Phuong Dinh Waste Incineration Plant (Hanoi), which started operation in 2015 with a capacity of 300 tons/day; and a project for waste incineration in Thai Binh

Province (capacity of 300 tons/day (5 MW/day)), i.e., the Nam Sown incineration plant.

5.9.2.4 Recycling of Municipal Solid Waste

The informal sector collects the majority of the recyclable and reusable waste in urban areas. A 1996 survey of the informal sector in Hanoi estimated that 18–22% of all waste was being diverted from the landfill by the informal sector (World Bank 2004). Recycling activities have been actively performed for many years but are mainly carried out by the informal sector. The most recycled material is plastic, followed by paper and metals (World Bank 2004). Approximately 90% of waste such as paper, plastic, metal, etc., is made into recycled products and approximately 10% is recycled (MONRE 2011). The recycling rate of organic waste and recycling of paper, plastic, glass, and metals such as iron, copper, lead, and aluminum account for only approximately 8–12% of collected urban MSW (MONRE 2011). In 2009, waste recycling was 348 tons per day in Hanoi, 86.5 tons/day in Hai Phong, 56.7 tons/day in Danang, and 16.9 tons/day in Hue (JICA 2011).

Recycling of waste paper, plastic, metal, etc., in Vietnam is mostly implemented by private and craft villages. Although this activity is spontaneous, it is very developed, bringing economic benefits to the people. However, this is usually carried out on a small scale, with outdated technology; thus, it produces more waste during the recycling process. Waste from the recycling facility is normally not treated and is instead discharged directly to the environment, causing severe environmental pollution. Alternatively, the waste, without treatment, is disposed of in a landfill site together with MSW.

Recycling villages only occupy 90 of the 450 trade villages. Certain recyclable trade villages are experiencing serious environmental problems, such as Minh Khai plastic recycling village (Hung Yen), Dong Mai lead recovery village (Hung Yen), and Duong O paper recycling (Bac Ninh). In addition, many recycling facilities are not located in the village but are located in urban areas. Ho Chi Minh City has 302 recycling facilities located in the city, mainly in District 11, of which 67 are plastic recycling facilities, 15 are glass recycling facilities, 9 are metal recycling facilities, 7 are establishments that recycle paper, and 2 are rubber recycling facilities (MONRE 2010).

Generally, the recycled products are of low quality and there is no transparent market for them. In addition to that, there is a lack of appropriate and effective policy framework and support for recycling industries.

5.9.2.5 Recycling of Construction and Demolition (C&D) Waste

The typical flow of C&D waste in Hanoi is presented in Fig. 5.75. This figure shows where waste is generated/stored, who collects that waste, where the waste is transferred, how the waste is transported, what technology is used, and finally how the waste is treated and disposed of (Thanh and Matsui 2011). There are three main sources of C&D waste generation: private, public, and industrial/commercial projects. The two main collection actors are URENCO and private companies. The transfer destinations include on-demand landfills, designated district landfills, and outer city landfills. C&D waste recycling is implemented only for concrete, which is used for aggregate applications such as road sub-bases and in the foundation of new buildings.

Normally, C&D waste is generated during building construction, demolition activities, renovation of buildings, road maintenance and improvement, and excavation. C&D waste in Vietnam in 2013 was composed of soil and gravel (36%), brick and masonry debris (31%), concrete (23%), steel (5%), and a small amount of plastic and wood (Nguyen et al. 2018; Tong et al. 2013).

C&D waste generation was approximately 1.9 million tons in 2009 (Tong et al. 2013). By 2015, the government of Vietnam had a plan to replace 23 old buildings (4 or 5 stories each on average), which equates to approximately 1 million m^2 floor per building that needs to be demolished and rebuilt in Hanoi (Lockrey et al. 2016; Tong et al. 2013). In addition, the Ministry of Transport and Hanoi City authorities have proposed eight rail routes to serve as the backbone

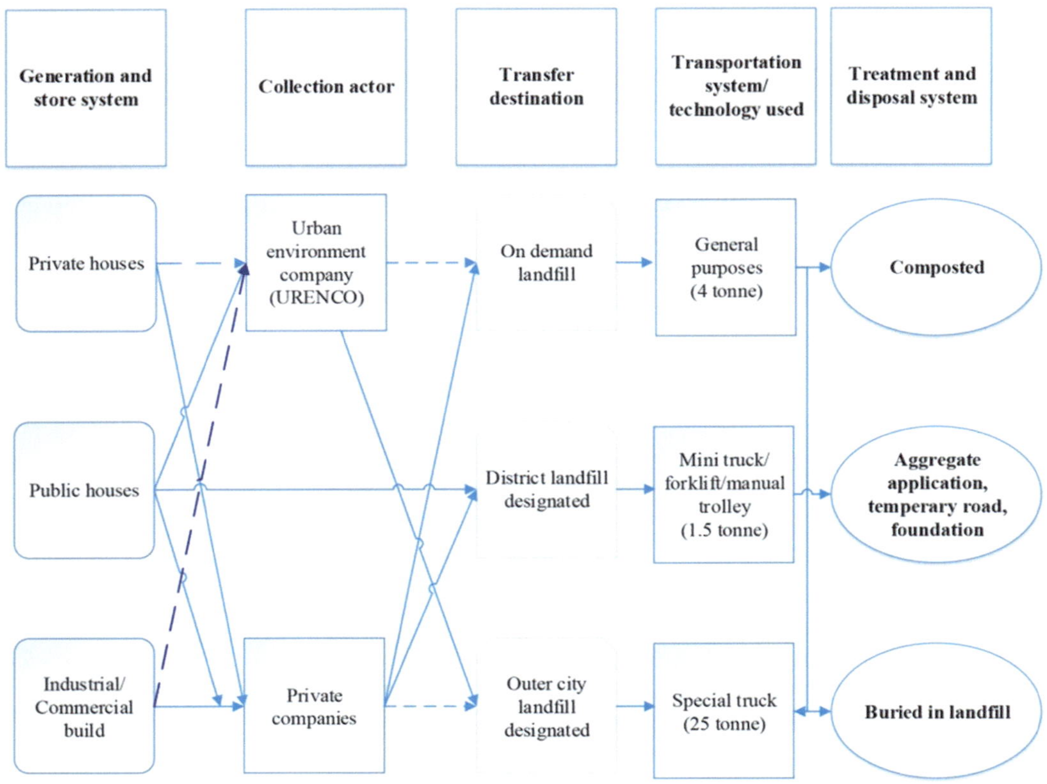

Fig. 5.75 Typical C&D waste concrete flow from waste generation to treatment and disposal (Redrawn from Lockrey et al. 2016)

of the public transportation system in Hanoi, in support of the government's policy to expand the role of mass transit in Hanoi. These routes are (1) Yen Vien–Ngoc Hoi (25 km), (2) Hanoi–Ha Dong (15 km), (3) Bac Co–Hanoi Station–Voi Phuc–Nhon (16 km), (4) Hanoi–Noi Bai (25 km), (5) Daewoo–Trung Kinh–Hoa Lac (32 km), (6) Giap Bat–South Thang Long (19 km), (7) Buoi–Dong Anh–Soc Son (24 km), and (8) Co Bi–Gia Lam–Kim No (26 km). Because of the difficulty in expanding the existing public transportation (buses) capacity of road networks, the development of an urban railway system is considered the most appropriate method for the mitigation of traffic congestion. However, underground and tunnel constructions in Hanoi are predicted to create a large amount of C&D waste. Therefore, utilization of C&D waste is an important and challenging issue for Hanoi.

Recently, the utilization of C&D waste for road foundations such as subgrade layers and pavement has been implemented in Vietnam (Tong et al. 2013). Tong et al. (2013) conducted extensive experiments to evaluate properties such as maximum dry density, optimum water content, compressive strength, and splitting strength for C&D waste mixture with and without cement in order to utilize C&D waste as a road foundation. The results showed that C&D waste from concrete and masonry can be utilized as a natural material for all subgrade, base, and sub-base layers of road foundations without cement mixing. In case of using recycled fine aggregate (RFA) for the lower base layer of high-performance pavement structures, it should be reinforced with 7.5% cement for RFA from concrete debris and 10.5% cement for RFA from masonry debris. In addition, when using RFA from concrete and masonry debris for the upper layer of high-performance pavement structures and bituminous pavements, it should be reinforced with 10.5 and 12.5% cement, respectively.

The Science and Technology Research Program (SATREPS) project has been developed by JICA and the Japan Science and Technology Organization (JST). In Vietnam, the project is chaired by the National University of Civil Engineering for a period of 5 years (2018–2023) ("Establishment of Environmentally Sound Management of Construction and Demolition Waste and Its Wise Utilization for Environmental Pollution Control and for New Recycled Construction Materials in Vietnam"). The objectives of the project are (1) to establish guidelines necessary for environmentally sound C&D waste management, (2) to develop quality standards for recycled materials produced from C&D waste, (3) to develop new technologies utilizing recycled materials produced from C&D Waste, and (4) to propose strategic business models designed to promote C&D waste recycling in Vietnam and examine their effectiveness and feasibility through on-site pilot projects (Nguyen et al. 2018).

A SATREPS's laboratory was established in Hanoi in 2019. Many laboratory tests have been carried out to evaluate the mechanical properties and environmental safety of concrete aggregates and recycled bricks for road foundations. In addition, a mill line was installed in early 2020 to perform quality control of recycled aggregate from the site and pilot research on the production of recycled materials processing is promoted. In addition, a Vietnamese standard "Technical requirements—Methods Test for recycled materials from construction materials for road foundation" is being built.

5.9.2.6 Recycling of Municipal Solid Waste Incinerator Ash (MSWIA)

Currently, MSWIA can be used as construction material in Vietnam (Nguyen et al. 2021). MSWIA samples were collected from the Dan Phuong incinerator plant and their physical, chemical, and mechanical properties such as grain size distribution, mineral composition, compaction, and shear strength were evaluated according to the ASTM standards. Experimental results revealed that the MSWIA samples are well-graded materials and MSWIA exhibits a good bearing capacity with its high frictional angle, and it would be suitable in geotechnical applications such as road embankments (Nguyen et al. 2021).

5.9.3 Waste Landfill in Vietnam

5.9.3.1 Landfilling and Open Dumping

The average number of landfill sites for one urban is 1, except for Hanoi and HCM City, each of which has approximately 4–5landfill/waste treatment facilities. Although the majority of MSW in Vietnam is directly disposed of by landfilling, the ratio of engineered and sanitary landfills was only 16% in 2004 (centralized landfill sites) and 26% in 2017, as indicated in Table 5.16, and 85% of urban areas do not have sanitary landfill sites (MONRE 2010). It was also estimated that approximately 50% of collected solid waste (not only MSW) is disposed of by sanitary landfills (MONRE 2010).

Figure 5.76 shows the composition of waste accepted in Nam Son landfill, which is the biggest landfill in Hanoi. This landfill accepts organics, paper, plastics, rubber, glass, metal, soil, sand, etc. The main composition of this landfill is organics, which account for 50–60% of the total volume. Another research reports that MSW in Hanoi has a water content ranging from 40 to 70% and wet densities varying from 0.39 to 0.5 t/m^3 depending mainly on the season (Nguyen et al. 2013).

5.9.3.2 Typical Structure and Design of Waste Landfills

In Vietnam, the majority of waste is not pretreated and is open-dumped, which seriously affects the surrounding environment (MONRE 2011). Pollution of the environment due to leachate leakage from waste landfills is a well-known and widespread problem. Bottom liners are essential for controlling the intrusion of

Table 5.16 Number of landfills with different types

	Total	Engineered and sanitary landfill		Open and control dumps (unsanitary landfill)		Year of data	Reference
		Number	Percentage	Number	Percentage		
Number of cities and Province having landfill sites	61	12	20%	49	80%	2004	World Bank (2004)
Centralized Landfill sites	98	16	16%	82	84%	2004, 2011	World Bank (2004), MONRE (2010)
Number of landfill sites (>1 ha)	458	121	26%	337	74%	2017	MONRE
Landfill area (ha)	1883.5	997.3	54%	836.2	46%	2017	MONRE

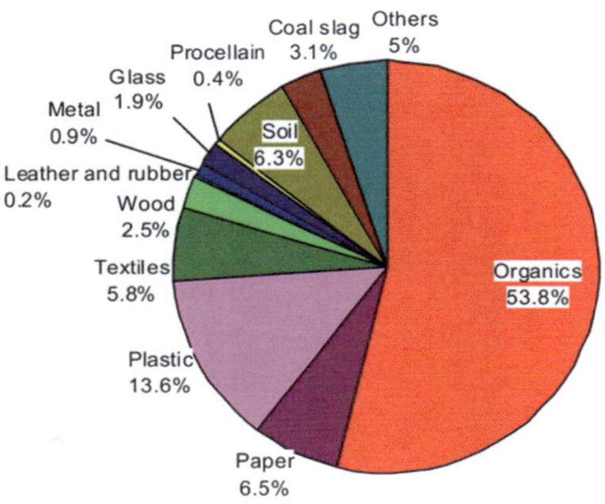

Fig. 5.76 Composition of waste accepted in Nam Son landfill (Hanoi) (Nguyen et al. 2013)

groundwater into landfill facilities and to prevent leached contaminants from being released to the surrounding environment. The regulations on the specifications of bottom liners in waste landfills are, therefore, recently being strengthened significantly in many countries.

There are two technical standards for landfill design and construction in Vietnam including:

- TCVN 6696 (2009): Solid wastes – Sanitary landfill General requirements for environmental protection
- TCXD 261 (2001): Solid waste landfills Design standard.

According to TCXD 261, liners must be installed at both the bottom and slope of waste landfills, and they must have hydraulic conductivity lower than 1×10^{-9} m/s and thickness greater than 60 cm. A typical cross-section of waste landfills in Vietnam is shown in Fig. 5.77. Hydraulic conductivity is a property of soil that describes the ease with which a fluid can permeate through the soil media. Therefore, lower hydraulic conductivity means higher barrier performance for barrier materials.

5.9.4 New Challenges and Perspectives in Landfilling in Vietnam

By applying various new technologies and systems being used in other countries, waste disposal and landfilling can be expected to be sophisticated. In this section, some possible new challenges and perspectives in Vietnam are introduced.

5.9.4.1 Fukuoka Method (Semi-aerobic Landfill)

The Fukuoka method is a semi-aerobic system developed more than 40 years ago at Fukuoka University, Japan. Generally, this method can be explained as a system where the leachate and gas are continuously removed from the waste mass using leachate collection and gas venting

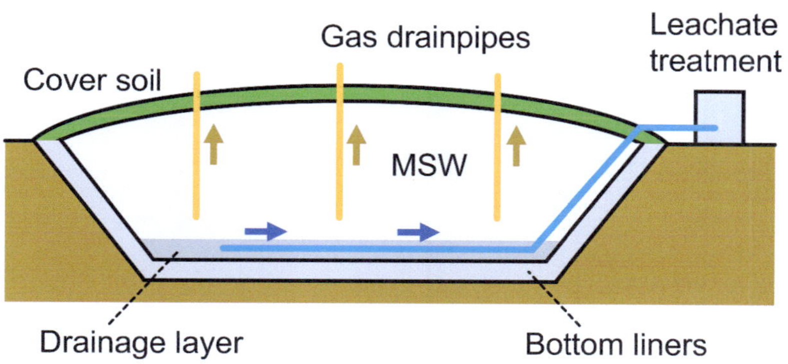

Fig. 5.77 Cross-section of a typical landfill in Vietnam

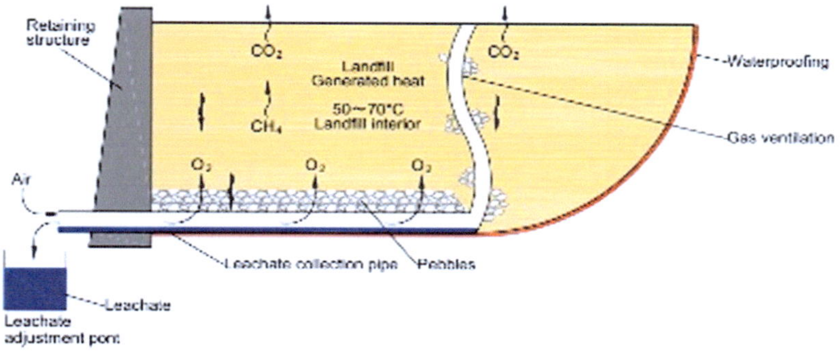

Fig. 5.78 Concept of Fukuoka method, semi-aerobic landfill (Redrawn from Chong et al. 2005)

systems (Fig. 5.78). The Fukuoka method of semi-aerobic landfill can be applied in Vietnam because the method can improve the waste stabilization process and improve the leachate quality due to the enhancement of the micro-organism activities in the waste body (Chong et al. 2005).

5.9.4.2 Bioreactor Landfill

Landfilling is the primary disposal method for MSW in many countries including Vietnam (Nguyen 2004). The predominant forms of waste disposal facilities in Vietnam are open and controlled dumping sites. Most of them do not have a necessary liner system or a cover system (Nguyen 2004); thus they cannot ensure landfill stability and pose various environmental problems such as potential groundwater contamination and emission of uncontrolled landfill gases. Recently, the use of bioreactor landfill has increased due to its many advantages such as an increase in rates of gas production, increase of waste stabilization, increase in rate of landfill settlement, and decrease in leachate disposal and post-closure cost (Hettiarachchi et al. 2009; Reinhart et al. 2002). In contrast to the traditional landfill approach, a bioreactor landfill rapidly transforms and degrades the organic components of the waste mass. Figure 5.79 shows a schematic view of a bioreactor landfill. A bioreactor landfill is one in which liquids are added to help bacteria break down the waste. By adding liquid and air to the waste body, an increase in waste degradation and stabilization is accomplished owning to enhanced microbial processes. In an aerobic bioreactor landfill, leachate is collected from the bottom and re-circulated into the landfill in a controlled manner. Air is injected into the waste mass using vertical or

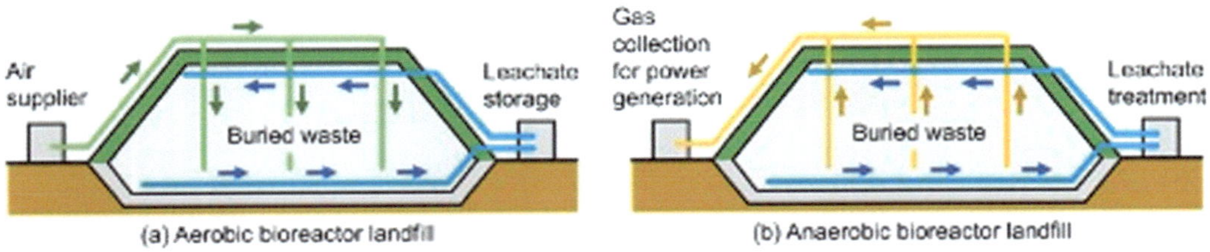

Fig. 5.79 Concept of bioreactor landfills

horizontal wells to promote aerobic reactions and accelerate waste stabilization. In an anaerobic bioreactor landfill, biodegradation occurs in the absence of oxygen and produces landfill gas. Landfill gas, primarily methane, can be captured to minimize greenhouse gas emissions and can be used for energy projects. Therefore, bioreactor landfills can be expected to serve as an alternative to address landfill management problems in Vietnam (Nguyen et al. 2013).

5.9.4.3 Coastal Landfill

In Japan, due to inland limitation, the locations of landfills have been diversified to coastal areas (Fig. 5.80). The role of coastal landfill sites has been increasing steadily (Tanaka et al. 2005). In Japan, coastal landfill sites account for only 2% of the total number of sites, but 28% of landfill volume due to their large capacity. In coastal landfill sites, natural clay deposits of more than 5 m in thickness and a hydraulic conductivity not exceeding 1×10^{-7} m/s play the role of the bottom barrier. In the future, coastal landfills can serve as a necessary solution for coastal cities of Vietnam such as Hai Phong city, Quang Ninh city, and Danang city.

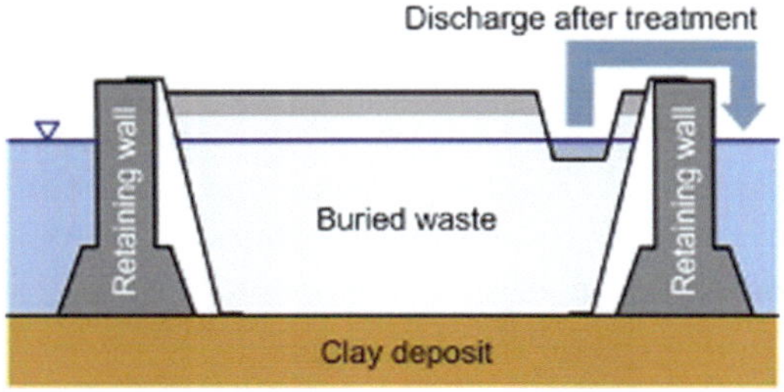

Fig. 5.80 Concept of a coastal landfill

5.9.5 Summary

The current status of MSW in Vietnam shows that MSW generation has been increasing over the years and is forecasted to drastically increase in the future. However, the targets for the year 2015 in the National Strategy on Municipal Solid Waste Management 2009 were not met, except for the collection rate in urban areas. Moreover, the management of MSW in rural areas is still poor and MSW source separation is mostly not implemented across the country. In addition, landfilling is the major method of treatment and disposal, in which very few landfill sites meet the environmental standard. The landfill site is a significant source of environmental pollution in urban areas. A shift of waste treatment and disposal from direct landfilling to other environment-friendly technologies can have some positive results, as shown by increasing rates of composting, recycling, and waste-to-energy projects; however, such efforts are still very limited. Biogas recovery technology, which can be a suitable potential candidate for MSW characterized by high organic content, has not been considered. In order to meet the new targets of the revised national strategy in 2025, MSW treatment capacity and investment should be increased and public awareness of the 3Rs should be enhanced.

Column 8

Motorbike and Hanoi

When I was a new hire at work, a Swiss economic professor told me that the reason he

Fig. 5.81 Transportation in Hanoi. Minh Khai street, photo by Phuong Nam Nguyen, Sept 4, 2019

came to Vietnam was to see again a peaceful place with bicycles. This reminded me of my childhood when my parents took me along the city on a bicycle, passing the Red river to our hometown. We traversed along almost the whole city on a bicycle. Then came the time when I bought my own motorbike. At that point in time, almost every adult in the city had a motorbike. My friends and I used to travel long distances, even to other provinces, on motorbike for leisure. Driving a motorbike … feeling the toughness of the wind on the face while looking at the beautiful landscape was a sense of our youth.

Hanoi just operated its first line of sky trains in the city in November 2011. While we do have buses for public transportation, private vehicles still dominate the roads. More than 80% of vehicles are motorbikes (Fig. 5.81). The average traffic growth is more than 6%/year, while the increase in the land area for transport is smaller than 1%/year. Besides the high volume of vehicles, busy streets with stop-and-start traffic as well as traffic jams worsened the magnitude of air pollution. This is in spite of the emission standards as well as fuel quality standards being continuously updated over the past two decades. While this has had some effect in the form of lower growth in the emission of volatile organic compounds, CO, and NO_x compared to the growth in private vehicles, it cannot be denied that emission from the transport sector is still high and more actions need to be taken. Discussion about whether motorbikes should be phased out soon is a hot topic in public. However, the decision is not easy as without motorbikes, some families may simply lose their livelihoods, while for others, no motorbike means less flexibility and even less freedom.

Currently, several regulations are being developed to control air pollution and reduce the emission of greenhouse gases from transportation in Vietnam. With these efforts, Hanoi may no longer be the city with the highest rate of motorbikes in the world. I personally hope that the spirit of freedom that motorbikes once brought us does not die but converts into other environment-friendly vehicles such as bicycles or skateboards.

Acknowledgements Section 5.4: This research was funded by the GSGES seeds research (Kyoto University, Japan) and the Vietnam Ministry of Education and Training under project number B2021-DNA-14. We are sincerely thankful for the support.

References

ACCCRN (Asian Cities Climate Change Resilience Network) (2016) Resilience Strategy for Da Nang City, Vietnam. 100 Resilient Cities Project Report. Acdcessed 12 Dec 2021 from https://resilientcitiesnetwork.org/downloadable_resources/Network/Da-Nang-Resilience-Strategy-English.pdf

Adamson M, Ejdeholm LM (1999) At the Heart of Hue: Assessment of the Public Spaces along the Son Huong [Academic Dissertation]. Department of Architecture, Lund Institute

Almeida MC, Butler D, Friedler E (1999) At-source domestic wastewater quality. Urban Water 1(1):49–55. https://doi.org/10.1016/S1462-0758(99)00008-4

Anh TNQ, Harada H, Fujii S, Anh PN, Lieu PK, Tanaka S (2016a) Preliminary analysis of phosphorus flow in Hue Citadel. Water Sci Technol 73(1):69–77. https://doi.org/10.2166/wst.2015.463

Anh TNQ, Harada H, Fujii S, Lieu PK, Hieu DV, Tanaka S (2016b) Sewer discharge characteristics and water balance in dry season in Hue, Vietnam. J Sci Technol 54(2A):265–272

Balasubramanian R, Gao X, Hatakeyama S, Hwang J, Tsai C-J (2017) Overview of the Special Issue "PM2.5 in Asia" for 2015 Asian Aerosol Conference. Aerosol Air Qual Res 17:351–355. https://doi.org/10.4209/aaqr.2017.01.0042

Balica SF (2012) Applying the flood vulnerability index as a knowledge base for flood risk assessment [Doctoral thesis]. Delft University of Technology

Brandes K, Schoebitz L, Nguyen VA, Strande L (2016) SFD promotion initiative: Hanoi, Vietnam: final report. Eawag/Sandec

Bui HN (2015) Current situation and policies on wastewater management in Viet Nam, Retrieved May 1, 2024 from https://www.pseau.org/outils/ouvrages/ministry_of_natural_resources_and_environment_current_situation_and_policies_on_wastewater_management_in_vietnam_2017.pdf

Butler D, Friedler E, Gatt K (1995) Characterising the quantity and quality of domestic wastewater inflows. Water Sci Technol 31(7):13–24. https://doi.org/10.1016/0273-1223(95)00318-H

Chen F, Zhang X, Zhu X, Zhang H, Gao J, Hopke PK (2017) Chemical characteristics of PM2.5 during a 2016 winter haze episode in Shijiazhuang, China. Aerosol Air Qual Res 17:368–380. https://doi.org/10.4209/aaqr.2016.06.0274

Chong TL, Matsufuji Y, Hassan MN (2005) Implementation of the semi-aerobic landfill system (Fukuoka method) in developing countries: a Malaysia cost analysis. Waste Manag 25(7):702–711. https://doi.org/10.1016/j.wasman.2005.01.008

Chuang M-T, Chou CC-K, Lin N-H, Hsiao T-C, Lin T-H, Fu JS, Kumar S, Lu Y-R, Yang T-Y (2017) A simulation study on PM 2.5 sources and meteorological characteristics at the northern tip of Taiwan in the early stage of the Asian haze period. Aerosol Air Qual Res 17:3166–3178. https://doi.org/10.4209/aaqr.2017.05.0185

Cohen DD, Stelcer E, Santos FL, Prior M, Thompson C, Pabroa PCB (2009) Fingerprinting and source apportionment of fine particle pollution in Manila by IBA and PMF techniques: A 7-year study. Xray Spectrom 38:18–25. https://doi.org/10.1002/xrs.1112

Cohen DD, Crawford J, Stelcer E, Bac VT (2010) Characterization and source apportionment of fine particulate sources at Hanoi from 2001 to 2008. Atmos Environ 44(3):320–328. https://doi.org/10.1016/j.atmosenv.2009.10.037

Danang GSO (Danang City Statistics Office) (2020) Da Nang City Statistical Yearbook 2019. Statistical Publishing House

Danang International Airport (2021) Welcome to Danang International Terminal. Accessed 10 Oct 10 2021 from https://danangairportterminal.vn

Danang People's Committee (2019) Danang Aims to Drive Growth with Smart City Plan. Accessed 24 Oct 24 2019 from https://mic.gov.vn

Danang People's Committee (2020) Report on the Natural Disasters in Da Nang City from 1998 to 2020

Dao CA, Con PM, Khai NM (2010) Characteristic of urban wastewater in Hanoi City—nutritive value and potential risk in using for agriculture. VNU J Sci: Earth Environ Sci 26(1):42–47. https://js.vnu.edu.vn/EES/article/view/1773

Derrible S, Truong TTM, Pham HT, Nguyen QH (2021) Learning from Hanoi to imagine the future of water distribution. NPJ Urban Sustain 1:9. https://doi.org/10.1038/s42949-020-00001-x

Dessai S, Hulme M (2004) Does climate adaptation policy need probabilities? Climate Policy 4(2):107–128. https://doi.org/10.1080/14693062.2004.9685515

Dey S, Di Girolamo L, van Donkelaar A, Tripathi SN, Gupta T, Mohan M (2012) Variability of outdoor fine particulate (PM2.5) concentration in the Indian Subcontinent: a remote sensing approach. Remote Sens Environ 127:153–161. https://doi.org/10.1016/j.rse.2012.08.021

Do TVH, Nagasawa R (2014) Potential flood hazard assessment by integration of ALOS PALSAR and ASTER GDEM: a case study for the Hoa Chau commune, Hoa Vang District, in Central Vietnam. J Appl Remote Sens 8:1–12

Do DH, Van Langenhove H, Walgraeve C, Hayleeyesus SF, De Wispelaere P, Dewulf J, Demeestere K (2013) Volatile organic compounds in an urban environment: a comparison among Belgium, Vietnam and Ethiopia. Int J Environ Anal Chem 93(3):298–314. https://doi.org/10.1080/03067319.2011.620708

Do DH, Walgraeve C, Amare AN, Barai KR, Parao AE, Demeestere K, van Langenhove H (2015) Airborne volatile organic compounds in urban and industrial locations in four developing countries. Atmos Environ 119:330–338. https://doi.org/10.1016/j.atmosenv.2015.08.065

Eckstein D, Hutfils M-L, Winges M (2020) Global climate risk index 2019. German Watch

Feyissa G, Zeleke G, Gebremariam E, Bewket W (2018) GIS based Quantification and mapping of climate change vulnerability hotspots in Addis Ababa. Geoenviron Disasters 5:14. https://doi.org/10.1186/s40677-018-0106-4

Friedler E, Butler D, Brown DM (1996) Domestic WC usage patterns. Build Environ 31(4):385–392. https://doi.org/10.1016/0360-1323(96)00008-X

General Statistics Office (n.d.) Number of administrative units as of 31 December by province. Accessed 12 Aug 2022 from https://www.gso.gov.vn/en/px-web/?pxid=E0101&theme=Administrative%20Unit%20and%20Climate

General Statistics Office (2021) Result of the Viet Nam Household Living Standards Survey 2020 (Issue December). Statistical Publishing House

Hai CD, Kim Oanh NT (2013) Effects of local, regional meteorology and emission sources on mass and compositions of particulate matter in Hanoi. Atmos Environ 78:105–112. https://doi.org/10.1016/j.atmosenv.2012.05.006

Hanoi Statistical Office (2011) Hanoi Statistical Yearbook 2011. HSO

Harada H, Dong NT, Matsui S (2008) A measure for provisional-and-urgent sanitary improvement in developing countries: septic-tank performance improvement. Water Sci Technol 58(6):1305–1311. https://doi.org/10.2166/wst.2008.715

Harada H, Strande L, Fujii S (2016) Challenges and opportunities of faecal sludge management for global sanitation. In: Katsumi T, Hashimoto S (eds) Towards future earth: challenges and progress of global environmental studies. Kaisei Publishing, Tokyo, pp 81–100

HEPCO (Hue Urban Environment and Public Works State Limited Company) (2013). Drainage system in Hue city [unpublished data]. Thua Thien Hue Province, Vietnam

Hettiarachchi H, Meegoda J, Hettiaratchi P (2009) Effects of gas and moisture on modeling of bioreactor landfill settlement. Waste Manag 29(3):1018–1025. https://doi.org/10.1016/j.wasman.2008.08.018

Hien PD, Loc PD, Dao NV (2011) Air pollution episodes associated with East Asian winter monsoons. Sci Total Environ 409:5063–5068. https://doi.org/10.1016/j.scitotenv.2011.08.049

Hien PD, Hangartner M, Fabian S, Tan PM (2014) Concentrations of NO_2, SO_2, and benzene across Hanoi measured by passive diffusion samplers. Atmos Environ 88:66–73. https://doi.org/10.1016/j.atmosenv.2014.01.036

Hien PD, Bac VT, Tham HC, Nhan DD, Vinh LD (2002) Influence of meteorological conditions on PM2.5 and PM2.5−10 concentrations during the monsoon season in Hanoi, Vietnam. Atmos Environ 36(21):3473–3484. https://doi.org/10.1016/S1352-2310(02)00295-9

Hien PD, Bac VT, Thinh NTH, Anh HL, Thang DD, Nghia NT (2021. A comparison study of chemical compositions and sources of PM1.0 and PM2.5 in Hanoi. Aerosol Air Qual Res 21:210056. https://doi.org/10.4209/aaqr.210056

Hinkel J (2011) Indicators of vulnerability and adaptive capacity: towards a clarification of the science policy interface. Glob Environ Chang 21(1):198–208. https://doi.org/10.1016/j.gloenvcha.2010.08.002

Hoang TTP, Kato T (2016) Measuring the effect of environmental education for sustainable development at elementary schools: a case study in Da Nang City, Vietnam. Sustain Environ Res 26(6):274–286. https://doi.org/10.1016/j.serj.2016.08.005

Hoang TTH, Van DA, Nguyen TTH (2017) Driving variables for eutrophication in lakes of Hanoi by data-driven technique. Water Environ J 31(2):176–183. https://doi.org/10.1111/wej.12231

Hoàng TT (1999) Tâm thức người Việt và nhà vườn xứ Huế (Belief of Vietnamese people and Hue garden house) [Master's thesis]. University of Architecture Ho Chi Minh City

Hoornweg D, Thomas L, Otten L (2000) Composting and its applicability in developing countries (Urban Waste Management Working Paper Series, 8). The World Bank, Washington, DC. Accessed 6 Oct 2021 from https://documents.worldbank.org/en/publication/documents-reports/documentdetail/483421468740129529/composting-and-its-applicability-in-developing-countries

Hopke PK, Cohen DD, Begum BA, Biswas SK, Ni B, Pandit GG, Santoso M, Chung Y-S, Davy P, Markwitz A, Waheed S, Siddique N, Santos FL, Pabroa PCB, Seneviratne MCS, Wimolwattanapun W, Bunprapob S, Vuong TB, Duy Hien P, Markowicz A (2008) Urban air quality in the Asian region. Sci Total Environ 404(1):103–112. https://doi.org/10.1016/j.scitotenv.2008.05.039

Huynh LT, Harada H, Fujii S, Nguyen LPH, Hoang THT, Huynh HT (2021) Greenhouse gas emissions from blackwater septic systems. Environ Sci Technol 55(2):1209–1217. https://doi.org/10.1021/acs.est.0c03418

Hy NT, Bang D, Dang NV (1999) Vietnam urban areas during Nguyen dynasty. Thuan Hoa Publisher, Thua Thien Hue

Imamura K, Maeda Y, Takenaka N, Pham HV, Tran TNL (2007) Investigation on air pollution in Vietnam: volatile organic compounds in Hanoi and Ho Chi Minh. In: Annual Report of FY 2006, The Core University Program between Japan Society for the Promotion of Science (JSPS) and Vietnamese Academy of Science and Technology (VAST), pp 47–54

IPCC (2014) Climate Change 2014. The Fifth Assessment Report (AR5) of the Intergovernmental Panel on Climate Change on Climate, Copenhagen

ISET (Institute for Social and Environmental Transition), Danang City Government (2009) Report on Climate Change Impact and Vulnerability Assessment for Da Nang City. Project on the "Asian Cities Network resilient to climate change (ACCCRN)", 100 Resilient Cities—Pioneered by The Rockefeller Foundation

Japan Sewage Works Association (2013) Design standard for municipal wastewater treatment plants, 2nd edn. Tokyo

JBIC (Japan Bank for International Cooperation) (1999) Urban Development and Housing Sector in Viet Nam. Japan Bank for International Cooperation, Tokyo

JICA (2011) The study on urban management in Vietnam

JICA (2018). The Project for Capacity Development on Integrated Management of Municipal Solid Waste in Vietnam Final Report. Accessed 10 Sept 2021 from https://openjicareport.jica.go.jp/618/618/618_107_12304358.html

Kieu TK, Singer J, Gannon TJ (2016) Education for sustainable development in Vietnam: lessons learned from teacher education. Int J Sustain High Educ 17(6):853–874. https://doi.org/10.1108/IJSHE-05-2015-0098

Kim Oanh NT, Upadhyay N, Zhuang YH, Hao ZP, Murthy DVS, Lestari P, Villarin JT, Chengchua K, Co HX, Dung NT, Lindgren ES (2006) Particulate air pollution in six Asian cities: Spatial and temporal distributions, and associated sources. Atmos Environ 40:3367–3380. https://doi.org/10.1016/j.atmosenv.2006.01.050

Kumokawa S, Shirakawa Y, Flamand P (2019) Characteristics of domestic wastewater and estimation of required johkasou capacity for buildings in Japan. Water Practice Technol 14(3):738–747. https://doi.org/10.2166/wpt.2019.061

Le HN, Ly BT, Thai KP, Pham GH, Ngo IH, Do VN, Le TT, Vuong NL, Dang SH, Nguyen TYL, Pham HD, Vu VT (2021) Assessing the impact of traffic emissions on fine particulate matter and carbon monoxide levels in Hanoi through COVID-19 social distancing periods. Aerosol Air Qual Res 21:210081. https://doi.org/10.4209/aaqr.210081

Lockrey S, Nguyen H, Crossin E, Verghese K (2016) Recycling the construction and demolition waste in Vietnam: opportunities and challenges in practice. J Clean Prod 133:757–766. https://doi.org/10.1016/j.jclepro.2016.05.175

Ly BT, Matsumi Y, Nakayama T, Sakamoto Y, Kajii Y, Nghiem TD (2018) Characterizing PM2.5 in Hanoi with new high temporal resolution sensor. Aerosol Air Qual Res 18:2487–2497. https://doi.org/10.4209/aaqr.2017.10.0435

Ly BT, Kajii Y, Nguyen TYL, Shoji K, Van DA, Do TNN, Nghiem TD, Sakamoto Y (2020) Characteristics of roadside volatile organic compounds in an urban area dominated by gasoline vehicles, a case study in Hanoi. Chemosphere 254:126749. https://doi.org/10.1016/j.chemosphere.2020.126749

Ly BT, Matsumi Y, Vu VT, Sekiguchi K, Nguyen TT, Pham CT, Nghiem TD, Ngo IH, Kurotsuchi Y, Nguyen TH, Nakayama T (2021) The effects of meteorological conditions and long-range transport on PM2.5 levels in Hanoi revealed from multi-site measurement using compact sensors and machine learning approach. J Aerosol Sci 152:105716. https://doi.org/10.1016/j.jaerosci.2020.105716

McCarthy JJ et al (eds) (2001) Climate change 2001: Impacts, adaptation, and vulnerability—contribution of working Group II to the third assessment report of the intergovernmental panel on climate change. Cambridge University Press, Cambridge

MONRE (Ministry of Natural Resources and Environment) [Bộ Tài Nguyên Và Môi Trường] (2010) Báo cáo môi trường quốc gia 2010—Tổng quan môi trường Việt nam (National environment report 2010—overview of Vietnam environment)

MONRE (2011) Báo cáo môi trường quốc gia 2010—Chất thải rắn (National state of environment report 2011—solid waste)

MONRE (2015) Báo cáo môi trường quốc gia giai đoạn 2011–2015 (National state of environment report period 2011–2015)

MONRE (2016) Báo cáo môi trường quốc gia 2016—Môi trường đô thị (National state of environment report 2016—urban environment). Accessed 13 Dec 2022 from http://vea.gov.vn/Documents/bao%20cao%20moi%20truong%20quoc%20gia/BCMTQG2016_BTNMT-nen.pdf?csf=1&e=fli306

MONRE (2018) Báo cáo hiện trạng mô! trường quốc gia năm 2018—Môi trường nước các lưu vực song (National state of environment report—Water environment of river basins. Accessed 13 Dec 2022 from http://dwrm.gov.vn/uploads/download/files/bao-cao-htmt-quoc-gia-2018-moi-truong-nuoc-cac-lvs_signed.pdf

MONRE (2019) Báo cáo môi trường quốc gia 2019-Quản lý chất thải rắn sinh hoạt (National state of environment report 2019—waste management)

MONRE & UNEP-RRC.AP (2010) Viet Nam national environmental performance assessment report. GMS Environment Operations Center

Naing W, Harada H, Fujii S, Hmwe CSS (2020) Informal emptying business in Mandalay: its reasons and financial impacts. Environ Manag 65:122–130. https://doi.org/10.1007/s00267-019-01228-w

Newfoundland Labrador (n.d.) Drinking water quality index. Accessed 13 April 13 2023 from https://www.gov.nl.ca/ecc/waterres/quality/drinkingwater/dwqi/

Nga TTV (2008) Arsenic contamination in Hanoi City, Vietnam. In: Takizawa S (ed) Groundwater management in Asian cities: technology and policy for sustainability. Springer Japan, pp 273–299

Nghiem TD, Nguyen TTT, Nguyen TTH, Ly BT, Sekiguchi K, Yamaguchi R, Pham CT, Ho QB, Nguyen MT, Duong TN (2020) Chemical characterization and source apportionment of ambient nanoparticles: a case study in Hanoi, Vietnam. Environ Sci Pollut Res 27:30661–30672. https://doi.org/10.1007/s11356-020-09417-5

Nghiem TD, Mac DH, Nguyen AD, & Lê NC (2021) An integrated approach for analyzing air quality monitoring data: a case study in Hanoi, Vietnam. Air Qual Atmos Health 14:7–18. https://doi.org/10.1007/s11869-020-00907-6

Ngoc DTN, Duyen LT, Sakamoto Y, Kajii Y, Ly BT (2018) Source apportionment of VOCs by receptor modelling in an urban site in Ha Noi. Vietnam J Sci Technol 56:88–95

Ngoc BAP, Delbarre H, Deboudt K, Dieudonné E, Tran DN, Thanh SL, Pelon J, Ravetta F (2021) Key factors explaining severe air pollution episodes in Hanoi during 2019 winter season. Atmos Pollut Res 12:101068. https://doi.org/10.1016/j.apr.2021.101068

Nguyen TY (2004) Bio-waste reuse program: the contribution of a positioning campaign: the case of Hanoi. Wageningen University, Vietnam

Nguyen LH, Tran VNT (2020) Evaluating future water quality of urban rivers in Hanoi under effect of urbanization and climate change—the application of WEAP model for Cau Bay River. Vietnam J Sci Technol 58(3A):195–202

Nguyen NT, Kobayashi H et al (2015) Transformation and conservation of traditional garden houses in Hue Citadel. Thuan Hoa Publishers, Hue, Vietnam

Nguyen KA, Liou Y-A, Li M-H, Tran TA (2016) Zoning eco-environmental vulnerability for environmental management and protection. Ecol Ind 69(2016):100–117. https://doi.org/10.1016/j.ecolind.2016.03.026

Nguyen LC, Nguyen HH, Nguyen AT, Vu AT, Ngo HD (2021) Static and cyclic properties of municipal solid bottom ash in Vietnam. Int J GEOMATE 20(80):152–158

Nguyễn HT (2008) Nhà vườn xứ Huế (Garden houses in Hue). Văn Nghệ Publishers, Ho Chi Minh

Nguyen NT, Kobayashi H (2014) Spatial transformation of traditional garden houses in Hue citadel, Vietnam. In: Proceedings of international conference on vernacular heritage, sustainability and earthen architecture VerSus2014, pp 543–549

Nguyen ML Vuong CC (2016) Solid waste typology and management in Hanoi. IMV report. Accessed 10 Oct 2021 from https://umr-selmet.cirad.fr/content/download/4052/29638/version/3/file/IMV_REPORT_WASTETYPOLOGY_HANOI.pdf

Nguyen LC, Toru I, Takeshi K (2013) Effect of biodegradation on mechanical properties of municipal solid waste in bioreactor landfill in Vietnam. In: Paper presented at the 7th international joint symposium on problematic soils and geoenvironment in Asia, Kyushu, vol 9

Nguyen VT, Tong TK, Dang TTH, Tran TVN, Nguyen HG, Nguyen TD, Isobe Y, Ishigaki T, Kawamoto K (2018) Current status of construction and demolition waste management in Vietnam: Challenges and opportunities. Int J GEOMATE 15(52):23–29. https://doi.org/10.21660/2018.52.7194

Nguyen NT (2007) Transformation of Hue Traditional Garden Houses in Hue citadel Area, Vietnam [Master's thesis]. Chiang Mai University

Nguyen VA (2010) Why DEWATS is still not popular in Viet Nam? Water Practice Technol 5(4):wpt2010117. https://doi.org/10.2166/wpt.2010.117

Nguyen NT, Kobayashi H et al (2010) A study on spacial transformation of hue traditional garden houses in the citadel area, Vietnam. In: Proceedings of the 8th international symposium on architectural interchanges in Asia, B-32, pp 527–532

Nguyen NT, Kobayashi H et al (2012) No. 9 Ngo Thoi Nham Street: the evolution of a traditional garden house in Hue, Vietnam. Sansai: Environ J Global Community 6:65–84

Nhung CTH, Dung NT (2012) Determination of the probability distribution of air quality data monitored at Hanoi. Vietnam J Sci Technol 50:83–89

OECD (2011) Doing better for families. OECD, Paris

Ouma YO, Tateishi R (2014) Urban flood vulnerability and risk mapping using integrated multi-parametric AHP and GIS: methodological overview and case study assessment. Water 6(6):1515–1545. https://doi.org/10.3390/w6061515

People's Committee of Thuan Thanh Ward (2015) Population data of Thuan Thanh ward [unpublished data]. Thua Thien Hue Province

Pham NA, Harada H, Fujii S, Nguyen VA, Huynh TH, Tanaka S (2014) Accumulation rates and characteristics of sludge in septic tanks of Hanoi. J Sci Technol 52(3A):133–139

Pham NA, Harada H, Fujii S, Tran VQ, Hoang H, Tanaka S, Kunacheva C (2012) Effect of septic tank management on septage composition: a case study in Danang, Vietnam. Vietnam J Sci Technol 50(1C):138–144

Reinhart DR, McCreanor PT, Townsend T (2002) The bioreactor landfill: Its status and future. Waste Manag Res 20(2):172–186. https://doi.org/10.1177/0734242X0202000209

Saaty TL (2008) Decision making with the analytic hierarchy process. Int J Serv Scis 1(1):83–98

Sakamoto Y, Shoji K, Bui MT, Phạm TH, Vu TA, Ly BT, Kajii Y (2018) Air quality study in Hanoi, Vietnam in 2015–2016 based on a one-year observation of NO_x, O_3, CO and a one-week observation of VOCs. Atmos Pollut Res 9(3):544–551. https://doi.org/10.1016/j.apr.2017.12.001

Satapathy S, Porsché I, Rolker D, Bhatt S, Tomar S, Nair S (eds) (2014) A framework for climate change vulnerability assessments. Deutsche Gesellschaft für Internationale Zusammenarbeit (GIZ)

Shen R, Schäfer K, Shao L, Schnelle-Kreis J, Wang Y, Li F, Liu Z, Emeis S, Schmid HP (2017) Chemical characteristics of PM2.5 during haze episodes in spring 2013 in Beijing. Urban Clim 22:51–63. https://doi.org/10.1016/j.uclim.2016.01.003

Statistics Bureau of Japan (n.d.) e-Stat. Accessed 13 Aug 2022 from https://www.e-stat.go.jp/

Tanaka N, Tojo Y, Matsuto T (2005) Past, present, and future of MSW landfills in Japan. J Mater Cycles Waste Manag 7(2):104–111. https://doi.org/10.1007/s10163-005-0133-6

Tchobanoglous G, Burton FL, Stensel HD (2003) Wastewater engineering: Treatment and reuse, 4th edn. McGraw-Hill

Thành D (2021) Cả nước có thêm 33 nhà máy xử lý nước thải đi vào hoạt động (Vietnam has 33 more wastewater treatment plants in operation). Báo kinh tế & đô thị online (Economic and Urban Online Electronic media). Accessed 11 Sept 11 2021 from https://kinhtedothi.vn/ca-nuoc-co-them-33-nha-may-xu-ly-nuoc-thai-di-vao-hoat-dong-409508.html

Thanh NP, Matsui Y (2011) Municipal solid waste management in Vietnam: status and the strategic actions. Int J Environ Res 5(2):285–296. https://doi.org/10.22059/ijer.2011.314

Thanh NP, Matsui Y, Fujiwara T (2011) Assessment of plastic waste generation and its potential recycling of household solid waste in Can Tho City, Vietnam. Environ Monit Assess 175(1–4):23–35. https://doi.org/10.1007/s10661-010-1490-8

Thuy NTT, Dung NT, Sekiguchi K, Yamaguchi R, ChauThuy P, Nam DT, Bang HQ, An TT (2016) Seasonal variation of concentrations and carbonaceous components of nanoparticles at a roadside location of Hanoi, Vietnam. In: Proceedings in international conference "Environmental engineering and management for sustainable development", pp 81–86

Thuy NTT, Dung NT, Sekiguchi K, Yamaguchi R, Thuy PC, Bang HQ (2017) Characteristics of elemental and organic carbon in atmospheric nanoparticles at different sampling locations in Vietnam. Vietnam J Sci Technol 55(3):305–315. https://doi.org/10.15625/2525-2518/55/3/8820

Tong TK, Le TT, Phung VL (2013) Recycling construction demolition waste in the world and in Vietnam. In: Proceedings of the international conference on sustainable built environment for now and the future. Hanoi, pp 247–256

Tran RV, Masumoto S, Yonezawa G (2017) Application of multi-parametric AHP for flood hazard zonation in coastal lowland area of central Vietnam. Int J Geoinf 13(1):23–34

Tran NH, Hoang L, Nghiem LD, Nguyen NMH, Ngo HH, Guo W, Trinh QT, Mai NH, Chen H, Nguyen DD, Ta TT, Gin KY-H (2019) Occurrence and risk assessment of multiple classes of antibiotics in urban canals and lakes in Hanoi, Vietnam. Sci Total Environ 692:157–174. https://doi.org/10.1016/j.scitotenv.2019.07.092

Tran DA, Tsujimura M, Ho HL, Dang DH, Vo PL, Doan TH, Nguyen TTT, Le CC, Pham TBT, Dang TD, Batdelger O, Nguyen TV (2021a) Groundwater quality evaluation and health risk assessment in coastal lowland areas of the Mekong Delta, Vietnam. Groundwater Sustain Dev 15:100679. https://doi.org/10.1016/j.gsd.2021.100679

Tran TA, Raghavan V, Nguyen VL, Saizen I, Tsutsumida N (2021b) A GIS-based Approach for Flood Vulnerability Assessment in Hoa Vang District, Da Nang City, Vietnam. IOP Conf Ser: Earth Environ Sci 652(2021):012003. https://doi.org/10.1088/1755-1315/652/1/012003

Truc VTQ, Kim Oanh NT (2007) Roadside BTEX and other gaseous air pollutants in relation to emission sources. Atmos Environ 41:7685–7697. https://doi.org/10.1016/j.atmosenv.2007.06.003

United Nations (2018) World urbanization prospects: the 2018 revision (Online Edition)

Urban and Rural Planning Institute [Viện Quy Hoạch Đô Thị Và Nông Thôn Quốc Gia] (2004) Thuyết Minh Tông Hợp Quý Hoạch Chi Tiết Phường Phú Hiệp-Phù Cát (General Report for Master Planning in Phu Hiep and Phu Cat Wards). Ministry of Construction, Hanoi

VietNamNet Bridge (2015) Number of registered bikes exceeds 2020 vision. http://english.vietnamnet.vn/fms/society/123272/number-of-registered-bikes-exceeds-2020-vision.html

Vo VM, Nguyen MH, Chau P (2013) survey results and suggestion of climate change education model for students in urban areas of Da Nang. Institute for Social and Environmental Transition (ISET), Danang

Watanabe R, Harada H, Yasui H, Tuan LV, Fujii S (2021) Exfiltration and infiltration effect on sewage flow and quality: a case study of Hue, Vietnam. Environ Technol 42(11):1747–1757. https://doi.org/10.1080/09593330.2019.1680739

2030 Water Resources Group (2017) Vietnam hydro economic framework for assessing water sector challenges

Weier J, Herring D (2000) Measuring vegetation (NDVI & EVI). NASA Earth Observatory, Washington DC

WHO (2006) WHO air quality guidelines for particulate matter, ozone, nitrogen dioxide and sulfur dioxide: global update 2005: Summary of Risk Assessment. WHO

WHO (2010) WHO guidelines for indoor air quality: selected pollutants. WHO Regional Office for Europe

WHO (2017) Evolution of WHO air quality guidelines: past, present and future. WHO

Wilcox LV (1948) Classification and use of irrigation waters. U.S. Department of Agriculture, Washington DC

Winkel LH, Pham TKT, Vi ML, Stengel C, Amini M, Nguyen TH, Pham HV, Berg M (2011) Arsenic pollution of groundwater in Vietnam exacerbated by deep aquifer exploitation for more than a century. Proc Natl Acad Sci 108:1246–1251. https://doi.org/10.1073/pnas.1011915108

World Bank, MONRE, CIDA (2004) Vietnam environment monitor 2004. Accessed 10 Oct 2021 from https://documents1.worldbank.org/curated/

en/724701468308959503/pdf/331510rev0PAPER0VN0Env0Monitor02004.pdf
World Bank (2011) Urbanization review: technical assistance report. World Bank
World Bank (2019) Vietnam: toward a safe, clean, and resilient water system. World Bank
Zhang F, Chen J, Qiu T, Yin L, Chen X, Yu J (2013) Pollution characteristics of PM 2.5 during a typical haze episode in Xiamen, China. Atmos Clim Sci 3(4):427–439. https://doi.org/10.4236/acs.2013.34044

Toward Sustainable Development in Urban and Rural Areas

6

Shinya Funakawa, Jane Singer, Kinh Thi Kieu, Nhan Thi Hien Le, Noriko Nakamura, Miki Yoshizumi, Izuru Saizen, Hong Thi Nguyen, Akiko Iizuka, Ueru Tanaka, Duc Tan Ho, Thong Trung Ho, Tung Ngoc Nguyen, Hirohide Kobayashi, Phuong Hoang Truong, Misa Aoki, Shuhei Tanaka, Khac-Uan Do and Hidenori Harada

Abstract

Vietnam has been rapidly developing since the 1980s, with the start of the Doi Moi (*Đổi Mới*) innovation policy to urbanize many cities and grow many economic activities. The rampant urbanization and enormous economic growth in recent times have impacted the quality of life of people in cities as well as in rural villages negatively rather than positively. Furthermore, these processes have adversely affected the natural environment and local culture. Considering the recent situation in Vietnam, the present chapter introduces our challenging projects adopting community-based or grassroot approaches to sustainable development. The contents include various practices in our projects

S. Funakawa (✉) · I. Saizen · H. T. Nguyen · H. Kobayashi · S. Tanaka
Graduate School of Global Environmental Studies, Kyoto University, Kyoto, Japan
e-mail: funakawa.shinya.2w@kyoto-u.ac.jp

J. Singer
Department of Global Studies, Kyoto University of Foreign Studies, Kyoto, Japan

K. T. Kieu
University of Science and Education, The University of Danang, Danang, Vietnam

N. T. H. Le
Nonprofit Organization FuDO (Fundament Design Organization), Nagasaki, Japan

N. Nakamura
NIJI Commercial and Development Joint Stock Company, Quang Tri, Vietnam

M. Yoshizumi
College of Gastronomy Management, Ritsumeikan University, Kyoto, Japan

A. Iizuka
Center for International Exchange, Utsunomiya University, Utsunomiya, Japan

U. Tanaka
Faculty of Agriculture, Setsunan University, Neyagawa, Japan

D. T. Ho
Department of Organization, Administration and Facilities Management, University of Agriculture and Forestry, Hue University, Hue, Vietnam

T. T. Ho
Faculty of Animal Science and Veterinary Medicine, University of Agriculture and Forestry, Hue University, Hue, Vietnam

T. N. Nguyen · P. H. Truong
Faculty of Architecture, University of Sciences, Hue University, Hue, Vietnam

M. Aoki
Faculty of Human Life and Environment, Nara Women's University, Nara, Japan

K.-U. Do
School of Chemistry and Life Sciences, Hanoi University of Science and Technology, Hanoi, Vietnam

H. Harada
Graduate School of Asian and African Area Studies, Kyoto University, Kyoto, Japan

S. Fujii et al. (eds.), *Livelihood and the Environment in Vietnam*, Sustainable Development Goals Series,
https://doi.org/10.1007/978-981-97-2289-1_6

related to environmental education, ecotourism with local resources, community development, and local cultural conservation.

Keywords

Sustainable development · Urban and rural areas · Local culture · Local resources · Community-based approach · Grassroot approach

6.1 Introduction

6.1.1 Outline of Section Topics on Sustainable Development

This chapter introduces our challenging projects and research on sustainable development in the background of the impact of urbanization and economic growth in urban and rural areas in Vietnam. Section 6.1 introduces the contents of this chapter, along with the essay entitled "Toward Sustainable Development in Rural Areas of Vietnam." Section 6.2 described two campus sustainability projects at the Hue University of Agriculture and Forestry and Danang University, one focusing on Education for Sustainable Development (ESD) and the other targeting operational sustainability on campus. Section 6.3 describes the "Building DanangEnvironment City" project and a case study of a community-based environmental education project realized through collaboration between the Danang local government, residents, Danang University, and Kyoto University. Section 6.4 summarizes research achievements regarding the possibility of tourism implementation through quantitative evaluation of local resources in the Dong Giang district of the Quang Nam province. Section 6.5 introduces a joint project titled "Enhancing Community Resilience and Livelihood Security to Cope with Natural Disasters in Central Vietnam" under the auspices of the JICA (Japan International Cooperation Agency) Partnership Program, implemented by the Kyoto University in collaboration with the Hue University of Agriculture and Forestry. Section 6.6 focuses on the traditional dzeng weaving activity of women in the Hong Ha commune. This activity provides a cultural dimension and offers an invaluable opportunity to preserve and use indigenous knowledge as one of the most precious local resources. Section 6.7 details projects on small-scale local livelihood activities for enhancing household economy and conserving ecosystems in two pilot villages near Hue city. Finally, Sect. 6.8 describes the reconstruction project of a traditional Cotu community house in the A Ka hamlet in the Nam Dong district, as the traditional architecture of ethnic minorities is disappearing in the modern context.

6.1.2 Toward Sustainable Development in Rural Areas of Vietnam

As outlined in Chap. 2, the production infrastructure that supports Vietnam's primary production may be divided into three regions, depending on climate, geology, topography, and soil conditions: (1) the northern mountainous area, characterized by generally steep topography and climates similar to temperate zones; (2) the central plateau, characterized by a relatively gentle topography and tropical savanna climate; and (3) alluvial plains in the lower reaches of rivers. Considering the sustainable development of rural areas in Vietnam, in addition to the above differences in the natural environment, the following three perspectives are essential: (1) establishing a primary industry adapted to the natural conditions of the region and producing a low environmental impact; (2) enabling people in the respective regions to build a livelihood based on the history of that region; and (3) developing a sustainable relationship with the urban areas of Vietnam or even the outside world. With the above perspectives in mind, we overview the direction of sustainable development in rural areas in the three regions with different production infrastructures.

6.1.2.1 Northern Mountainous Area

Spanning through Myanmar, Thailand, Laos, and northern Vietnam, the mountainous region of the north Indochina Peninsula is characterized by a steep terrain that primarily comprises sedimentary or metamorphic rocks and granite and a climate that gradually shifts from tropical savanna to warm temperate depending on latitude and altitude. From the viewpoint of not only the natural environment but also the ethnic distribution, people's livelihood, and history, the northern mountainous area can be seen as a transition zone or barrier between the southern part of China and the large river delta in the Indochina Peninsula. This area is mainly inhabited by ethnic minorities, who have partially migrated from South China. They have been cultivating upland rice, which initially originated in the lowlands of South China, as shifting cultivation since at least the nineteenth century. In the latter half of the nineteenth century, most of the Indochina Peninsula was occupied and divided by Britain and France. Nonetheless, the mountainous area acted as the border between major countries. Given this historical background, the region may be the most affected by wars, as Southeast Asian countries achieved independence and became involved in global politics under the Cold War structure after the two world wars of the twentieth century. One must be aware of the background of people's (sometimes immigrants') history as the minorities of each country, their unstable livelihoods, and their economic backwardness in the war as the basic knowledge for planning future regional development. On steep lands, which are generally unsuitable for agricultural production, the main question is "what is sustainable primary production that can improve the livelihoods of local people while maintaining the surrounding environment and that can be practiced by ethnic minorities who are not economically wealthy?".

6.1.2.2 Central Vietnam Plateau

The Central Vietnam plateau is characterized by a tropical savanna climate and extensive distribution of Oxisols with low fertility. Oxisols are known to show low land productivity under low input management because of their poor chemical fertility. Currently, however, Oxisols are rather advantageous for modern agriculture, in which unlimited chemical inputs and energy can be applied to achieve high production because of the relatively gentle terrain favorable for mechanized farming and suitable soil physical properties for crop production. The recent agricultural progress in Brazil is a good example of the utility of Oxisols. On the central Vietnam plateau, coffee cultivation, which has been developed since long, surged in the 1990s. Vietnam is now the second-largest coffee-producing country in the world. In this way, the central Vietnam plateau has historically been developed in terms of commercial agriculture compared with the northern mountainous areas. However, problems threatening the sustainability of agriculture in this region have become apparent in recent years. The forest area in this region has decreased due to the expansion of agricultural land despite its general increasing trend at the nationwide scale. Global coffee prices have declined, and monoculture agriculture is becoming unstable. Further, environmental deterioration due to the indiscriminate use of pesticides and chemical fertilizers is severe, and changes in and shrinkage of suitable cultivation areas due to global warming have been anticipated. Therefore, intensive and diversified agriculture that is economically beneficial while conserving forest resources must be established.

6.1.2.3 Alluvial Plains

Lowland rice cultivation developed in the middle and lower reaches of rivers has long supported many people and played a leading role in Vietnamese agriculture. Due to the warm climate, two- and three-stage crops are possible, which remains the driving force for Vietnam's agricultural export. As introduced in Chap. 4 of this book, coarse-textured soil is widely distributed in Central Vietnam, where relatively short and steep rivers flow. In contrast, fine-textured soils are more common along the Song Coi River in the north and the Mekong River in the south. Breeding crop varieties and optimizing agricultural land management optimal for each region in response to such differences in land productivity across regions and variations in climate peculiar to Vietnam, which is long from the north to south, have become issues in agricultural technology development. Meanwhile, the emission of greenhouse gases, such as methane and nitrous oxide, from lowland areas, including paddy fields, has become a global concern. Therefore, establishing technical measures to mitigate these emissions from paddy fields is a common issue in humid regions of Asia. Furthermore, since most of the lowland rice-growing areas are adjacent to populated urban areas, simultaneous management across urban and rural areas in terms of the use of water resources and emission of pollutants as well as the use of energy and logistics are expected to become essential for the sustainable development of this region.

6.2 Collaboration on Campus for Tertiary ESD in Central Vietnam

6.2.1 Introduction

Based on the idea that universities should be models for society and "living laboratories" for approaches and technologies to combat climate change, campus sustainability (CS; see Fig. 6.1) has become a priority for universities worldwide. CS refers to universities' operational efforts in adopting policies and strategies for reducing energy use, maximizing energy efficiency, reducing waste and CO_2 emissions, sustainably sourcing goods and services, and minimizing disaster risk (Alshuwaikhat and Abubakar 2008). It also refers to the promotion of sustainability research and education on campus as well as outreach to the local community and businesses to promote awareness and behavior change (Savelyeva and McKenna 2011).

Many universities in Japan and other developed countries, including Kyoto University, have established offices to promote sustainability

Fig. 6.1 Campus sustainability components

and have instituted ambitious sustainability targets and long-term plans. Until now, however, few universities in developing countries such as Vietnam have had the funding or capacity to institutionalize sustainability efforts or promote multi-stakeholder collaboration between the campus and communities, even though regions such as Central Vietnam are regarded as being highly vulnerable to the impact of climate change, pollution, and natural disasters. By involving students and the community in sustainability activities, research, and education, these universities can maximize their limited funding as well as educate government, residents, and the private sector.

This section will briefly describe two CS efforts implemented since 2013 in Central Vietnam by Kyoto University researchers through collaboration with faculty at two Vietnamese universities, Hue University of Agriculture and Forestry (HUAF) and The University of Danang-University of Science and Education (UD-UED). The first project focused on Education for Sustainable Development (ESD), while the other targeted operational sustainability on campus. Both produced measurable outcomes and were intended to initiate broader efforts to engage students, faculty, administrators, and community in raising awareness regarding sustainability issues, changing attitudes, and promoting behaviour change for sustainability.

6.2.2 Localized Tertiary Education for Sustainable Development

In 2012, a team of Kyoto University researchers led by Associate Professor Tracey Gannon initiated a Japanese government-funded project known as the Tertiary ESD Initiative. The aim was to create a model ESD course for undergraduate students and to assess the effectiveness of the ESD course in enhancing students' eco-literacy and fostering environmental advocacy. After the course was implemented in two consecutive years for students at Kyoto University, the researchers tested its replicability by offering the course to undergraduates at HUAF in 2013. The Kyoto researchers first collaborated with Hue faculty to tailor course content for a Vietnamese context and to address the sustainability challenges affecting central Vietnam, such as poverty alleviation, waste management, chronic flooding, and other disaster risks. The course featured short lectures by pairs of international and Vietnamese faculty; field visits on campus, to rural villages, an open food market, and an open disposal site for municipal solid waste; and frequent student engagement in group discussions and classroom exercises.

Several methods were used to gauge the effectiveness of the course in Vietnam, including pre- and post-course surveys, class observation, student reflection, staff observations of fieldwork and group work, and follow-up interviews conducted one year after the course to learn if students were continuing to practice pro-sustainable behavior. Students were also asked if they felt they had acquired or strengthened skills associated with sustainability, such as systems thinking—the ability to analyze complex systems across different domains and scales—and interpersonal skills, such as communication, leadership, and empathy. Students' progress in improving knowledge, motivation, and action was assessed by coding their survey and interview responses using NVivo software. The results for both Kyoto and Hue students in terms of self-assessed competency development are shown in Fig. 6.2. Hue students showed particular improvement in communication competencies and the ability to listen well to others, to make links, and to draw ideas together.

The Hue course was shown to be especially successful in facilitating behavior change (Gannon et al. 2016). Before the course, students reported that they engaged in common sustainability behaviors such as recycling and energy conservation. After the course, however, the range of behavior was broader, ranging from reduced motorcycle use to organic agriculture for household consumption. One student even sold his harvested organic vegetables online. As their knowledge increased, students were motivated to not only behave sustainably themselves but to promote such behavior among their families, friends, and classmates.

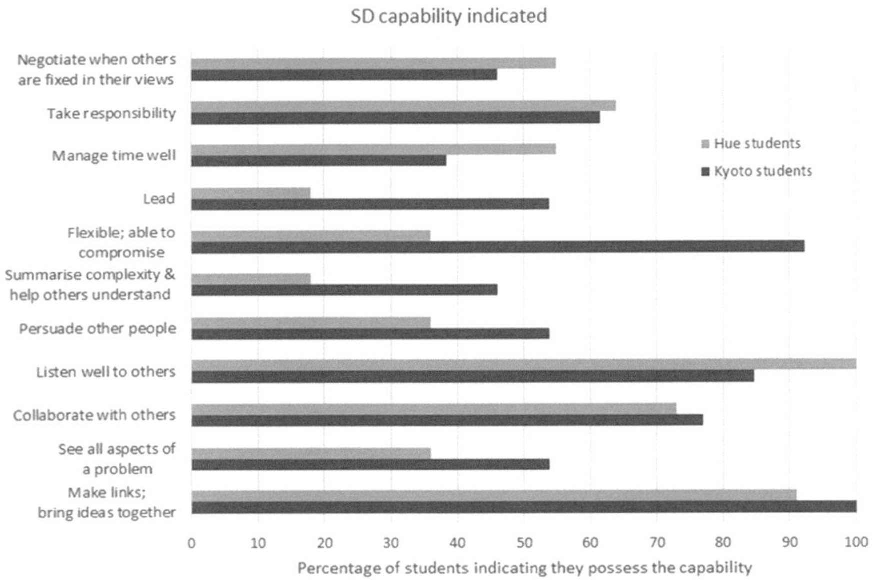

Fig. 6.2 Self-assessed competency gains in Kyoto and Hue

6.2.3 Student-Led Initiatives Promote Green Campuses in Vietnam

6.2.3.1 Introduction

In Vietnam, university students regularly conduct several types of sustainability-linked activities, such as litter campaigns, public awareness raising, environmental advocacy campaigns, and sustainability-related training courses. The Youth Union, a mass socio-political organization with branches on every university campus and nearly universal membership in Vietnam, is mainly in charge of mainstreaming environmental activities on university campuses among students and promoting greater awareness. Yet, university campus administration and faculty have undertaken very few CS initiatives. To broaden awareness and institutionalize green practices, students, faculty, and administrators at UD-UED, a university, in Danang with the support of Kyoto University, conceived and implemented several initiatives to promote CS.

6.2.3.2 Methods and Results

Workshop participation and investigation

In November 2016, two representatives of UD-UED participated in a large-scale CS workshop organized at HUAF with three visiting administrators and faculty from Kyoto University with experience and expertise in operational sustainability and ESD. The visit was held under the auspices of a research fund for Asian development administered by the Graduate School of Global Environmental Studies. The UD-UED staff learned about Kyoto University's efforts to improve the sustainability of its built capital and innovative campus practices such as an environmental tax. Subsequently, UD-UED organized its own workshop with experts from KU to identify how the university can apply campus sustainability.

Table 6.1 Plan to conserve energy and to reduce and recycle municipal solid waste at UD-UED

No.	Activity	Period	Expected outcome
1.	Reduce and recycle solid waste at student dormitory	15/11–25/12/2016	
1.1	Action 1. Investigate student's awareness regarding solid waste management (SWM) at UD-UED	15–17/11/2016	Assessment of student awareness of SWM
1.2	Action 2. Train students about SWM waste reduction and recycling	18–20/11/2016	40 students will be trained in SWM
1.3	Action 3. Segregate, reduce, and recycle SW	20–25/11/2016	10 student rooms at dormitory will be involved in SWM Utilize organic waste for composting
1.4	Action 4. Evaluate students' awareness change after participating in the program	25/11–25/12/2016	Report on change in student awareness of SWM
2.	Campus promotion of energy saving	15/11–25/12/2016	
2.1	Action 1. Install sensor systems at halls, meeting rooms, and some lecture rooms	15–30/11/2016	2 meeting rooms, 2 lecture rooms, and 1 hall (UD-UED main building)
2.2	Action 2. Observe and evaluate energy consumption	1–25/12/2016	Report on energy consumption and student awareness before and after sensor installation

Action plan formation and implementation

To follow-up on the information provided at the workshops, UD-UED launched an action plan in December 2016 to promote CS. A student working group was formed to determine which activities they should implement in order to green the campus. Based on consultation with the university faculty and administrators, an action plan was drafted and approved by the higher administration (Table 6.1). This plan targeted enhancing awareness and motivating lecturers, staff, and students to save energy and to reduce and recycle solid waste by carrying out several short-term activities at UD-UED.

Results

A group of 20 student volunteers residing in 10 rooms at a dormitory on campus were trained in solid waste segregation and reduction (see the photographs in Figs. 6.3, 6.4, 6.5 and 6.6). This group helped train other students to separate solid waste before disposing it in campus trash bins labeled "Recyclables" and "Other waste." An average of 30 kg of waste a day was collected, of which 26.6 kg, or 89%, was organic waste. The organic waste was utilized for composting, producing 70 kg of compost over a three-month period. The composted material was used as fertilizer for campus gardens and trees. Recyclable waste (7%) was sold to a local recycling company.

The remaining 4% of the waste was collected by URENCO, the company responsible for the collection and disposal of municipal solid waste in Danang.

Installation of infrared sensors and LED bulbs to save energy

After investigating the current lighting system on campus, a volunteer group suggested that the Department of General Administration replace the bulbs in public areas on campus from 250 W halogen bulbs and 36 W fluorescent lamps to LED bulbs and install four infrared sensors to automatically shut off the lights when the classroom emptied in order to conserve energy. Initial results showed that investment expenses were reduced by 21%, while energy costs were reduced by 67% (Figs. 6.7, 6.8 and 6.9).

Fig. 6.3 Student volunteer training. Photo by Nguyen Van Khanh (2017)

Fig. 6.4 Solid waste segregation at dormitory. Photo by Nguyen Van Khanh (2017)

Both the ESD and CS initiatives have raised awareness and fostered commitment by the universities to expand their efforts. At UD-UED, for example, the administration intends to invest in additional LED bulbs and a group of interested students is now training local households in composting techniques. It has been shown that students can serve as effective models and instructors for local residents in promoting sustainable behavior (Singer et al. 2019). The researchers are also planning to link the Vietnamese universities to an Asian network of universities active in CS (Asian Network for Campus Sustainability) for further exchange of information and strategies.

Fig. 6.5 Solid waste segregation at campus. Photos by Nguyen Van Khanh (2017)

Fig. 6.6 Compost. Photos by Nguyen Van Khanh (2017)

6.2.4 Promoting Campus Sustainability Through Research

In Vietnam, research on CS remains very limited. Previously published studies mainly focus on describing the construction of a set of assessment criteria (Kieu 2020), recommendations for a CS model (Phạm 2017; Phan 2020), and reporting implementation of CS at specific universities (Le 2018; Ngo and Trinh 2016; Waller et al. 2014).

In 2019, the Ministry of Education and Training (MOET) approved a research grant for UD-UED to build a CS assessment framework for Vietnamese higher education institutions.[1] Notably, this is the first CS research funded by the Ministry of Civil Affairs in Vietnam, although Vietnam National University, Hanoi, received funding for a self-initiated study on promoting a "Green University" in 2016. At the end of 2020, the research generated a set of criteria issued to

[1] Grant number B2019-DNA-02.

Fig. 6.7 Installation of infrared sensor. Photos by Nguyen Van Khanh (2017)

conform to the unique context of Vietnam. In this study, one of the authors conducted a survey at four universities: UD-UED (regional university), Hue University University of Science (regional university), Quang Nam University (provincial university), and Ho Chi Minh University of Agriculture and Forestry (Gia Lai campus). The survey results unearthed certain "bright spots" in CS activities at Vietnamese universities:

1. The role of student clubs and the Youth Union is crucial in promoting regular and continuing volunteer activities on topics of sustainable development. In addition to the activities that take place on campus, there are also activities that take place in the community, which help create connections with the community and enhance the student experience and thinking about sustainable development. It is noteworthy that the cost of these activities is very low.
2. Sustainable development content has been introduced in some courses in formal education syllabi. This mainstreaming also helps increase students' knowledge and attitudes on sustainability. Most of the courses are related to climate change, environment, gender equality, and biodiversity conservation.

Fig. 6.8 Installation of new bulbs

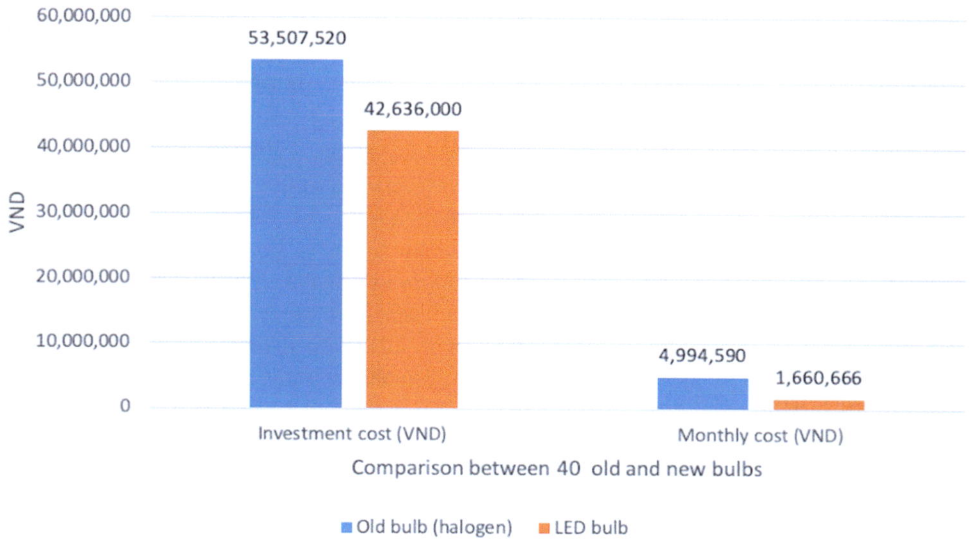

Fig. 6.9 Comparison between old and new systems

3. Vietnam's public universities are moving toward autonomy; thus, the "hard" operational aspects of a CS approach, particularly in campus operations, are of great interest, as they will help cut costs for the university.

The initial survey results revealed certain constraints in the implementation of CS at Vietnamese higher-education institutions (HEIs), as noted below:

1. Omission of CS issues in university development policies
2. Lack of awareness among university heads
3. Low awareness of staff and students about CS
4. Lack of resources to implement CS activities
5. Lack of coordination between faculties and departments in the university
6. Lack of systematic study and practical guidelines for implementation
7. Implementation of current initiatives depends heavily on the individual efforts of a few officials and lecturers who are interested in CS.

Overall, compared to other countries in the region such as Malaysia, Indonesia and Thailand, CS in Vietnam is only in the early stages. Some managers think that implementing CS and organizing the assessment will require extensive resources and see CS as an "add-on" task. In the context of higher education reform in Vietnam, most universities are trying to accredit their training programs according to a set of domestic criteria (Circular 04 issued by the Ministry of Education and Training (MOET)) and the region's quality assurance system (ASEAN University Network Quality Assurance). It should be noted, however, that most public universities are now tasked with achieving increasing autonomy. Hence, research on CS is important for guiding HEIs to embed CS quickly and efficiently and without putting undue pressure on administrators.

6.2.5 Recommendations

6.2.5.1 Combining Higher Education Reform Targets with CS

Higher education reforms in Vietnam focus on the following main issues: improving the quality of education, enhancing university autonomy, strengthening the research and publication capacity of lecturers, increasing international cooperation, and increasing its influence on society (Lâm 2013; MOET 2006; Pham 2012; Tran and Villano 2017; Tran 2014). These goals can easily integrate with CS initiatives:

- First, CS promotes systemic change, beginning with managing consumption and emissions, thereby contributing to reducing the university's operating costs. This has important implications for financial autonomy at HEIs. In fact, the first two universities in Vietnam to apply the GreenMetric criteria to evaluate CS were motivated by budget cuts (Le 2018; Waller et al. 2014).
- Second, CS promotes research on sustainability science and ESD, fostering student competence-based learning, an emerging global educational theme. Infrastructure or operational reforms can be included in living laboratory initiatives or when using community research sites as learning venues. This will help improve the quality of higher education and increase community outreach, which is currently very limited in Vietnam.
- Third, the topics of sustainable development and CS are among the pressing subjects of research, particularly given the impetus provided by the United Nation's 17 sustainable development goals (SDGs). Promoting CS, therefore, affords an opportunity to promote international cooperation and research in the field, generating academic publications as well as novel findings.

As shown above, the implementation of CS can help achieve grassroots reform in Vietnamese higher education, yet it will require ample time and resources. However, examination of the long-term benefits suggests that CS can generate proven benefits in university educational quality as well as sustainability.

6.2.5.2 Promotion of Leadership Consensus and Institutionalization

In order for CS activities to be sustainable, it is necessary to raise the awareness of HEI leaders about the importance of CS and to empower leaders and staff to catalyze and implement new paradigms. This will help ensure that CS permeates the entire university system, which includes the following inter-linked elements: education, research, campus operations, community outreach, assessment, and reporting (Lozano et al. 2013: 11). Literature on campus sustainability suggests that the impetus for CS can be top-down, centering on administrators, or bottom-up, from student-led initiatives (Brinkhurst et al. 2011). In many Vietnamese HEIs, student sustainability activities are quite innovative and sustainably led by youth organizations. Hence, the engagement of leaders and staff to institutionalize CS is a prerequisite to ensure that CS can be sustained.

6.2.5.3 Taking Advantage of Available Resources

In developing countries such as Vietnam, low-cost solutions that require few resources are urgently needed. For instance, the experience from developed countries about CS will help reduce uncertainty among Vietnamese universities when building CS. Application of information technology (i.e., mobile apps and web-based sharing information) can promote systematic assessment to reduce data collection and reporting time.

Universities should motivate faculty to integrate sustainability thinking into their specialized studies for more CS initiatives. In addition, universities are encouraged to seek external funding sources to implement CS initiatives, such as the Green fund introduced by the US-based organization AASHE,[2] funding from research collaboration with local government or private sectors, or even by fund-raising activities.

6.2.5.4 International Cooperation

UD-UED can be a successful showcase of low-cost but effective international cooperation in promoting CS. The cooperation between

[2] https://www.hub.aashe.org/browse/types/greenfund/.

UD-UED and Kyoto University was conducted through:

- Capacity building: a lecturer from UD-UED visited the Graduate School of Global Environmental Studies (GSGES) at Kyoto University. Through this opportunity, the CS models at Kyoto University were introduced directly and effectively to the UD-UED staff. Lessons learned from CS at Kyoto University were crucial for replication at UD-UED.
- Participating in networking seminars or workshops: UD-UED, HUAF, and Kyoto University participated in an experience-sharing workshop and this can be seen as a starting point for those institutions to show interest and later strengthen their commitment to CS. UD-UED staff was introduced to the CS network of Asia, ASCN, and they learned from the experience of developing countries in the region such as Malaysia and Thailand.
- Funding support through SEEDs research fund: the first CS activities at UD-UED and HUAF were funded by a SEEDs grant from GSGES, Kyoto University. Cooperation with a professor from Kyoto University was also an important factor in encouraging UD-UED staff to apply for a MOET research grant (2019–2020).

International cooperation can be an important catalyst for promoting CS at HEIs in Vietnam. Through international cooperation, universities will be able to leverage different resources to implement initiatives and models quickly and sustainably.

6.2.6 Conclusion

CS is an appropriate approach to help Vietnam achieve dual goals: reforming higher education and achieving the SDGs. Vietnam can capture many opportunities for international integration through research and implementation of CS. With the support of resources and funding from Kyoto University, two universities in central Vietnam, UD-UED and HUAF, launched the first CS activities in operations (waste management, energy saving) and education (a non-formal ESD course at HUAF). Notably, research on CS was promoted after initial achievements at UD-UED and HUAF.

Universities need to capitalize on these early gains to further develop CS. Some of the solutions suggested in this section are: (i) combining higher education reform targets with CS; (ii) promoting leadership consensus and institutionalization; (iii) taking advantage of available resources; and (iv) international cooperation.

6.3 Developing Next-Generation Environmental Leaders Through the Japan-Vietnam Joint Environmental Education and Research Projects

6.3.1 Introduction

The Sustainable Development Goals (SDGs) entered into force on 1 January 2016. They set 17 targets to address the challenges not adequately addressed by the Millennium Development Goals and the escalating environmental problems discussed at Rio + 20. The SDGs are based on the recognition that strategies to end poverty and improve, education, health, social protection, and employment, especially tackling climate change and protect the environment while addressing a wide range of societal needs need to be achieved without missing out on multiple targets. Education for Sustainable Development (ESD) is positioned as a way to develop human resources capable of implementing the SDGs. According to the Japanese National Commission for UNESCO, the human resources ESD aims to develop are capable of recognizing their relationship with others, society, and the natural environment, respecting their 'involvement' and 'connection,' and can deal with various issues related to sustainable development, such as the environment, peace, and human rights, not only in individual fields but also in the environmental, economic and social fields. The ability to take an interdisciplinary and comprehensive approach from environmental, economic, and social perspectives is required to deal with issues related to sustainable development, such as the environment, peace, and human rights. Education to foster such capabilities is needed to develop the next generation of leaders.

One of the issues that have arisen in recent years in the field of community development is that each local entity is often working individually. While stove-piped administration within the government is a problem, there is also a need for more cooperation between local community organizations. In particular, many local community organizations are established under the leadership of a few "conscious" people, and it is sometimes difficult for leaders to cooperate. Furthermore, many of the members of local community organizations are active older adults with time to spare, and only a few young people from the local community participate in these activities.

On the other hand, even within the same university, problems of sectionalism in different research fields arise, which significantly detriment the development of the next generation of leaders who can play an active role in society. As well as research, lectures at universities are provided by specialists in the respective research fields, which inevitably leads to a bias in the research field and results in students learning only one side of an issue.

In these issues arising in university education and community development, initiatives to develop next-generation leaders to create a sustainable society were conducted in Danang and Hue, Vietnam, through collaboration between actors within the region and universities.

6.3.2 Community-Based Environmental Education Project in Danang City, Vietnam

6.3.2.1 Overview of Danang City

Danang City is one of the most rapidly urbanizing cities in Vietnam, and its economy is following the same trajectory. The city's gross domestic product (GDP), in terms of 1994 prices, rose by a remarkable average of 11.56% annually between 1997 and 2006 and 10.87% between 2006 and 2012. Overall, the GDP growth rate for the whole period 2002–2012 was 12.53%. GDP per capita aims to reach 4,500–5,000 USD by the end of 2020. The city's import–export turnover grew at 13.2% to nearly 3,200 million USD while the trade surplus hit 413 million USD in 2020. However, Danang has been growing at the price of several environmental problems. To address this issue, in 2008, Danang City launched the "Building DanangEnvironment City" project, wherein several initiatives were undertaken to address the environmental issues.

This section introduces the "Building Danang—Environment City" project and a case study of a community-based environmental education project in Thuan Phuoc Ward of Danang City, Vietnam, which was conducted by Danang local government, residents, Danang University, and Kyoto University.

6.3.2.2 "Building Danang—Environment City" Project

"Building Danang—Environment City" project was approved by the municipal people's committee in 2008, in order to make Danang an environment-friendly city. The project aims to develop quality standards for air, water, and land to ensure safety and health of local residents, investors, and tourists by 2020. The table shows specific targets in each period (Table 6.2).

Table 6.2 Specific targets of "Building DanangEnvironment City" for each period (Danang People Committee 2008)

Period	Targets
2008–2010	• Garbage leakage from Khanh Son garbage, Dam Rong Lake, and lower section of Phu Loc River; polluted seawater in the coastal: Thanh Binh, My Khe, Tho Quang lock area, and Cu De river mouth • Build wastewater treatment systems in industrial zones: Hoa Khanh, Danang aquiculture service, Hoa Cam, etc. • Handle air pollution from factories and enterprises, especially from steel, cement, aquiculture processing factories; decrease dirt, socialize environment tasks, and look after and protect trees • Collect household garbage (95%) and dangerous garbage (100%); strengthen solid waste matter by over 30%
2011–2015	• 90% wastewater quality of industrial and export processing zones meets environment standard • 90% household water quality in inner city is collected and handled as per environment standard • Control potential wastewater resources and implement suitable sanitation • Form and develop waste recycle industry to reuse; strive to reuse over 50% of the collected waste that is recycled • 90% people in inner city and 70% people in suburb should be able to use pure water • Control air pollution: from potential resources including road traffic, industrial, and urban waste; ensure that air pollution index (API) is less than 100 • Develop urban green spaces with suitable density and species of trees; strive to achieve 3–4 m^2/person • Preserve forest diversified biology in the city
2016–2020	• Continue to perfect the targets in 2007–2015 period; ensure achievement of the following criteria: 100% industry and household wastewater is handled, 70% solid waste is recycled, and 25% amount of water is reused • Complete the database to evaluate and promulgate the environment city after achieving all criteria laid above

6.3.2.3 Community-Based Environmental Education Project in Thuan Phuoc Ward, Danang City

In terms of urban population, Danang is the 4th largest city in Vietnam with the highest rate of people living in urban areas, making up 87.7% in 2018. The population was 1,134,310 in 2019. Thuan Phuoc Ward People's Committee developed a community environmental management project in partnership with Danang Department of Natural Resource and Environment (DONRE), schools, and Danang University of Technology (DUT) in 2006. In the area, water pollution in lakes and rivers and garbage-strewn streets and rivers pose serious environmental problems. The project was developed to tackle these problems.

The objective of the project was to raise awareness among residents and school children to foster an understanding and take initiatives toward solving the environmental pollution issues in Danang, namely pollution in the river and streets. In order to ensure a participatory approach, various local organizations, including local government, schools, and NGOs, were identified for project implementation.

A model community-based environmental management project is being conducted in Thuan Phuoc Ward, Danang since 2006. Through questionnaire surveys, interviews, and discussions in local workshops, action plans were developed and implemented in Danang (Table 6.3 and Fig. 6.10).

6.3.2.4 Achievements and Issues of the Project

There are several achievements of the project. First, teachers and students of the Vo Thi Sau School were considerably interested in

Table 6.3 Project activities of the local community (left) and the primary school (right)

Local community	Primary school (Vo Thi Sau)
Identify environmental issues	Planting and observation of trees
Establish a regulation on environmental protection	Drawing contest on environmental activities
Sunday clean-up	Games and tests on environmental activities
Publish posters on environmental actions	Eco-camping
Improve water pollution	Training for school teachers
Environmental monitoring	Publish environmental educational materials

Fig. 6.10 Activities in the model project conducted in Danang city

the environmental education program; they thus actively conducted the programs. The participants were satisfied with the activities and wanted to continue them further. Second, the school established an environmental room wherein it exhibited the outcomes of the environmental education programs (drawings, writings, environmental maps, etc.) to share with other teachers, students, and the local community. Third, Danang city government gave a high evaluation to the environmental education program and promoted training workshops as well as publication of a guideline of environmental education to disseminate the program in Danang city. Thuan Phuoc ward was awarded for its community-based environmental activities in Danang. Fourth, the environmental program in Danang was disseminated all over Vietnam through a national newspaper (Nhan Dan).

However, several issues were also identified. First, conflicts between the local community and the government emerged. The community requested the government to solve the water pollution first. Some participants complained that it was the people living in street restaurants and a local market who polluted the lake and river. Second, while senior citizens took part in the activities, the response from young and middle-aged people was lukewarm due to a lack of understanding about the project and pre-occupation with livelihood work. Third, although people living around the lake and river areas participated actively, other residents in the same watershed area were more hesitant to participate as they did not bear any direct negative impact from the problem. Finally, it was found that it is difficult to solve water pollution in the short term and that it is difficult to make the initiative sustainable.

6.3.3 The Challenge to Contribute to Environmental Education in Vietnam Based on Learning from the University's Environmental Education Project

6.3.3.1 Internship Program in Danang

During the project, environmental education was not well recognized among the citizens, and environmental protection activities mainly included only cleaning. Many participants were less concerned with environmental issues and did not understand what environmental education entailed. A survey regarding environmental problems among the residents who participated in the project and those who did not was also conducted. It was found that residents who participated in the project were more interested in environmental problems than those who did not. Therefore, there is a need to find a method of getting more residents involved. The community house was considered as a key link for connecting the project activities to the residents. In Vietnam, there is one community house in each ward that is used for residents' meetings and activities and residents visit there often. Therefore, with the cooperation of the Thuan Phuoc ward people's committee, posters introducing environmental education activities in Vo Thi Sau School were posted, and a contest on handmade environmental protection products was held at Thuan Phuoc Community House through the project. Community participation is an effective way for making new networks and for changing people's "attitude" toward environmental conservation and sustainable society into "action". Both "attitude" and "action" and their relationship constitute one of the goals of environmental education (Fig. 6.11).

6.3.3.2 Current Status of Environmental Education in Vietnam

In Vietnam, which has developed even further in the last 10 years, compared to when the project was conducted, the number of people who enjoy traveling or camping has increased, and at the same time, an increasing number of people are interested in environmental issues, especially youngsters. However, older people are still less concerned with protecting their surrounding environment. For example, they often throw garbage out of their house, which children tend to follow. In the past, garbage was mostly biodegradable and returned to the soil after a few years, but that is not the case in modern times, especially with plastic. Although, children study public health and environmental conservation in their schools through subjects such as "Morality," "Wellness," and "Nature and Society," their content has not yet been systematically integrated and teachers are not aware of them. In addition, there are few opportunities to learn through experience (Fig. 6.12).

6.3.3.3 Expectations for Future Environmental Education in Vietnam

Now that interest in environmental issues is increasing, it is necessary to introduce a new framework of environmental education programs involving multiple stakeholders, similar to the aforementioned project conducted

Poster introducing the school activities

Contest of handmade environmental protection products

Fig. 6.11 The activities conducted in Thuan Phuoc Ward Community House in 2009

Fig. 6.12 "Nature and Society" textbooks describing environmental protection

in Danang City. If residents in the community can also participate in activities, for example, to teach children or if children are allowed to describe to their elders what they have learned, their awareness of environmental issues will increase. In addition, according to the Vietnamese government information, Ministry of Education and Training (MOET) and Ministry of Natural Resources and Environment (MONRE) have been working together for developing the content of environmental education suitable for each subject and all levels of education and training for teachers associated with environmental protection in the period of 2019–2025. It is expected that the new educational content will be attractive and effective. A protected environment leads to a greater number of tourists, which enables further development of the economy in Vietnam.

6.3.4 Environmental Human Resource Development Through the Development of Environmental Education Programs in Japan and Vietnam

6.3.4.1 Lesson Learned from the Experience Through the Community-Based Project in Danang

As part of the "Building Danang—Environment City" plan over 2008–2020, a community-based environmental management project was launched in 2006 in Thuan Phuoc ward of Danang City to address pollution in lakes and rivers of the area.

The pollution level in Dam Rong Lake in Thuan Phuoc ward was severe as people living around it discharged wastewater and threw garbage directly into the lake. The ward government thus promoted the project in order to raise awareness among local residents living around the lake. To do so, children from nearby schools were enrolled to foster their understanding and leadership toward environmental issues in their neighborhood. The project stressed on the importance of participatory approach by including local people as well as local government, schools, and NGOs based in the ward.

There were many meetings and discussions carried out among the stakeholders before and during the implementation of the project activities. It was found that the final agreement on activities such as cleaning up every Sunday, promoting an environmental poster, or establishing an environmental monitoring system by local residents involved many valid arguments from all sides and required compromises from all stakeholders. An approach regarding how to work with different groups of beneficiaries was thus generated through these project activities.

At the school, the activities included tree planting and observation, environmental picture drawing, school teacher training on environmental issues, and publication of environmental materials. One of the biggest achievements of the whole project was the active participation of school teachers and students in environmental activities and their continuance of environmental education activities even after the project finished. It was learned that community-based projects should include local schools, whose role was seen to be vital in promoting activities in the long term. The importance of educating kids from small age and instilling in them a sense of protecting their surrounding environment in Danang is an initial step toward further implementation of similar activities in Hue and Japan.

6.3.4.2 Development of Activities in Hue and Japan

Under a Japan International Cooperation Agency (JICA) Partnership Program in Thua Thien Hue Province, environmental and disaster education among locals was identified as one of the priorities within the project. It was considered that education regarding such issues is indispensable for younger generations. Three schools in three communes of the province joined in activities aiming (1) to enhance pupils' awareness and knowledge regarding the environment and environment-related disasters and (2) to influence other people in the community through children's actions. The preparatory steps before implementing activities at schools included identification of environmental problems at sites, study of environmental and disaster activities at schools, discussions with school teachers and pupils about their needs and capacity, and, finally, establishment of environmental clubs at schools.

Two main concepts were applied in the project. First, the approach centered on children. Thus, children played the primary role in implementing activities while the other components such as parents' association, project members, communities, schools, etc. were used to support children. The second concept involved the schools' initiative, especially in organizing activities, an initiative from school teachers took to get ideas and give feedback to pupils and relevant stakeholders.

Direct beneficiaries of the activities were 142 pupils and 10 teachers from 3 schools. There were certain activities organized regularly such as 5 times of the Sunday Green at one school and 12 times of every Sunday Garbage Collection at another school. Moreover, 130 recycled products were made, 50 trees planted, 87 papers for the eco-journal contributed, 300 beer cans collected, and 168 pictures drawn by the children of the 3 schools (Fig. 6.13).

It was learned that since children were key actors, activities should be prepared and implemented by children based on their physical and mental abilities. The importance of selecting members was another key finding. With regard to that, it is recommended not to choose members of the last grade as they are generally very busy with their examinations. In addition, before implementing any activity, school teachers and pupils should take time to discuss and make a detailed schedule. Finally, as stated by a school teacher, it is important to report to higher

Environmental Protection Club

Tree planting and caring

Wall paper on environmental protection theme

Presentation on environmental protection topics

Fig. 6.13 Environmental and disaster education activities conducted in Hue city

relevant departments, such as the Department of Education and Training of the district, so that they can help disseminate the activities later.

Based on the activities conducted in Danang and Hue under the JICA Partnership Program, environmental education and disaster mitigation education were studied under a Master's course on Global Environmental Studies at Kyoto University. It involved a comparative study between Japan's education system and that of Vietnam. Environment and disaster mitigation education in Maiko High School in Japan and disaster mitigation education trial activities at two schools (Phan Sao Nam and Chu Van An Secondary Schools) in Vietnam were examined as a case study. The study was conducted with three main objectives, stated below:

(1) To characterize the education system in Japan and in Vietnam;
(2) To point out the weaknesses as well as strengths of the education system of the two countries to their education for disaster risk reduction in schools; and
(3) To find the potential to integrate education for disaster risk reduction into the current education systems in Japan and Vietnam.

The study pointed out that both Japan and Vietnam were suffering from annual disasters, which have only increased in frequency and severely in recent decades. Education on disaster risk reduction is therefore "a must" for these countries as one of the ways to tackle disasters. It was found that there had been efforts in both countries toward mainstreaming education for disaster risk reduction in their current educational system. Based on the three schools, it was found that Japan is a step ahead while Vietnam is only in its initial stages.

However, the current education system in both countries had some strengths and weaknesses that supported and hindered the promotion of education on disaster risk reduction.

Students' study trip to Rokko Mountain

Training on firefighting skills at Kobe Fire Academy

Hazard map making by Maiko high school students and kids of a nearby primary school

Students' volunteer work in Tohoku in April 2011

Fig. 6.14 Activities under environment and disaster mitigation education in Maiko High School in Japan

Specifically, students' awareness of daily prevention against disasters was enhanced by the general education in Japan as it valued physical training, group work, and moral manners. Japanese students practiced them on a daily basis and got accustomed to them, making them more resourceful in case of disasters. Moreover, education on disaster risk reduction started quite early in Japan, as early as kindergarten and normally from primary school. This consolidated kids' awareness of disaster preparedness and formed a culture of safety from a young age. Besides, the case of Maiko High School despite of its uncommonness showed that there was a way for education for disaster risk reduction to be bloomed in the current education system (Fig. 6.14).

In contrast, in Vietnam, a lack of formal education on disaster risk reduction in the entire education system is yet to be addressed. Both Japan and Vietnam have a large teaching curriculum, making it difficult for teachers to find the time and motivation to integrate a new education concept. The current curriculum in Vietnam, which still puts heavy emphasis on theory, also hinders any education on disaster risk reduction. In Japan, however, students' life is centered around schools with a considerable focus on examinations, which weakens their role as members of communities where education on disaster risk reduction is especially vital.

Despite the abovementioned problems, there are ways to integrate education on disaster risk reduction into the current educational system in both the two countries. In Vietnam, local lessons as well as extra-curricular activities offer flexible time slots for local schools to conduct their own activities based on the local issues outside of the prescribed syllabus. Hence, schools can use these classes for education on disaster risk reduction. In Japan, the case of Maiko High School could be used as an example on the design of a course on disaster risk reduction. Moreover, the "Period for Integrated Study" and model of elective subjects can provide room for disaster risk reduction education as a subject. Schools in Vietnam could also learn and apply this method in their schools, where education on disaster risk reduction is emphasized and students have a choice in enrolling for disaster risk reduction classes.

In short, although disaster risk reduction is a new subject, its importance cannot be understated. School education on disaster risk reduction needs to be realized as early as possible to make disaster risk reduction a tangible reality.

6.3.5 Conclusion

Implementing the above field-based learning programs could be realized in Vietnam, through the collaboration of the Graduate School of Global Environmental Studies, Kyoto University with The University of Danang, and Hue University of Agriculture and Forestry. The programs have also enabled the university to collaborate with various local community groups that are involved in local environmental activities. The advantage for the university is that through the field-based education program, students can learn what the problems are in the field and why they are occurring, regardless of their field of study. The advantages for the local community are that the exchange of opinions with research experts and students through field-based learning stimulates the emergence of new perspectives and ideas and provides a forum for sharing information on the activities of various local actors.

As can be seen from the above activities, it is necessary to create a space where various local actors and all generations can work together to discuss and promote the creation of a sustainable society. By collaborating with universities, it may be possible to break through the rigidity of the discussion forum and develop the next generation of leaders.

6.4 Local Resources for Introduction of Ecotourism

6.4.1 Introduction

Proper utilization of local resources is an important step toward achieving sustainable development of rural areas (Biswanath 2017). Tourism is a typical example of successful local resource utilization, such as landscape, food, and culture (Uchiyama and Kohsaka 2016). It can provide benefits to local community through increased job opportunities and by establishing a market for local products. In practice, community-based tourism, which is a sustainable form of tourism in which beneficiaries are involved by mobilizing their own resources (Li 2004), has been widely implemented and has considerably benefited forest-dependent communities.

This section introduces the possibility of tourism implementation through the quantitative evaluation of local resources in Dong Giang district of Quang Nam province, Vietnam. Although tourism activities here have not been designed officially by the local government, there are certain spontaneous activities, such as sightseeing and tourist photography, especially for weddings, being carried out by youngsters from neighboring cities. In the near future, however, due to limited number of tourism products, tourism in Dong Giang may face major challenges in ensuring its competitiveness in the tourism market. Thus, diversifying tourism products is an important step in strengthening the competitive advantage of a destination (Sasana 2017) so that this section diversifies local resources through incorporating natural and cultural values.

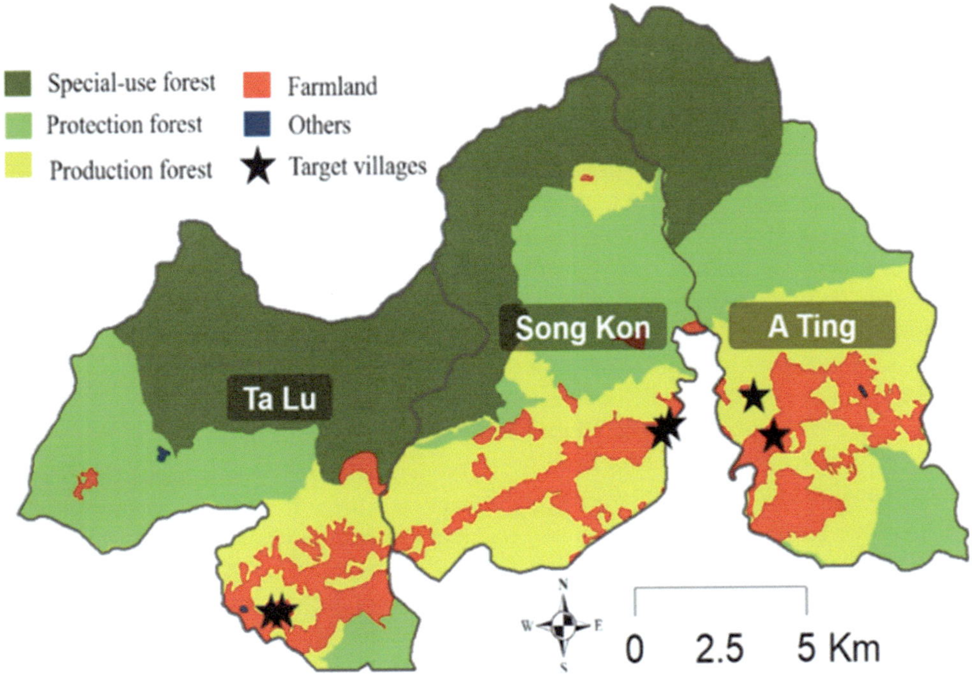

Fig. 6.15 Villages surveyed in Don Gian district

6.4.2 Methods of Identification of Local Resources with Natural and Cultural Values[3]

Field surveys were conducted from May to June 2017 at six target villages in Ta Lu, Song Kon, and A Ting communes (Fig. 6.15). Twelve meetings with sixty local participants collecting forest ecosystem services for their livelihoods were organized.

Twelve participatory mapping exercises and semi-structured questionnaires were undertaken to determine sites with outstanding natural beauty, such as hot water springs, waterfalls, hills, and natural lakes (Fig. 6.16). A base map of topographic characteristics was used for a mapping exercise, in which the participants were asked to place colored beans on the map to indicate the locations of the sites. Descriptions of the surrounding landscape and preliminary assessment of the mentioned sites were also discussed in the meetings. Many types of waterfalls were mentioned (Bătinaş 2010) so that classifications of waterfalls were shown to the participants to hear their description of the sites.

Assessing the possibility of local resources with natural value to be utilized for tourism development is often based on landscape quality and conditions for the enjoyment of visitors (Priskin 2001). A tourist destination should consist of the following main elements: (1) presence of attractions; (2) accessibility, facilities, and services related to tourism activities; (3) political stability for conducting tourism development activities; and (4) support for destination products from tourism stakeholders (Boniface et al. 2016; Manhas et al. 2016). Herein, three criteria

[3] This subsection is reproduction of the article by Nguyen Thi Hong, etc., one of the authors of this section and has been deposited in the repository of Kyoto University at https://repository.kulib.kyoto-u.ac.jp/dspace/handle/2433/259084. It does not violate any policy of the repository.

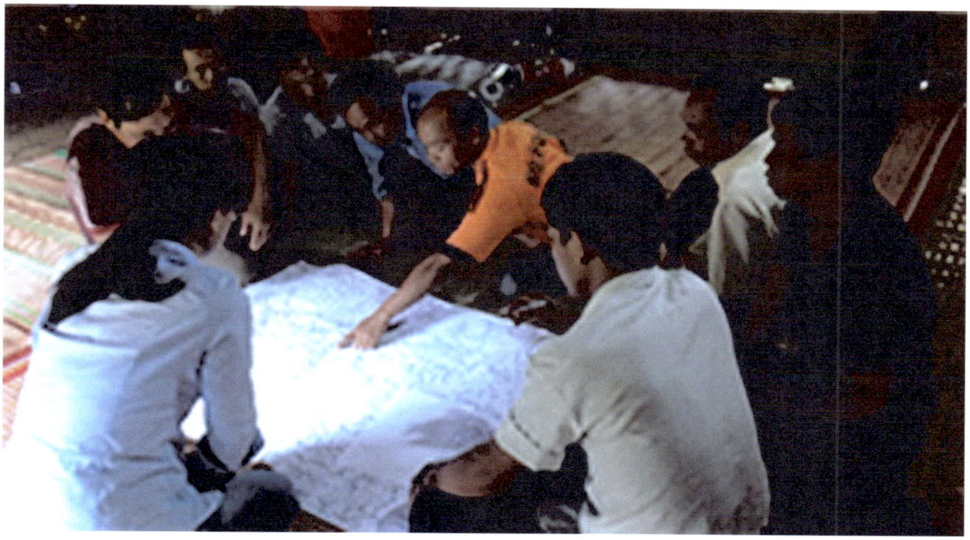

Fig. 6.16 Mapping of local resources with natural value by local participants. Photo by Nguyen Thi Hong

were considered, (1) attraction, (2) accessibility, and (3) adaptation to local legal conditions.

Tourists often expect to have access to diverse, rare, or unique destinations. The diversity of landscape, the possibility to be a tourism destination, and topographic characteristics were considered in the attraction criterion. Then, sub-elements in attraction criterion of nature-based tourism assessment were prepared and scores from 0 to 5 were assigned to the elements based on the number of sub-elements they contained 9. Three types of datasets, consisting of land use map, digital elevation model data, and the information from the meetings with the local participants, were used for counting the number of sub-elements and scoring (Table 6.4).

Accessibility refers to the ease of being able to physically access the destination (Priskin 2001). This criterion was divided into five ranges of distance from a main road to the sites mentioned by the participants in the meetings: "0" (more than 20 km), "1" (15–20 km), "2" (10–15 km), "3" (5–10 km), and "4" (0–5 km). The study unit's score was calculated by averaging the scores of all sites distributed within the study unit.

The adaptation to local legal conditions criterion was considered with respect to the local legal documents on forest management and usage: "0" (special-use forest), "1" (protection forest), "2" (production forest), and "3" (farmland and others), based on the community's access to different forest types. Generally, the communities' access is determined by the function of forests. In particular, a "special-use forest" is designed to conserve the natural forest ecosystem and genetic sources so that the access there is limited completely. A "protected forest" is designed to prevent erosion and desertification and to maintain its function of regulating climate; thus, access to it is limited in certain cases. Compared to the special-use forest and protected forest, access to the "production forest" is least restricted because its main function is to produce and trade timber and non-timber resources.

Finally, the possibility of study units to be utilized for tourism development was delineated through a sum of scores from attraction, accessibility, and adaptation to local legal conditions. This possibility was classified into five classes, as shown in Table 6.5.

Six group discussions were held with 30 local managers to confirm the existence of the local resources with the cultural values of the Cotu people. Then, these local managers classified

Table 6.4 Criteria of attraction

Elements	Sub-elements
Landscape (based on land use map and participants' opinions)	Forest scenery/View
	Rice field scenery/View
	Plantation scenery/View
	Garden/field scenery/View
	Waterbody (lake, river, etc.)
Tourism objective distribution (based on land use map and participants' opinions)	Natural beauty (forest, plantation, etc.)
	Natural phenomenon (Cave, crater, etc.)
	Waterbody (lake, waterfall, river, etc.)
	Cultural attraction
	Historical heritage
Uniqueness of resources (based on land use map and participants' opinions)	Forest ecosystem
	Karst ecosystem
	Landscape scenery/View
	Hot water spring/Waterfall/Lake/River
	Cultural and historical heritage
Value (based on participants' opinions)	Ecological value
	Knowledge value
	Medicinal value
	Economic value
	Belief, cultural, and historical value
Possible tourism activities (based on participants' opinion)	Research/Education
	hiking/Tacking/Climbing
	Photography
	Enjoying scenery
	Viewing cultural/Historical heritage attraction
Altitude	>2,000 m (high mountain)
	1,000–2,000 m (mountain)
	500–1,000 m (high hills)
	100–500 m (hills)
	15–100 m (lowland)
Slope	>45 (very steep)
	25–45 (steep)
	15–25 (wavy/undulating)
	8–15 (rather flat)
	0–8 (flat/level)

these confirmed resources into different categories. The classification criterion was the popularity of these confirmed resources in the area, which was considered based on the number of local households whose members have preserved these local resources. The four categories were "very popular" (more than 80% of local households with family members preserving the local resources); "popular" (50–80% of the households); "slightly popular" (less than 50% of the households); and "extinct" (none of the households). In addition, the possibility of reintroducing "extinct" local resources and reviving "slightly popular" local resources were also explored through these local managers. Based on these results, the detected resources with cultural values available in the area were classified into "very popular," "popular," "slightly

Table 6.5 Scores for classification of study units by criteria

Criteria	Classes of classification and scores				
	Very low	Low	Moderate	High	Very high
Attraction	0–7	7–14	14–21	21–28	28–35
Accessibility	0–0.8	0.8–1.6	1.6–2.4	2.4–3.2	3.2–4.0
Adaptation to local legal conditions	0–0.6	0.6–1.2	1.2–1.8	1.8–2.4	2.4–3.0
Suitability	0–8.4	8.4–16.8	16.8–25.2	25.2–33.6	33.6–42.0

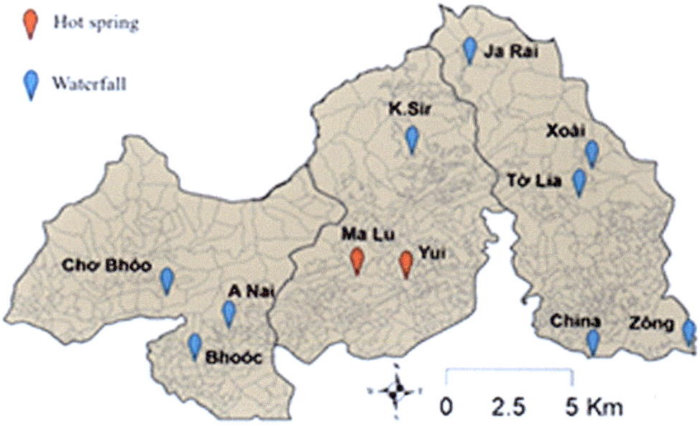

Fig. 6.17 Names and locations of sites mentioned by the local participants

popular," and "extinct." Among these local resources, the latter two showed the possibility of being reintroduced or revived.

Demographic characteristics of local people who have continually preserved the local resources were determined through 158 face-to-face interviews using structured questionnaires. The structured questionnaires were employed to obtain information regarding: (1) socio-demographic characteristics of respondents such as age, gender, and local unions to which they belonged; and (2) the local resources with cultural values preserved by the respondents.

6.4.3 Local Resources for Tourism Development

6.4.3.1 Description and Assessment of Local Resources with Natural Values

Eleven sites with outstanding natural beauty were mentioned and mapped by the local participants in the meetings. These sites were digitized and a map was generated to show their locations (Fig. 6.17). In particular, two types of local resources, waterfalls and hot water springs, were mentioned by the participants (Table 6.6). These types have been deemed significant in tourism by tourism planners and managers because of their high attraction to visitors 10).

Figure 6.18 shows the classification based on the abovementioned criteria. Among the five classes of the attraction criterion, only "moderate" and "high" classes were observed with six and five units, respectively (Fig. 6.18a).

While some of the sites exhibited low adaptation to local legal conditions on forest management and usage as shown in Fig. 6.18b, tourism activities are still planned in the protected and production forests. Ecotourism in protected forests can be carried out if the activities meet certain requirements for forest conservation,[4] although it is much easier to launch ecotourism

[4] Vietnam Government, 2015. Decision No. 17/2015/QD-TTg on Regulations on protective forest management.

Table 6.6 Description of the sites mentioned by the local participants

Type		Local name	Description
Waterfall	Fan	*Thác Chơ Bhóo*	The fall looks like a fan, with the base much wider than the top
	Curtain	*Thác Bhoóc*	Wide breadth, thin flow, and its height is taller than the width
	Cascade	*Thác Tờ Lia*	The fall descends gradually on sloping rocks, and a series of small steps in quick succession is observed
	Slide	*Thác Xoài* (another name is *Chơ Run*)	The fall is thin and descends on a smooth, inclined surface
		Thác A Nai	
	Horsetail	*Thác* China	The fall is very high; looks like a horsetail, with a spreading tail
	Tiered	*Thác Zông*	The height of flow decreases with the stratification of rock
		Thác K.Sir	
	Punchbowl	*Thác Ja Rai*	The fall is short, small, and then spreads out in a vast basin
Spring	Hot water	*Suối nước nóng Yui*	Since the spring links to a freshwater spring, water temperature is safe for bathing
		Suối nước nóng Malu	Since the spring links to a freshwater spring, water temperature is safe for bathing. It stems from Malu mountain, a very high mountain, where Cotu people believe that the mountain gods are dwelling

in a production forest when the planned activities do not have a negative impact on timber and non-timber production or the soil environment.[5] Furthermore, most identified sites were within 5 km to the road connecting to Danang and Tay Giang districts, and thus had very high evaluation scores for accessibility (Fig. 6.18c).

Figure 6.19 shows a map that was built by overlaying all the assessed scores. A significant percentage, 72.7% (8/11), of the study units that contain the mentioned sites were classified as "high possibility" in terms of being utilized for tourism development. The distribution of these units (with "high possibility") is spread widely throughout the area, which could boost the local economy of many villages if tourism is developed.

6.4.3.2 Description and Discussion of Local Resources with Cultural Values

This section presents the local resources with cultural values detected through the group discussions with the local managers and discusses the possibility that these detected resources could be utilized for tourism development through literature reviews. Seventeen local resources with tangible and intangible cultural values were detected, where tangible cultural values are handicrafts (knitting, traditional musical instruments, weaving, jewelry), local cuisines (lam rice, toasted buffalo meat, ants and their eggs, buffalo horn-shaped cakes), and traditional architecture (community housesGuol

[5] Vietnam Government, 2016. Decision 49/2016/QD-TTg on Regulations on production forest management.

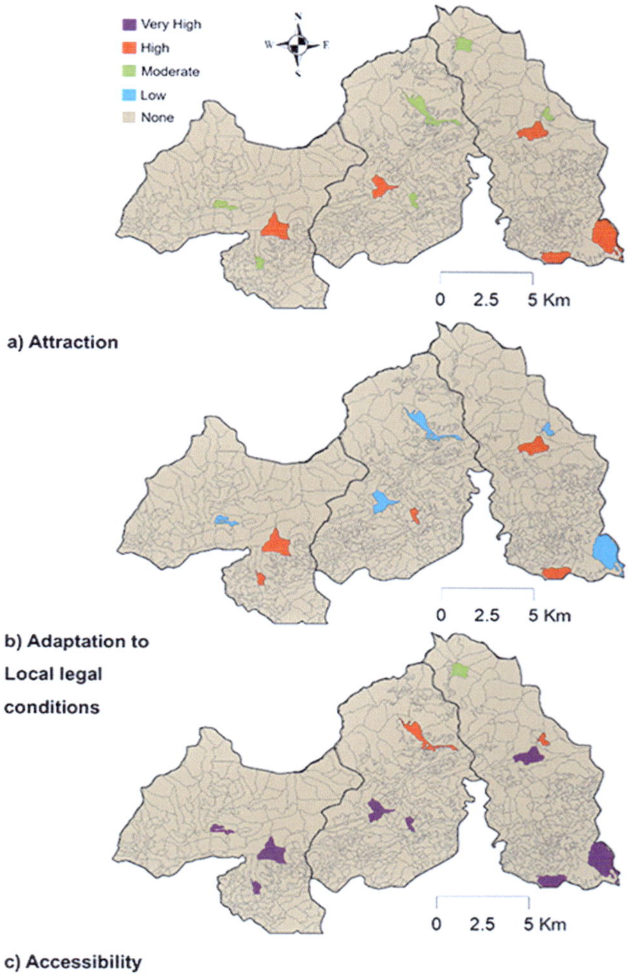

Fig. 6.18 Maps showing the results of classification based on criteria

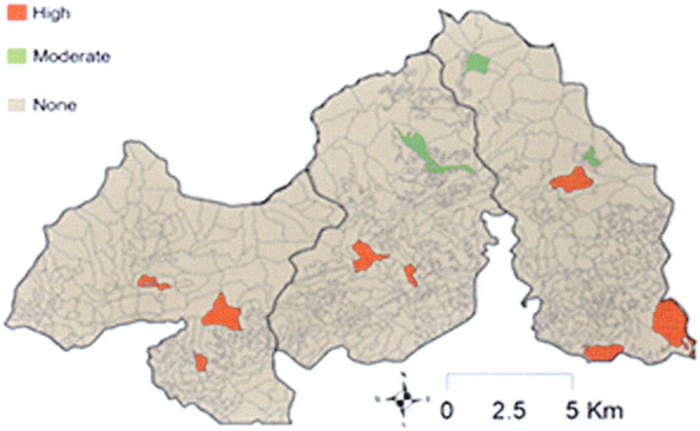

Fig. 6.19 Map showing the results of overlaying the criterion maps

Table 6.7 Local resources with tangible values and socio-demographic characteristics of local people preserving the resources

Local resources	Brief description	Popularity	Socio-demographic characteristics		
			Gender	Age	Union
Traditional handicraft					
Knitting	Making papooses and baskets using forest products	Popular	Male	51 ± 7	Farmer, elderly
Musical instruments	Making local musical instruments using forest products and animal horns	Slightly popular	Male	57 ± 6	Farmer, elderly
Weaving	Weaving threads into cloth and making traditional clothes	Popular	Female	33 ± 8	Women
Jewelry	Making jewelry using beads and wild animal fangs	Slightly popular	Female	56 ± 7	Women, elderly
Local cuisine					
Lam rice	Made by filling bamboo tubes with local rice, covering the heads of tubes with banana leaves, and cooking using firewood	Very popular	Female	35 ± 10	Women
Toasted buffalo meat	Made by mixing buffalo meat with local spices and toasting using charcoal	Slightly popular	Female	37 ± 10	Women
Ants and their eggs	Various foods made from local ants (kiến chua) and their eggs	Popular	Female	43 ± 1	Women
Buffalo horn-shaped cake	Made from local rice (proong) and local spices, and covered with leaves of a local plant (đót)	Very popular	Female	36 ± 9	Women
Traditional architecture					
Community house (Guol)	The floor has a rectangle/oval shape and is made with a wooden frame. The roof is arched at two gables. There is a big pillar at the house's center, which is decorated by sculptural and painting artwork. All used materials are collected from the forest	Very popular	Male	Over 18	Farmer, youth, and elderly

(*Gươl*) house) (Table 6.7). The local resources with intangible cultural values are festivals (new rice festival, twinning festival), rituals (killing buffalo and tung tung ya ya dance in the new rice festival; food festival and tung tung ya ya dance in the twinning festival), and ceremonial occasions (wedding, funeral, etc.) (Table 6.8, Fig. 6.20).

Generally, the local resources with tangible cultural values had a variety of references ("very popular," "popular," or "slightly popular"). Indigenous cultures change in accordance with economic and social changes. Since the 2000s, infrastructure development, such as roads connecting the district center to remote villages and nearby districts, in the area has led to the discontinuation of nomadism (Århem 2009). The investment provides an opportunity for the emergence of modern materials and goods produced by the Kinh people, which have limited the preservation of the local resources classified as "slightly popular." In addition, legal barriers on the usage of forest resources also contribute to this limitation. The results of the discussion

Table 6.8 Local resources with intangible values and socio-demographic characteristics of local people preserving the resources

Local resources	Brief description	Popularity	Socio-demographic characteristics		
			Gender	Age	Union
Festivals and rituals					
New rice festival	Conducted in October or November to thank local gods for supporting local people's lives and crops				
Killing buffalo	The ritual lasts two days and takes place in front of community houses. On the first day, local people sing and dance around a buffalo tied to a pillar. On the second day, the village patriarch prays and invites local gods by singing special songs. Healthy and prestigious men dance around the buffalo and puncture it while villagers cheer for them	Very popular	Male	Over 40	Farmer, elderly
Community dance	The Tung Tung Ya Ya dance	Very popular	All genders	All ages	All local unions
Twinning festival	Conducted out of crop seasons; to strengthen relationships between villages				
Food festival	Villagers gather, enjoy local foods and drinks, and then sing and dance	Very popular	All genders	All ages	All local unions
Community dance	The Tung Tung Ya Ya dance	Very popular	All genders	All ages	All local unions
Ceremonial occasions					
Wedding	Representatives of families of the bride and groom sing rhyming songs about love and happiness. The two families exchange gifts comprising forest and agricultural products. The bride and groom drink local wine, pray to their ancestors, and receive gifts from relatives. They perform the Pàzùm ceremony (wherein the new couple is covered by a blanket), which officially recognizes their marriage	Slightly popular	All genders	Over 18	All local unions
Funeral	Traditional music is used during the funeral. The relatives pray to local gods and ancestors for the deceased by singing special songs. Relatives and villagers prepare offerings for the deceased and bury them at a charnel house	Popular	All genders	All ages	All local unions

groups with the local managers showed that there is a possibility of reviving "slightly popular" resources such as musical instruments and jewelry, which are related to the spirituality of Cotu people. The Cotu men often use musical instruments in community gatherings, such as festivals and unplanned gatherings after work. Meanwhile, the Cotu women favor the jewelry. The participants in the group discussions mentioned that "we do not want to lose the resources because they are an essential part of who we are." On the contrary, most local resources with intangible cultural values were being well preserved in the villages (classified

Fig. 6.20 Detected local resources with cultural values

as "very popular" or "popular") (Table 6.8). The good preservation of the festivals and rituals can stem from the Cotu people's strong belief in the power of local gods. Through worship and rituals in festivals, the Cotu people communicate with and thank the local gods for their support to the lives and crops of local communities (Århem 2009).

There has been recently an increasing demand for tourism development based on culture, which is often referred to in literature reviews as "cultural tourism" or "culture-based tourism." This form of tourism is also identified and defined as "a type of tourism activity in which the visitor's essential motivation is to learn, discover, experience and consume the tangible and intangible cultural attractions/products in a tourism destination." Tourists tend to visit destinations with specific cultural attractions that are different from their normal places of residence.

The culture of the Cotu people is different from those of Kinh people (main ethnic) and others in Vietnam; thus, its uniqueness is highly evaluated by domestic and foreign visitors. Thus, local resources with cultural values of the Cotu people could become significant resources for tourism.

6.4.3.3 Socio-Demographic Characteristics of Local People Who Are Familiar with or Preserving the Local Resources

The participants in the exercise for detecting the local resources with natural values were local people who collected the ecosystem services for their livelihoods. More than half of the participants were aged between 38 and 49 years (61.7%, n = 37), followed by 50–59 years (20.0%, n = 12), and 18–37 years (18.3%, n = 11). These local participants were men belonging to the farmer union.

Table 6.7 shows the socio-demographic characteristics of the local people who were preserving local resources with tangible cultural values. The knitting and musical instruments were preserved by men who were 51 and 57 years old on average, respectively, and belonging to the farmer or elderly unions. Women belonging to women's or elderly unions played a major role in preserving other handicrafts. In particular, women who were 33 years old on average were preserving weaving while those older than 56 years old on average were preserving jewelry. In addition, women who belonged to the women's union played a key role in preserving the local cuisine. In contrast, men were mainly involved in the preservation of community houses (Guol house). According to the rules of the village, men older than 18 years old and younger than 60 years old, belonging to the youth or farmer union, were responsible for practicing the construction and restoration of the community houses. Meanwhile, older men, belonging to the elderly union, who were better acquainted with house architecture, history, and functions were responsible for managing the construction, usage, and restoration of the community houses. On the contrary, diversity in gender and local unions was observed in the preservation of local resources with intangible cultural values, except for the ritual of killing buffalo in the new rice festival (Table 6.8). Only men over 40 years old perform this ritual. Among them, the village patriarchs, who belonged to the elderly union, performed worships (praying and inviting local gods by singing special songs), and others (those belonging to the farmer union who were healthy and prestigious) performed the rest of the ritual activity, including dancing around the buffalo and puncturing it.

6.4.4 Values of Local Resources for Tourism Development

This section explores the local resources with natural and cultural values that were recognized and are being preserved by the local communities. The resources with natural values are highlighted at eleven sites with outstanding natural beauty, such as waterfalls and hot springs. In addition, eight study units that contain some of these sites were assessed as having a high possibility of utilization for tourism development in terms of attraction, accessibility, and adaptation to local legal conditions criteria. In addition, seventeen local resources with tangible and intangible cultural values were pointed out. While most local resources with intangible cultural values were being well preserved in the local communities (classified as "very popular" or "popular"), the resources with tangible cultural values had a variety of references (classified as "very popular," "popular," or "slightly popular"). Based on the discussion, these local resources with cultural values could provide to be significant resources for tourism development.

Indigenous values that are only saved in local communities and not managed by any public agency may gradually disappear because of the impact of economic and social development (Dlamini 2017). The local resources with natural and cultural values may also face threats if they are not detected, valued, and managed. Considering tourism assessment, previous research studies, and practices in tourism development, local resources could be utilized for tourism development to some extent. Meanwhile, as revealed in previous studies, the livelihood activities in the highland areas of Central Vietnam have faced certain limitations of local conditions, e.g., the lack of water irrigation for agricultural activities in the dry season (Nguyen et al. 2015) and small allocated areas of forestland to local households (Nguyen et al. 2016). Thus, tourism development could be an alternative livelihood option for the forest-dependent communities in the area. Because tourism development is a complicated activity that overlaps with different sectors of the society and economy (Inskeep 1991), there is a need for further discussions between multiple stakeholders, such as local communities, local governments, professional consulting organizations in tourism, and governmental organizations that are responsible for culture and forest conservation. The findings from this study could be a base for further discussion of tourism introduction in an area as well as the construction of appropriate regional planning in the future.

6.5 Enhancing Community Resilience and Livelihood Security in Three Diverse Communes in Central Vietnam

6.5.1 Background

More than 30 years have passed since Vietnam's government first implemented wide-ranging economic and social reforms under the Doi Moi (*Đổi Mới*) innovation policy in 1986. While significant economic growth has been achieved in this period, poverty still remains a pressing problem, particularly in the country's mountainous areas. Although Vietnamese government and foreign donors have tried to alleviate poverty, the gap between rural and urban areas remains wide and there is a limited number of foreign projects addressing the needs of ethnic minorities. This section introduces a joint project in Vietnam implemented by Graduate School of Global Environmental Studies (GSGES), Kyoto University in Japan in collaboration with Hue University of Agriculture and Forestry (HUAF) in Vietnam. An overview of the project, including project objectives, a description of the three sites, and project activities, is shared.

6.5.2 Universities' Initiative[6]

After three years of exchange of scholars and students between GSGES, Kyoto University in Japan and HUAF in Vietnam, a project called "Enhancing Community Resilience and Livelihood Security to Cope with Natural Disasters in Central Vietnam" was developed by GSGES in Japan in collaboration with HUAF, under the auspices of JICA (Japan International

[6]This subsection is reproduction of the article by Akiko Iizuka, etc., one of the authors of this section and has been deposited in the repository of Kyoto University at https://repository.kulib.kyoto-u.ac.jp/dspace/handle/2433/108247. "It does not violate any policy of the repository".

Cooperation Agency) Partnership Program (JPP). JICA is an implementation agency that administers Japan's efforts to provide official development assistance (ODA) in developing countries. JPP supports projects implemented by a variety of Japanese organizations and institutions in developing countries. In this case, GSGES, Kyoto University is the implementing organization in partnership with HUAF, and JICA Osaka center is in charge. Ministry of Education and Training of Vietnam approved the project by issuing Decision 702 on February 7, 2007.

This three-year developmental project ran from September 2006 to August 2009. During the project, the presidents of GSGES and HUAF signed a Memorandum of Understanding on September 16, 2008, at Hue University. This document expresses the commitment of both institutions to further academic exchanges and cooperation at the university as well as at the departmental level. Its purposes were to: (1) enhance community capacity to cope with natural disasters effectively, (2) conserve the natural environment which enriches people's lives and production activities, and (3) improve local people's livelihoods by enhancing their economy. Underlying each objective, and thus the project itself was the goal of enhancing the overall welfare of the people living in the three target communities where the project was carried out.

6.5.2.1 Project Members

GSGES, Kyoto University implemented the JPP with HUAF as its local counterpart. GSGES has worked together with HUAF through seminars, partner research, and student exchanges since communication with the Center for Agricultural Forestry Research and Development (CARD) at HUAF first began in 2003. CARD, established in 1993, implements a number of research projects with an aim to improve the livelihood of local communities and better protect natural resources and environments currently being degraded by economic development in central Vietnam. The JPP was the first project to be implemented by the two academic institutions. Its focus—to work toward improvement in people's livelihoods at the grassroots level through small-scale activities—represented a marked departure in terms of aim from the collaborative research and exchanges conducted in the past between GSGES and HUAF.

Preparation for the project began in 2005 when GSGES drafted a project proposal with the help of HUAF and JICA. The three-year project was initiated in September 2006. This project was implemented by project members from various disciplines of both universities. For the duration of the project, a researcher resided in Hue full-time to manage the project and coordinate efforts between Vietnamese and Japanese team members when the latter visited Hue to contribute ideas and experience to the project. On the Vietnamese side, a full-time project coordinator oversaw the entire project, while five full-time project assistants and approximately 15 part-time researchers facilitated project activities and provided local expertise. The Japanese team was staffed by researchers of GSGES. The Vietnamese project team included both researchers and administrative workers at Hue University.

The role of the project members was to help local people analyze their real situation so they were able to find a suitable solution for their own development. In addition, the project members took the role of facilitating and creating suitable conditions for the local people to access information and knowledge outside the community and support their efforts to implement developmental activities. Greater participation for local people in the decision-making process; an emphasis on the use of local, as opposed to outside, resources; and new efforts to reappraise the role of indigenous knowledge were all emphasized as starting points for future project activities.

6.5.2.2 Project Objectives

The project's goal, pursued by all project members, was defined as: to improve community resilience, environmental protection, and living conditions for better welfare and human security. This mission statement was understood in terms of three main objectives:

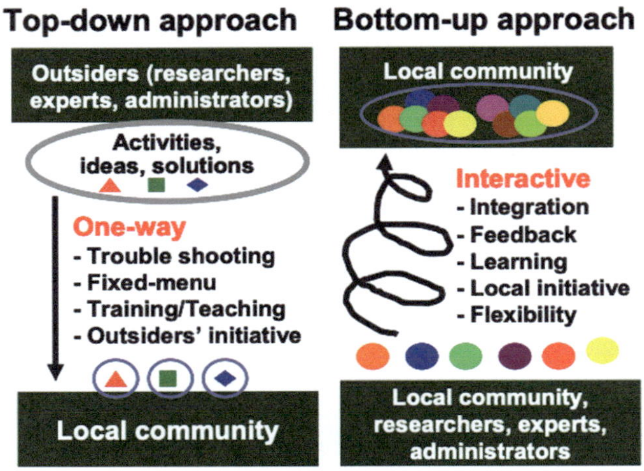

Fig. 6.21 Top-down and bottom-up approaches

- To enhance community resilience.
- To enhance environmental protection.
- To enhance household economy.

Underlying each objective, and thus the project itself, was the goal of enhancing the overall welfare and security of the people living in the three communities targeted by the project. Overlap between these objectives required that they must be pursued in concert for the overall success of the project. This is perhaps best shown by example. Villagers in Hong Ha commune, one of the project sites, engaged in illegal logging in protected forests because they did not have stable jobs and needed an immediate income. Clearly, people confronting the harsh realities of poverty cannot always afford to prioritize the potential future scarcity of natural resources over imminent needs. It followed that the problems faced by the community could not be easily solved by conventional single-focus initiatives. Schemes to plant trees or provide environmental education promised potential benefits, but these benefits would be reduced if pursued in isolation. Project members ascertained that a more holistic approach to community welfare would be required: one that enabled people to gain a stable income at the same time as conserving natural resources. A holistic perspective promises to incorporate economic, social, and environmental goals into welfare and human-security oriented activities.

6.5.2.3 Project Concepts

Project concepts are common notions or approaches shared by all the stakeholders participating in a given project or activity (in this case, members of the local community, project members, and any persons related to the project activity). Two concepts summarized below are particularly important to the project under discussion here.

One of the most important concepts of the project is the participatory method. It is mainly based on the bottom-up approach, departing from the top-down approach frequently employed in project works in Vietnam. Figure 6.21 shows a schematic diagram of the "top-down" and "bottom-up" approaches in community-based rural development. Generally, the top-down approach (left in Fig. 6.21) employs "trouble finding and trouble shooting," whereby outsiders identify problems and possible solutions, prepare activity options in advance, and then suggest them to the communities. Project guidelines are sometimes uniformed by knowledge of a local system or culture under consideration and there is often little scope for local people to share their opinions with project managers during the planning and implementation stages of a project. Typically, activities introduced by outsiders are associated with "training/teaching workshops," aimed at increasing awareness and bringing forward local experts among the local population to facilitate a certain participatory activity. This approach is efficient in mobilizing a number of participants, disseminating techniques and know-how, and thus in achieving project goals in a given period.

On the contrary, the bottom-up approach (right in Fig. 6.21) offers a method to realize and bring out the potential strengths—not only the problems—of the assisted areas. The participatory approach used in the GSGES-HUAF project attempts to encourage the involvement of local people in every aspect of the project's management. Local people are comprehensively involved from the initial stage of each activity. Potential strengths include a wider range of ecological, social, cultural, technical, and human resources that exist in the area. The bottom-up approach is characterized by flexibility, local initiative, and a learning process carried out in an interactive manner among the community members and outsiders. The project members believe that the bottom-up, or so-called participatory approach, is therefore particularly useful for fostering a sense of ownership among the local people that leads to self-empowerment in a sustainable manner. The local people living in the three communes targeted by the project are the main actors in the planning and implementation of the project work. They are assisted by project members, whose main role is to facilitate the participation of the local people by finding solutions they deem acceptable.

The project's implementation promoted local people's participation and encouraged them to enhance their potential by using indigenous knowledge and techniques. In order to do so, understanding the local situation was considered vital to both the planning and implementation of this project. Identifying and planning effective project activities depends on a deep understanding of local realities and the needs and potential of the location. Before any specific project activities were undertaken, the first critical step of GSGES-HUAF joint work in the project sites was therefore to understand the locality. The information used to determine what activities are suitable for a given locality was collected in a variety of ways, including statistics, reports, interviews, and in-depth discussions. Information was collected for each project site by means of (1) a baseline questionnaire survey, (2) a series of group discussions, and (3) field observations. These efforts to understand the local context continued long after a given activity was implemented. In this manner, project activities remained flexible when new issues or concerns arose over the course of their implementation.

6.5.2.4 Project Sites

The project targeted sites located in varied surroundings in upland, midland, and lowland areas of the Bo river watershed in Thua Thien Hue province, Central Vietnam, as shown in Fig. 6.22. The locations of the targeted communes for the project activities are: (1) Hong Ha, an upland commune; (2) Huong Van, a flood plain; and (3) Huong Phong, a coastal commune (Fig. 6.22).

Firstly, Hong Ha is a mountainous commune in A Luoi district, located 45 km south west of Hue city. The commune is divided into five villages—Con Tom, Pa-hi, Can Sam, Paring, and Arom— which together provide home to 1,332 people (Thua Thien Hue Statistical Office 2007), mainly Cotu and Taoi people and also Kinh people since 1975. People of Kinh ethnicity number as few as 15 families, most of whom are engaged in various businesses. Hong Ha is one of the 16 poorest communities in A Luoi district and among the 1,581 poorest communes nationwide, according to national poverty criteria (Hoang and Le 2006: 161). The household survey was conducted for all households in Hong Ha in August and September 2006, group discussion in December and January 2006, community walking in January 2007, and the feedback workshop on January 19, 2007. The workshop identified six activities, but one of these, a proposal to create a local market for the commune, was not implemented because it was considered unfeasible. Table 6.9 shows five project activities that were implemented in Hong Ha as a result of the household surveys, group discussions, and field observations: traditional community house construction, traditional

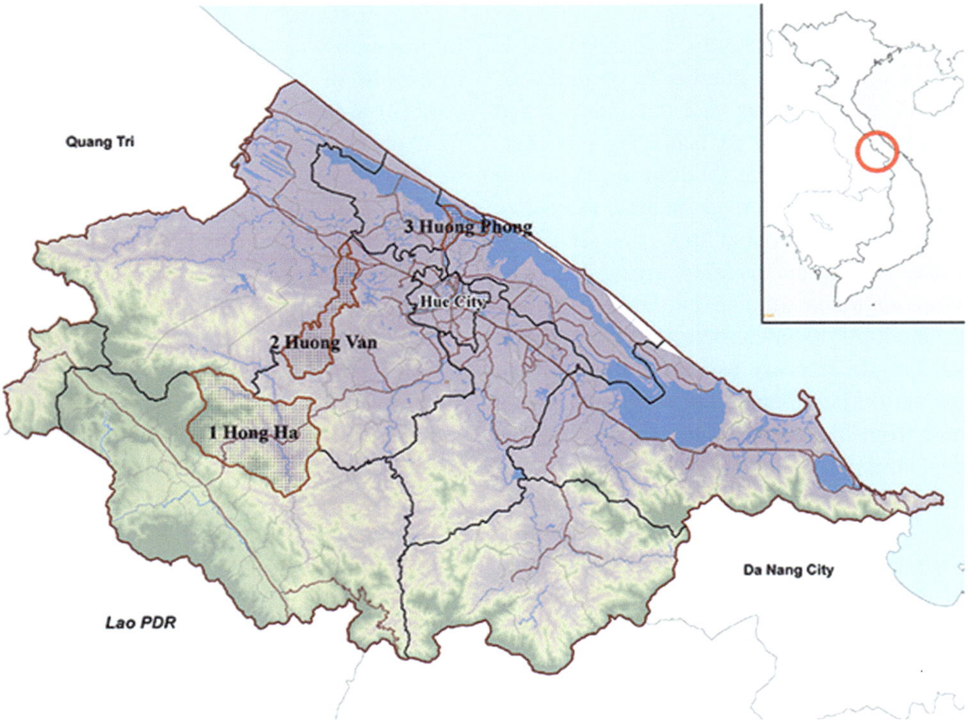

Fig. 6.22 Maps of the three project sites in Thua Thien Hue province

Table 6.9 List of activities in three sites

Hong Ha	Huong Van	Huong Phong
– Traditional community house construction – Traditional textile weaving for women – Traditional music instruments and folk-son class – Goat raising – Introduction of new cropping	– Establishment of community resource center – Livestock association (training, two-floor pig pan, and development of feed for pigs) – Cooperative biogas system – Safer community planning and vulnerable mapping	– Poly-aquaculture – Mushroom production – Capacity building on disaster management – Small businesses for poor households headed by women – Biogas system
Common activities of three sites – Disaster/environment education – Kids' exchange camp		

textile weaving for women, traditional music instruments and folk-song class, goat raising, and introduction of new crops for diversifying income sources. The activity of traditional weaving for ethnic minorities is discussed in the next section.

Huong Van is a plain commune in the Huong Tra district, located 20 km west of Hue city. There are 6,892 people living in the five villages of Lai Thanh, Long Khe, Son Cong, Lai Bang, and Khe Trai (Thua Thien Hue Statistical Office 2007). Most residents in Khe Trai village are of

Kinh ethnicity, although a small portion of the minority group of Van Kieu ethnicity. People in Huong Van make a living by cultivating rice, vegetables and fruits, and raising livestock such as chickens and pigs. However, Huong Van's location renders it particularly prone to natural disasters: the commune is severely affected by floods between September and December every year. Massive floods in 1999, 2004, 2007, and 2009 devastated housing, social infrastructure, and livestock. The household survey in Huong Van was conducted for 300 randomly selected households in August and September 2006 and group discussion was held in March 2007. Consequently, four activities were implemented in Huong Van commune, namely the establishment of a community resource center and livestock association, building cooperative biogas for utilizing pig manure and saving the cost of firewood, and making safer community planning and vulnerable mapping to reduce the damage from annual floods (Table 6.9).

Huong Phong is a coastal commune in Huong Tra district, 12 km north of Hue city. There are 10,824 people living in the six villages of Thuan Phuoc, Thuan Hoa, Tien Thanh, Van Quat Thuong, Van Quat Dong, and An Lai (Thua Thien Hue Statistical Office 2007). The commune's residents, all of whom are of Kinh ethnicity, engage in aquaculture fishing, rice production, and animal husbandry. Livelihood depends on geographical location: many residents of Thuan Hoa and Van Quat Dong, located near Tam Giang lagoon, engage in aquaculture, whereas the inhabitants of other villages are more typically engaged in rice cultivation and animal husbandry. Huong Phong's coastal location makes it vulnerable to strong winds, flooding, and typhoons. In October 2007, two people died when their fishing boat was capsized by a strong whirlwind. The household survey was conducted for 100 randomly selected households in Huong Phong in March 2007 and group discussion was held in June 2007. Consequently, five activities were implemented in Huong Phong commune: Poly-aquaculture, mushroom production, capacity building on disaster management, small business for the poor women-headed households, and biogas system (Table 6.9).

As seen in Table 6.9, activities clearly vary according to the sites. Huong Van is a flood plain where people are affected by annual floods and more than 70% of the population raises pigs. Therefore, the community mainly focuses on raising livestock and disaster management. On the contrary, in Huong Phong, located along a river and lagoon, people engage primarily in rice cultivation and/or aquaculture. In this sense, diversification of income sources is an important way to enhance their livelihood to avoid risks or shocks, such as those from disasters. Poly-aquaculture activities involve culturing several kinds of aquaculture species rather than mono-aquaculture of shrimp. Mushrooms, which can be grown on the abundant rice straw left after rice cultivation, offer a ready secondary income for local people. In Hong Ha's upland area, most people belong to ethnic minority groups and cultural issues are important for them. For this reason, three of the activities carried out in Hong Ha related to the culture of ethnic minority people at their request. For example, traditional textile weaving is practiced by women of the Taoi minority and its products are used as clothing and gifts on occasions such as weddings, which was taught in a class run by motivated local women. Such cultural activities added a unique dimension not explicit in the rubric of conventional income-raising activities in the project, e.g., goat raising, poly-aquaculture, and mushroom production activities, which aim for increasing household income (Table 6.9). In short, the activities carried out across the three project sites varied because all of them were organized and implemented based on local needs.

6.6 Dzeng Weaving Activities for Ethnic Minorities in Central Vietnam

This section focuses on the implementation of the gender and cultural activities of the project. Previous sections provided an overview of the joint project by two universities in Japan and Vietnam. The activities across the three project sites typically focused on income-raising strategies designed to build markets, raise animals, and plant cash crops. Since the activities of environmental education, ecotourism, and livelihood activities in central Vietnam have been described in the other parts of the section as well as the other parts of this book, this section focuses on the *dzeng* weaving activity for women in Hong Ha. The *dzeng* weaving activity provides a cultural dimension not explicit in the rubric of other project activities and affords the women of Hong Ha an invaluable opportunity to preserve and use indigenous knowledge, which is one of their most precious resources.

6.6.1 Ethnic Minorities in Central Vietnam

The cultural aspect of Vietnam is diverse, officially comprising 54 ethnic groups, accounting for 13% of the population. Central Vietnam is home to 17 ethnic minorities who have lived in this area for a long time and have a strong attachment to this region (Dang et al. 2000). Despite the economic difficulties, the cultural aspects of ethnic minorities are diverse. Each ethnic minority has its own characteristics of living, such as housing, clothing, and eating. Ethnic minorities in Central Vietnam retain many of their typical traditional values, one of which is traditional weaving. The literature written by Howard, M. C. and Howard, K. B. provides an analysis of the textiles of the central highlands of Vietnam, with description of the textiles, including materials, looms, and decorative techniques. Furthermore, their study also comprehensively summarizes the textile traditions of each group (Howard, and Howard 2002). For instance, among Malayo-Polynesian-speaking people, the Jarai group living primarily in Gia Lai province has a traditional textile made of cotton called *kpai,* and Rade group, who live primarily in Dak Lak province, prepare a similar textile as Jarai, but the Roglai group, who live in Ninh Thuan and Khanh Hoa province, have no tradition of weaving, but produce clothes made of bark and bast fiber (Howard and Howard 2002). In this way, traditional weaving, one of the cultural identities of ethnic groups, is complex and diverse.

However, the living conditions of ethnic minorities and their customs relating to clothing have undergone changes. Most ethnic minorities now wear Western clothes, and only some elderly women in remote areas wear traditional clothes. Taking the example of the Hong Ha commune, many young Taoi and Cotu women do not have traditional clothes, and only a few elderly women usually wear traditional skirts. Taoi people in some places near the border of Laos in Thua Thien Hue province still know how to weave; for instance, all Taoi women have traditional clothes and know how to weave in A Roang commune in A Luoi district, Thua Thien Hue province. However, the production of cotton thread is remarkably rare in central Vietnam at present. Some people used to produce their own cotton thread during the first half of the twenty-first century, but most started purchasing commercial threads from local markets due to the dislocation caused by the Vietnam War in the 1960 and 1970s and the easier access to external commercial goods (Howard and Howard 2002). One exceptional can be found in a very isolated area near the border of Laos in Quang Nam province. According to the interviews in Cong Dzong village in Zuoi commune, Nam Gian district in Quang Nam province from May 4 to May 7, 2009, some local Cotu women in Cong Dzong village in Zuoi commune, Nam Giang district, still grow cotton, process it with traditional tools, dye it with natural materials, and weave it on their own (Fig. 6.23). However, places such as Cong Dzong have been declining, and now it is the only place among Cotu groups to complete the whole process of making traditional clothes. Despite such exceptional cases, many traditional elements of ethnic minorities have suffered from neglect and erosion for a long time.

6.6.2 Taoi Group and Traditional Weaving, Dzeng

Taoi is one of the 21 ethnic minority groups belonging to the Mon Khmer language group in Vietnam (Dang et al. 2000). There are approximately 35,000 Taoi people in Vietnam, and most of them live in mountainous areas along the border with Lao in Huong Hoa district in Quang Tri province and A Luoi district in Thua Thien Hue province (Nguyễn et al. 2004). The traditional cloth and garments made from the cloth are called "*dzeng*" in Taoi. In the past, the cloth woven by Taoi women was highly admired by neighboring ethnic groups and bartered for valuable materials, such as livestock (Dang et al. 2000). The Pacoh group, classified as a subgroup of Taoi official nationality, do not weave but obtain their clothing from the Taoi group, who have a tradition of weaving. Both the Pacoh and Taoi groups live in the Quang Tri and Thua Thien Hue provinces. Cotu, who also live in Thua Thien Hue and Quang Nam province, have few cultural differences from Taoi, and their textiles are identical (Howard and Howard 2002).

6.6.3 Dzeng Weaving in Hong Ha

6.6.3.1 Overview of Hong Ha Commune

Hong Ha is a mountainous commune in A Luoi district, located 45 km southwest of Hue city. The commune is divided into five villages, Con Tom, Pa-hi, Can Sam, Paring, and Arom, which together provide home to 1,332 people (Thua Thien Hue Statistical Office 2007). Figure 6.24 shows the ethnic ratio of Hong Ha in 2009, mainly consisting of the Cotu, Pacoh, Taoi, and Kinh people. Cotu, Pacoh, and Taoi are ethnic minorities in Vietnam, but they make up the majority of the Hong Ha commune, contributing to the cultural nuances of ethnic groups. People of Kinh ethnicity—Vietnam's majority ethnic group—comprise less than 10%, most of whom are engaged in various businesses. Hong Ha is one of the 16 poorest communes in A Luoi district and among the 1,581 poorest communes nationwide, according to national poverty criteria (Hoang and Le 2006). The Cotu and Taoi ethnic groups share similarities in many aspects, such as the language group of Mon Khmer, living place, production, beliefs, housing, and family relations according to Vietnamese sociologists (Nguyễn et al. 2004; Hoang and Le 2006).

Life for many local people living in Hong Ha historically depended on slash-and-burn cultivation and the use of natural resources from the forest. These agricultural practices are discouraged by current government policies, which promote sedentary farming methods. As livelihoods in Hong Ha depend almost entirely upon cultivation and the forest, government sponsored changes in agricultural practices have a large impact on village life. Not all of these impacts are positive: a great deal of once invaluable indigenous knowledge is being lost, for example. That loss impacts upon many aspects of domestic life, such as food, housing, clothes, and family relations, for the commune's ethnic people.

The situation of Hong Ha's women gives a particular cause for concern. An overview of the research project named Community-based Upland Natural Resource Management in Hong Ha from 1998 to 2003 authored by two of its facilitators, Hoang and Le, cites previous research showing that while the women of Hong Ha commune spend 10 times as much time doing housework than men, they also spend more time doing farm-work than men (Hoang and Le 2006). The same overview highlights the fact that women have a lesser role in decision-making and less control over the allocation and use of natural resources in the commune. Commenting that "While men talk, it is often the women who do most of the work," Hoang and Le call for participatory approaches at the community level that enhance the capacity for Hong Ha's most disadvantaged people. These approaches, the authors emphasize, should "direct attention to the role and needs of women." The *dzeng* weaving activity described in the following section provides a good

Fig. 6.23 Women and *dzeng* weaving in Cong Dzong village. Photos by Akiko Iizuka, May 6, 2009

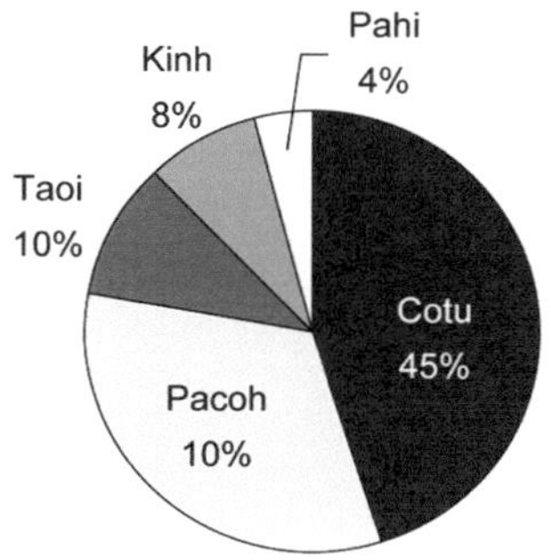

Fig. 6.24 Ethnic ratio of Hong Ha

example of the kind of capacity building proposed by Hong and An (Hoang and Le 2006).

6.6.3.2 Discussions and Planning of Weaving Class[7]

As one of the project activities introduced in the previous section, *dzeng* weaving activity was set up in April 2007 in response to the wishes voiced by women at meetings held in the Hong Ha commune in December 2006 between the project facilitators and villagers. Villagers made suggestions about the activities they wanted to pursue in their villages. Women of various age groups asserted that they were interested in learning how to weave *dzeng* textiles woven traditionally by women of the Taoi minority group. Project activities across the three project sites are mainly livelihood activities that increase income and/or diversify income sources using local resources. *Dzeng* weaving activities added a cultural aspect that is essential for ethnic minorities in Hong Ha.

The women explained that they were highly motivated to learn how to weave if a teacher could be found for them. They would also need equipment and materials. They listed both financial and cultural considerations for their high levels of motivation. Purchasing *dzeng* at A Luoi market located 15 km from Hong Ha commune is both time consuming and costly: the fabric needed to make a skirt costs between 15 and 30 USD, which accounts for a large part of a household's expenses in Hong Ha terms. By learning how to make *dzeng*, the women hope to reduce their expenses as well as restore tradition and cultural pride in their village.

The first task in meeting the women's wishes was to help them find a teacher. This task was facilitated by Ms. Lam, the leader of the local women's union, who introduced project members to Ms. Kan Tam, one of the few women in Hong Ha commune who knew how to weave. Reluctant at first to share her skills with the other women, Ms. Tam agreed to teach the weaving class after some discussion. Next, Ms. Lam oversaw the selection of five learners to participate in the weaving activities. Learners were sought to (1) participate throughout the learning process despite other demands on their time, (2) share their learning experiences with other women, and (3) attend classes without reimbursement. Over the course of several discussions with the women learners, their new teacher, and project members, Ms. Lam identified when, where, and how often classes should take place.

6.6.3.3 Implementation

The classes started in April 2007, with four learners making daily visits to Ms. Kan Hong's house, one of the participants, to study weaving. Later, in September, the place of weaving moved from Kan Hong's house to Hong Ha's newly built community house (Fig. 6.25). The community house's location at the center of the commune was easy to access for most participants. The activity received a great deal of attention from other women in the village, attracting large numbers every day to observe the class. Class numbers quickly expanded to allow six of the most eager observers to join. The teacher offered to teach them simultaneously as the first intake, on the grounds that the first group was already able to weave by themselves and able to provide supervision when they needed it. In this case, and on other matters such as changes to when

---

[7] This subsection is reproduction of the article by Akiko Iizuka, etc., one of the authors of this section and has been deposited in the repository of Kyoto University at https://repository.kulib.kyoto-u.ac.jp/dspace/handle/2433/108247. It does not violate any policy of the repository".

Fig. 6.25 Teacher and women in *dzeng* weaving class. Photos by Akiko Iizuka

and where classes would be held, the learners and the teacher were entirely self-organized, managing their classes capably, without the need for intervention from project members. The next section describes the steps taken to enable women to master weaving.

Their first task was to prepare warp wool for weaving. The teacher showed how to roll wool onto a large wooden lattice structure (Fig. 6.26). After transferring it to looms, the learners started to learn how to weave. Subsequently, the learners faced three basic stages in learning how to make *dzeng*:

Weaving with colors The first stage in *dzeng* weaving depends entirely on the use of intense colors (Fig. 6.26). Black, white, and red are the staple colors, although other colors, such as yellow, green, blue, and purple, may be added as weavers grow in skill. The width of the *dzeng* is approximately 40 cm across (the breadth of a loom). Lengths (approximately 150 cm) vary according to the height of the woman weaving. The size is usually determined using women's body modules. The fastest learners finished their first lengths of cloth within six weeks of starting to weave. Next, they proceeded to make their second *dzeng*. The second attempt was smoother and tighter as the wool broke less often the more proficiently it was woven.

Weaving with patterns and colors Mastering the second stage of proficiency in *dzeng* requires the learners to make patterns with colors (Fig. 6.26). Patterns in *dzeng* are largely traditional, with each pattern carrying a unique meaning, as shown in some examples in the previous section. Most of the learners in the activity were nervous in embarking on this stage, but soon mastered the skill, with explanations from the teacher.

Weaving with beads The third stage of *dzeng* weaving incorporates glass beads into the pattern (Fig. 6.26). Beads are not cheap and their use requires a great deal of skill; therefore, beaded *dzeng* are far more expensive to purchase than plain colored or patterned colored styles. The learners were told by their teacher that learning to weave with beads would take more time than any other stage in the learning process. Only skilled learners can progress to this stage.

Overall, the group progressed well. All women mastered the technique of weaving patterns with colors in the first six months of the activity. Eight of them were able to incorporate the use of glass beads into their weaving, a level of skill that tests even the proficient *dzeng* weaver, by January 2008.

6.6.3.4 Evaluation of Weaving Activity for Women

The impact of any project activity is often measured by its economic impacts, such as household income or amount of harvests, or physical gains, such as completed physical infrastructure; however, this cultural activity is intangible in nature and its impact cannot be quantified. Therefore, individual interviews were carried out to evaluate the impact of the activity. Individual interviews with 12 women who participated in the activities were conducted to listen to the opinions and voices of the women. The interviews were not only meant for evaluating the activity but for listening to the oral histories (experiences and lives) of the women in Hong Ha since the project activities are just a part of their lives. Twelve interview transcriptions between 20 and 30 pages per interview were completed between July 2007 and November 2010.

Women who took part in weaving activities worked on average 15 h a day, combining farmwork, foraging in the forest for fuel and raw materials, housework, childcare, and the day-to-day business of feeding their families. They took time they could not afford to lose to learn how to weave. The natural question is why? Why did they want to learn how to weave? What did they expect to gain from the activity? *Dzeng* are considered invaluable as they play the role of social and symbolic currency, and the process of learning to make it, meant many things to the women: tradition and financial stability, but also, by extension, identity and pride. "I believe my daughter is proud of me," Khach, one of the participants, says, simply. "I will teach my three daughters when they are old enough not to break the wool," explains E. As for At, forced to

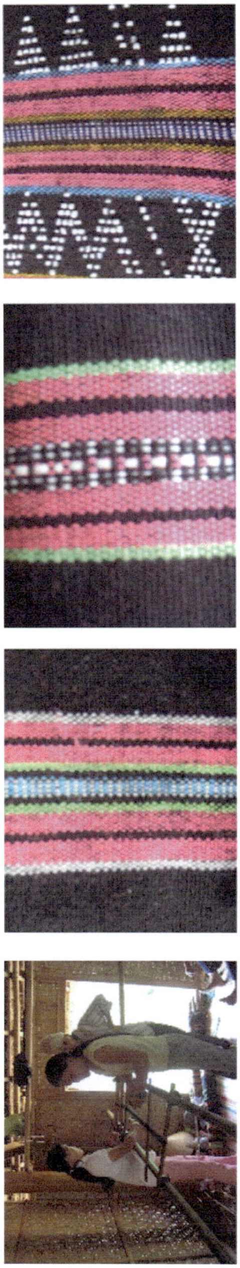

Fig. 6.26 Preparation and three stages to weave *dzeng*. (Photos by Akiko Iizuka

abandon her hopes of higher education because her family could not afford to pay her travel fees to the local high school in A Luoi, she simply enjoys having the opportunity to learn.

At, like many of the others in the group, comes to study "even when the teacher is sick." On other days, the women help their teacher in sowing, growing, and harvesting. Even after ending the training period, Nang says "learners still help each other, for example, with cultivation. It is our custom to help each other. We knew each other before and were friends in the class, so we often visit each other and help out when we can." Why? Oral testimonies have provided a vital context for these reasons. To sustain the conditions that allow them to come together for learning, for sure, but also for the simple joy of being together. Ngoat continues her weaving after the training period and prefers to weave at the community house together with the other learners rather than alone at home "because it is cheerful." Khach, who attended the class diligently, finished only four *dzeng* over the course of the weaving class. According to the other learner, "Khach comes to the community house to cook and chat with the group, not to weave." Nhac enjoyed so much during the weaving class and said "Having friends is happy… I like coming here, to work, eat and rest with the other women." As for E, "it comforts me to be around the other women."

Their motivation and enthusiasm for learning *dzeng* are recognized through their testimonies; however, three women, Bien, At, and Xuan, quit training in the middle. Bien quit weaving one month after the class started due to her assumed responsibility to take care of her family and her mother's illness. She says, "I like weaving, but because there is no one working in my family, I cannot weave." At quit the class when she became pregnant in the middle of the course. She was the best weaver at that time, just after she learned how to weave with beads. She deeply regrets quitting halfway to achieve the weaving skills and says, "First I quit academically at school and now I've quit weaving." After At quit, Yew, the younger sister of At, took At's place. Both sisters did not attend the class during the entire period, but they obtained weaving skills. Xuan, just 17 years old when she started to learn weaving, quit weaving after two months of the classes. She started weaving because her parents encouraged her to learn, but later she found it very hard, although she liked to attend the class and chat with the other learners. Truc, Xuan's sister-in-law, was also encouraged by Xuan's mother and took the second class. She had completed the training and gave *dzeng* to her mother-in-law. These three women quit weaving in the middle of the training course; however, each of them has their own reasons for consideration, and their hope and motivation are followed by some other people. At weaved one *dzeng* after quitting the class and hopes to weave again in the future, saying "if my conditions change." The progress of the learners thus varies greatly based not only on the ability of the individual but also on her personal circumstances.

The weaving class continued from April 2007 to June 2008. Some women continued to weave *dzeng* at their own pace. Ha, an accomplished weaver who made 13 *dzeng* during the training, continues to weave whenever she finds the time. When it rains, she weaves from 7 to 11 am and from 1 to 7 pm. When she is busy, she weaves only for 30 min. Ngoat finished 10 *dzeng* during the training and made another 10 by April 2009 on her own after the training. She enjoys weaving, giving, or selling at a low price to her friends. Both Ha and Ngoat helped the second weaving class that started from September 2008. Hong, the supervisor and learner of weaving activity, also helped a new learner who was stuck during the classes. Therefore, the second class went more smoothly than the first one, especially because the learners in the first class often stopped by the class and assisted in teaching.

The solidarity and cohesion this disparate group of women from five ethnic groups and five different hamlets are now showing is a powerful indicator of the value they place on the *dzeng* weaving activity. Their enthusiasm for learning is recognized and supported by many husbands and family members in the village, who encourage their wives, daughters, mothers, and siblings to attend the class. The opportunity to be and learn together is particularly meaningful for women for whom work, home, and family would more ordinarily define the day-to-day. What will the women of Hong Ha choose to do in the years to come? Will they uphold the fabric of rural life through their hard-won traditional knowledge and the fabric they have created by their own hands? Will they weave a future for Hong Ha that builds upon their village's distinctive and diverse ethnic past and present, and help shape a sustainable rural future for their families, neighbors, villages, and natural surroundings? Only time, or the women themselves, will perhaps tell.

6.7 Small-Scale Local Livelihood Activities for Enhancing Household Economy and Ecosystem Conservation

Central Vietnam, known to be an area of poverty along with the mountainous areas in the north and plateaus in the south, is located in a fragile environment where land resources and ecological environments are easily deteriorated by the expansion of human activities. Land use pressure is high since the land suitable for agriculture and habitation is limited to narrow flood plains, coastal plains, and gentle hills. In recent times, ever-increasing population and human activities have expanded the land use to steep mountains and hills. Shallow soils are susceptible to soil erosion under intensive rainfall and nutrient depletion through cultivation. Major livelihood activities such as agriculture, forestry, and fishery are small-scale activities that can no longer be sufficient income to sustain a livelihood. In search of better incomes and employment opportunities, there is an outflow of youth and adult population as migrant workers to Hanoi, Ho Chi Minh City, and nearby cities, and, consequently, the percentage of elders and women in the area is increasing. This trend may have a serious impact on maintaining local livelihood activities and land resources, as well as on ecosystem conservation. It must be noted that a "vicious cycle of poverty and ecosystem degradation" may easily occur in areas where local livelihood activities are dependent on surrounding natural resources, the youth and adult populations are declining, and human activity continues to expand.

Central Vietnam is also a natural disaster-prone area. For example, the frequency of typhoons landing in Vietnam between 1951 to 2010 is 0.7 times a year in the north and 0.2 times in the south, but it is 1.8 times in the central part (Tran et al. 2012). In addition, although recent dam construction eliminated severe floods during the rainy season, flooding above floor level still occurs three to four times a year. On the contrary, in a drought year, salt damage occurs due to the inflow of seawater into paddy fields in the coastal plains. Repeated natural disasters gradually erode people's household economy and the damage to vulnerable people such as elders, widows, handicapped, and economically poor is especially serious.

In certain selected areas near Hue city, Central Vietnam, we developed new livelihood activities and improved existing ones based on the following requirements: (1) must be manageable with low labor input, (2) must bring a higher value than existing products, (3) must be effective for conserving land resources and ecosystems, (4) must use local materials to save cost, and (5) must have room for participation of local people. In this section, the case of small-scale pig raising in areas prone to natural disasters and the case of small-scale livestock-based livelihood activities in mountainous areas are introduced (Hồ et al. 2010; Le et al. 2013).

6.7.1 Case 1: Small-Scale Pig Raising in Natural Disaster-Prone Area

Huong Van commune, where the trial of small-scale pig raising was carried out, is located approximately 30 km north of Hue city, Central Vietnam. The major livelihood activities here are rice cultivation in the flood plains

Fig. 6.27 Pig pen under small-scale pig raising

and acacia plantations in the hills. In order to increase household income, in the early 2000s, Hue University of Agriculture and Forestry and Women's Union of the commune started small-scale pig raising with 5–10 pigs (Fig. 6.27). With this trial, the household income, which traditionally relied on rice cultivation, became double. The activity, however, posed some challenges such as contamination of feed and low market price of pigs derived from seasonal flooding, unsanitary environment caused by foul odor and sewage from pig pens, and further increase of income. It was also necessary to encourage people who are willing to participate but cannot afford to input the initial cost.

6.7.1.1 Evacuation Terrace to Escape from Flooding

The annual floods are a major problem for small-scale pig raising. In the event of a flood, people prioritize their lives and property and cannot afford to evacuate pigs to flood-free areas. Thus, at the beginning of the rainy season, pigs are put up for sale one after another; due to this, the selling price in the market goes down. An "Evacuation Terrace" was devised as a measure against floods in small-scale pig raising. This is a terrace made of concrete or bamboo at the height of the roof. In case of a flood, pigs can be guided to the evacuation terrace and kept for several days until the water drains.

After the flooding season, Vietnam's Lunar New Year (*Tết*) begins soon and the price of pigs in the market rises; thus, selling pigs at this time will increase household income. The cost of the evacuation terrace can be recovered with this profit in a few years.

6.7.1.2 Cassava Silage for Pollution Control and Income Generation

Flooded water mixes with filth and manure, which may contaminate feed and forage crops. Pigs, once escaped from the floods, then face problems of shortage of feeds and pollution. As a countermeasure, we devised cassava silage, which is a fermented feed made from cassava, for pig raising during flooding period. Cassava (Fig. 6.28), a root crop native to South America, is usually used for human consumption, livestock feed, and extracting starch. There is a starch factory near the commune, and cassava is cultivated as one of the cash crops. The purchase price at the factory, however, fluctuates in tandem with the price of starch in the international market. In certain years, when the purchase price is low, cassava is left in the field without being harvested. In such cases, the cassava can be harvested, sliced, mixed with rice bran in some cases, packed and sealed in a plastic bag, and buried in the soil to make fermented feed (silage). The reason to burry cassava in the soil is to avoid being exposed to contaminated flood water. After the flood water is drained, the fermented cassava is dug up to feed to pigs without cooking. The nutritional value improves due to the increase in digestible sugars, amino acids, and proteins through fermentation. Some fermented cassava is dried and stored for home-consumption, or the surplus can be sold on the local market. This can increase household income and also help buffer the income decline associated with the cassava price plunge.

6.7.1.3 Wasted Acacia Bark Improving Sanitation and Health for Pigs

Acacia plantations spread out in the hills. Acacia is cut down once every six to seven years after

Fig. 6.28 Cassava utilized for fermented feed

planting and exported as a papermaking material. Since the bark is not suitable for papermaking, most of it is left as waste and only a small part is used as fuel. We tried charcoal and wood vinegar (a liquid of pyroligneous acid), which is produced from the bark, to apply to pig raising for the health and the improvement of sanitation (Hồ et al. 2016).

Examination revealed that the wood vinegar diluted to about 0.2% with water was effective in preventing diarrhea in piglets. In local pig raising, feeds containing antibiotics and growth hormones are often given to piglets for preventing diseases and promoting growth. The fact that the use of diluted wood vinegar can prevent diseases during the piglet period directly translates to a lower production cost. This also dispels concerns about residual agents such as antibiotics and hormones remaining in the meat. In addition, by sprinkling the charcoal powder and undiluted wood vinegar on the pig pen, it is possible to suppress the generation of foul odors and flies, which are a nuisance to the neighborhood and make the sanitary environment worse.

6.7.1.4 Support for People Who Cannot Participate

In the commune, there are people who cannot participate in the small-scale pig raising for various reasons, even if a certain approach or technique is excellent and rational. When we asked the members of Women's Union for ways to encourage the participation of such people, they laid a unique proposal, wherein those who can afford could lend a pregnant sow (female pig) to poor households free of charge. These households first build a simple pig pen on their own and take care of the sow every day, including feeding. Advice and encouragement come from women who are good at raising pigs. The sow is then returned to the owner after birth, the piglets remain in the poor household, and the pig raising activity is continued with it. This activity is thought to have derived from the traditional idea of mutual support that still remains among local people.

6.7.2 Case 2: Small-Scale Livestock Raising in Mountainous Area

About 30 km west of Hue city, there are communes of ethnic minorities in mountainous and hill areas. The major livelihood activity here is acacia plantation, which produces pulpwood for papermaking. Rice (both in paddy and upland fields), cassava, and vegetables are cultivated mostly for home-consumption. Acacia is harvested every six to seven years. Each time, soil on the slope land gets exposed and is washed away during the rainy season after the cutting. The recent expansion of acacia plantations has increased the chances of soil erosion.

On the contrary, the area of a single acacia plantation owned by each household is getting smaller due to the division of land from inheritance. The labor and materials costs for managing the plantation are also increasing, which makes it difficult to obtain satisfactory profits. Those who could not inherit the land earned a small income as wage labor working in the plantation and as seasonal migrant workers. Outflow of youth and adult populations from the commune as migrant labors causes labor shortage, thereby making maintenance of local livelihood activities even more difficult. This situation also makes it difficult to take the necessary measures for preventing land and ecosystem degradation as well as their recovery.

In response to the above situation and as an alternative to acacia plantation, we tried to seek certain livelihood activities that satisfy the following conditions: (1) to be manageable with less labor and materials input in order to enable the participation of vulnerable households, (2) to utilize locally available resources, (3) to provide sufficient income, (4) to provide employment opportunities for young people who do not have regular jobs, and (5) to have less burden to surrounding land resources and ecosystems. The livelihood activities thus selected included raising of indigenous mini-pig, raising of crossbred chicken with wild and ordinary poultry, and a beekeeping at home-garden.

Fig. 6.29 Indigenous mini-pigs

6.7.2.1 Raising of Indigenous Mini-Pig

In the commune, people raise cattle, buffaloes, pigs, and chickens. Since the land on flat or gentle slopes is narrow, fields for forage cultivation and cattle grazing are limited. In the past, grazing of buffalo in the forest was popular, and it was sold to the communes in flood plain areas as a draft animal for rice cultivation and pulling carts. Demand for buffalo is however declining due to mechanization in paddy fields, as well as a shortage of labor to take care of buffalo. Pigs are raised for cash income; however, the numbers are now decreasing because the selling price of local pigs has declined due to an increase in meat supply from large-scale pig farms, the shortage of labor necessary for feeding and cleaning pig pens, and the outbreak of foot-and-mouth disease as well as the so-called blue ear disease. Chickens are free-range and the production scale is small. Livestock raising is, however, attractive since the products (meat and eggs) have higher selling prices as compared to the cultivation of grain crops, vegetables, and fruits.

We paid attention to indigenous mini-pigs, as shown in Fig. 6.29, which have been inherited by ethnic minorities in the mountains of the Indochina region. In recent years, pig breeds that grow large in a short period of time have become common; thus, the indigenous mini-pigs that grow slowly and are small in size have almost disappeared. Indigenous mini-pigs, however, still retain many benefits for small-scale households. These include, for example, disease resistance, good meat quality, low labor input due to grazing, and feeding on locally available feeds such as fresh leaves of sweet potato and taro, chopped pseud-stem of banana and fermented cassava, and so on. Simple pig pens and fences can be made from locally available bamboo and acacia, and grazing space is available in acacia and rubber plantations (Fig. 6.30).

Raising one male and five female mini-pigs yields at least 25–30 piglets annually, which can

Fig. 6.30 Grazing space for indigenous mini-pigs and crossbred chicken under rubber plantation

be fattened and sold on the market within a year. Due to the rarity and the meat quality, as well as being free from antibiotics or growth hormones, they are sold at prices that are three to five times higher than those for regular pigs. If the selling price of one mini-pig fattened for one year is 100 USD (conversion from Vietnamese dong to U.S. dollar), including 20 USD for its production cost and assuming that 20 pigs are sold, the profit will be approximately 1,600 USD, which is equivalent to 58.9% of Vietnam's 2019 GDP per capita. In addition, when there are events such as weddings, small sized mini-pigs suitable for roasting are traded at even more favorable prices. Depending on the labor availability and motivation of the household, the number of mini-pigs raised can be increased to get more income. Therefore, raising of indigenous mini-pigs can be a suitable local livelihood activity for improving and stabilizing household income. In addition, using the above profits as a source of funds, sows bred with excellent traits can be selected or can be obtained from the communes of ethnic minorities in the mountain areas. It should be noted that the raising of indigenous mini-pigs itself also facilitates the conservation of genetic resources.

6.7.2.2 Raising of Crossbred Chicken with Wild and Ordinary Poultry

Wild chickens live in mountainous forests. Local people get crossbred chickens (Fig. 6.31) by mating a trapped male wild chicken with ordinary poultry, or mating naturally by free-ranging ordinary poultry hens on the edge of a natural forest for several days.

The crossbred chicken is kept in a simple poultry hut made with local materials such as bamboo. Some hens are used for breeding, and surplus chickens and eggs are sold. Since the traits of wild chickens disappear from the crossbred in about three to five years, at that time, the wild chicken crossbreds are obtained again by natural mating. The crossbreds are free-ranged in acacia plantations, rubber plantations, and fallowed agricultural lands. Locally available

Fig. 6.31 Crossbred chicken (female) with wild and ordinary poultry

materials such as rice bran, chopped pseud-stem of banana, and fermented cassava are given as feed. This method of keeping requires little labor input.

Compared to ordinary poultry, the crossbred is considered to be more resistant to diseases. The meat is thus recognized as being a safer and more secure product than that of ordinary chicken, due to non-use of antibiotics and growth hormones. Moreover, due to its rarity, it is sold at a high price of approximately 30–40% above usual prices. If the selling price of one crossbred kept for half a year is 20 USD (conversion from Vietnamese dong to U.S. dollar), including 5 USD for its production cost, and assuming 50 crossbreds are sold, the profit will be approximately 1,000 USD which is equivalent to 41.4% of Vietnam's 2019 GDP per capita.

Since this livelihood activity uses wild chickens that live in forests, it is premised on direct coexistence with forests and may incentive local people to make efforts for ecosystem conservation. In addition, the supply of wild chicken crossbred to the market can be expected to reduce poaching of wild animals, including endangered species, which is one of the major problems threatening ecosystem conservation in the mountainous areas of Vietnam.

6.7.2.3 Beekeeping at Home-Garden

In the vast areas of acacia plantations around the commune, beekeeping can be carried out using honey sources from the flowers and new shoots of acacia. Current beekeeping activities are carried out by beekeeping companies using Western honeybees, which are seasonal (April to July) and are limited to plantation areas along easily accessible paved roads (Fig. 6.32). Beekeeping is possible in other seasons, but very few households do such beekeeping on a small scale (Fig. 6.33). In other words, most beekeeping resources in acacia plantations and natural forests remain unused. In addition, some species, including subspecies, of wild honeybees inhabit the forests and acacia plantations, and it is possible to use them for beekeeping.

Beekeeping at home-garden is an activity that involves setting up beehives for Western or wild honeybees in the garden of a residential area adjacent to an acacia plantation that beekeeping companies cannot access. By setting up beehives on the premises of the house, the management, for example, cleaning the hives, controlling natural enemies such as wasps, and harvesting honey every two weeks can be done easily. One of the advantages of beekeeping is that honey can be collected within the first few months of installing a new beehive. This is attractive to those who want to quickly recover their initial investment. Since the labor input is lower than other livelihood activities, it can be introduced even in elderly households and poor households that are busy with their daily work. Due to the abundance of honey sources, honey can be harvested from the hives of Western honeybees

Fig. 6.32 Beehives installed by a beekeeping company

Fig. 6.33 Beehives at home-garden

once every two weeks. Western honeybees are obtained by purchasing a hive containing a swarm of bees from a beekeeping company, or by placing a queen bee raised by a farmer in the handmade hive to form a bee swarm. In the case of wild honeybees, the harvest is two to three times a year. A swarm of wild honeybees can be captured by a handmade hive partially coated with beeswax.

The price of honey in Vietnamese market fluctuates greatly from year to year and is not easy to estimate it; however, in our field demonstration, honey from 30 hives of Western honeybees brought 1,200 USD a year. This is equivalent to 44.2% of Vietnam's 2019 GDP per capita. Due to its rarity, honey of wild honeybees is traded in the domestic market at three to four times the price of Western honeybees, making it an attractive new source of cash income.

Beekeeping activities encourage ecosystem conservation as well. If the profits from beekeeping increase and the scale grows, the chance of soil erosion can be reduced by selectively planting flowering trees and fruit trees, as honey sources, on fragile steep slopes instead of acacia.

6.8 Reconstruction Project of Cotu Traditional Community House in Nam Dong District, Thua Thien Hue Province, Vietnam

6.8.1 Background

Cotu, one of the 53 ethnic minorities in Vietnam, mainly inhabits the mountainous areas of Quang Nam province, Thua Thien Hue province, and part of Danang city, Vietnam. Their population was approximately 92,000 in 2020 (Bh'Riu Liếc 2020: 9). The original Cotu village consists of several tens of houses, surrounding a traditional community house named Guol. The layout of Cotu villages can have oval, horse-shoe, or polygonal layout (Kobayashi 2008: 119; Nguyễn et al. 2004: 98). In Cotu village, Guol is normally located at the center of the village. It plays the role of an administrative, social, and cultural center of the village (Lưu 2007; Trần 2009). However, traditional Cotu village and Guol houses have gradually disappeared and deteriorated due to lifestyle changes.

From 2013 to 2015, a series of field surveys were implemented in different Cotu villages in Thua Thien Hue province (A Luoi, Nam Dong), Quang Nam province (Dong Giang, Tay Giang), and Danang city for exploring traditional architectural and cultural values of the Cotu. Therefore, construction of a Guol house based on traditional methods and processes was suggested. The purpose of these field surveys was to find potential sites for the construction of Guol. Finally, A Ka hamlet, Thuong Quang commune, Nam Dong district, Thua Thien Hue province was selected for the reconstruction project because A Ka villagers were found to show sufficient motivation and have knowledge of traditional construction as well (Fig. 6.34).

A Ka hamlet, one of the seven hamlets of Thuong Quang commune, has 107 households and 328 inhabitants, and all of them are Cotu (according to the statistical data in 2018). The hamlet is on the west, 12 km far from the center of Nam Dong district.[8] In the past, A Ka hamlet had a traditional community house, which was built in 2004 according to an interview with the hamlet leader. The house was leaky on the roof in 2010, mostly decayed in 2012, and it completely collapsed in 2013 (Fig. 6.35). Therefore, the villagers indeed needed to construct a new traditional community house. Besides, the patriarch and certain old villagers still knew the traditional method for the construction of Guol. Thus, the "Reconstruction of traditional community house of the Cotu in A Ka hamlet" project was established in 2016 (Nguyen et al. 2019).

After different field surveys, meetings, and discussions with local villagers and the local

[8] Nam Dong, a mountainous district in Thua Thien Hue province, is located on the southwest of Hue city. According to 2017 statistical data, its area is 64,777.9 km^2 with 11,945 Cotu inhabitants, which constitute 70% Cotu ethnic minority in Thua Thien Hue province.

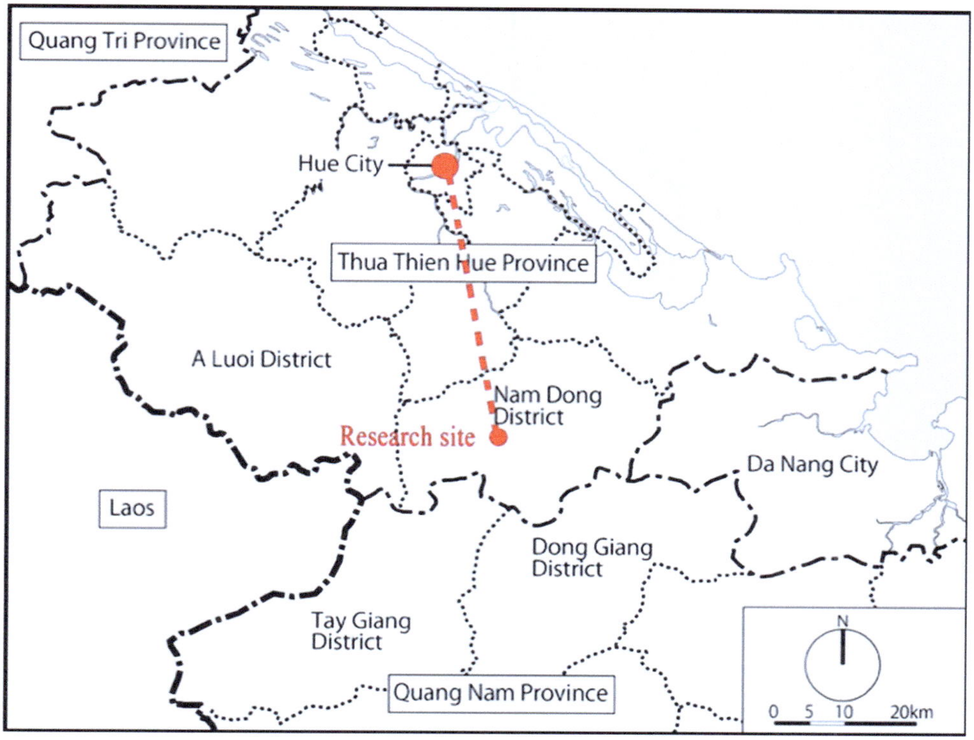

Fig. 6.34 Research site of the project

government (district, commune, and hamlet levels), an agreement among A Ka hamlet, local government, consultative board (consultants from Kyoto University—Japan, University of Sciences—Hue University, and Vietnam Institute for Cultural and Arts Studies—Hue Branch) was signed on March 2016. The following summarizes the major terms of the agreement:

- Implementation period: March 2016–October 2017 (the project was actually finished on August 2018).
- Budget: Labor of villagers and Kyoto University, Japan uses part of the research fund (95 million VND) to support part of labor (calculated budget was 245 million VND).
- Purpose: The construction of Guol was implemented based on traditional methods; local community was the main subject for the construction; the construction process was recorded as a useful source for maintaining unique cultural value of the Cotu.
- Responsibilities of each stakeholder: A Ka hamlet was responsible for materials (wood, leaves, etc.), implementation of outwork for Guol, construction of Guol, organization of necessary ceremonies, and preservation of the house after finishing of construction. The local government was responsible for support with administrative documents and facilitation of the process. The consultative board was responsible for partial financial support, consultation on construction techniques, traditional ceremonies, etc. Besides, the researchers were asked to record and preserve the process of reconstruction, interview and organize meetings, and conduct seminars to share knowledge of construction and preservation after completion.

Table 6.10 shows the progress of the project. First, the management board for reconstruction

Fig. 6.35 Deterioration process of Guol house in A Ka hamlet

Table 6.10 Progress of reconstruction of Guol house

No	Time	Content
1	March 2016	Statement agreement
2	March 2016–August 2016	Discussion, work division: Group 1-explotation of materials; Group 2-doing outwork; Group 3-construction technique; Group 4-ceremony organization
3	13 June 2016	Decision no 1279/QĐ-UBND of the Chairman of Thua Thien Hue province for promulgating reconstruction project for Guol house at A Ka hamlet
4	02–14 August 2016	Ceremony for choosing land Ceremony for disinfection Ceremony for wood exploitation
5	17 August 2016	Field survey in forest and mark selected trees
6	August 2016–March 2017	Exploitation of some main components for Guol house
7	April 2017	Transport
8	March 2018–May 2018	Performing outwork

was established. It was divided into four groups for carrying out the four main works. Then, the hamlet organized a ceremony to choose the land for constructing the Guol house, worship for disinfection, and other similar traditional customs of the Cotu. The wood was sourced and transported, and the outwork and components of the Guol house were prepared. Because of several reasons (weather, certain villagers passing away, policy of central government for closing forest, etc.), the project timeline was extended to May 2018.

6.8.2 Reconstruction Process

The process was implemented based on traditional methods with the participation of villagers. The villagers were divided into four groups, as shown in Table 6.10. Members of each group were decided depending on their work and capabilities as well as the specific time of reconstruction project. For example, villagers who have considerable experience in choosing good wood were employed for sourcing wood from forest, while the collection of roof leaves was allocated to ordinary villagers. The main steps of the construction process are described below.

6.8.2.1 Material Collection

Before the reconstruction, the hamlet selected the land of old Guol house for setting up new Guol house. A ceremony for choosing land, disinfection, and wood sourcing were organized on August 2016. Group 4 (leader is the patriarch of the village) was responsible for organizing and implementing traditional ceremonies and worship during the project. The new house faces the Northeast (same as the orientation of the former Guol house) because this is a good orientation as the Cotu's conception.

Material collection constituted the first stage of construction. Consultants and villagers agreed to implement this step based on traditional methods. Therefore, villagers did not buy wood from the market and instead collected it from the forest. However, exploiting wood from natural forest was forbidden by a policy promulgated by the Prime Minister's Office of Vietnam in 2016. Thus, the villagers sourced the timber from the community forest, which led to timber quality becoming a major concern. Therefore, Group 1 was divided into small sub-groups (each subgroup had 4–8 persons) who went to forest to evaluate wood quality and mark trees for cutting. According to local knowledge, the optimal time for cutting a tree is at the end of the month

Fig. 6.36 Wood transported by a buffalo

(as there is no moon at the end of month and termites will not eat the wood) and in dry season for more convenient transport.

After sourcing the timber, buffaloes were used to transport these trees out of the forest (Fig. 6.36). Then, delivery tricycles were used to carry them to the hamlet. This work was implemented over a period of more than one year. By March 2018, wood was sourced for all main components (pillars, beams, and purlins). The material for the roof, wall, and floor was sourced during the time for outwork and construction of the house.

6.8.2.2 Component Outwork

Component outwork, managed by group 3, constituted one of the most important works in the construction process. For implementing the outwork, villagers used traditional tools such as axe, knife, chisel, etc. However, for convenience and to speed up the process, the villagers also used several modern tools such as saw, hammer, etc. Cotu uses body-based units for the measurement and design of their dwellings (Kobayashi and Nguyen 2013; Truong et al. 2013; Nguyễn et al. 2014). According to surveys, Cotu uses 18 body-based units for measurement, design, and construction.[9] In case of A Ka hamlet, the villagers used 15 body-based units for the design and construction of the Guol (Fig. 6.37).

During the reconstruction, the formal hamlet leader, Mr. Ho Van Bang, headed group 3 and managed all the component outwork. He used his body size for designing and providing the size of the house's components (Fig. 6.38). Pillars, especially a center pillar, are critical for a stable Guol structure. Usually, pillars are

[9]According to previous surveys, Cotu ethnic group has 17 body-based units (Kobayashi and Nguyen 2013). However, based on the interview with 25 patriarchs in Nam Dong district in 2018, Cotu ethnic group was found to have one more unit for measurement, design, and construction.

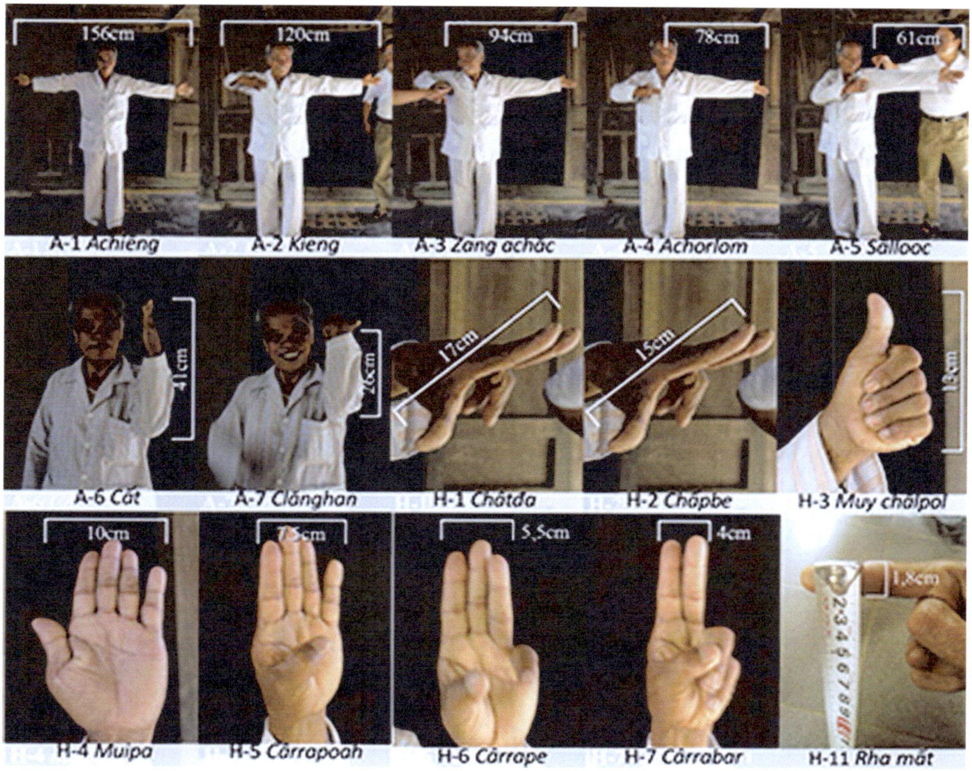

Fig. 6.37 Fifteen body-based units used for the reconstruction in A Ka hamlet

approximately 20–35 cm in diameter. This work took three weeks to complete (Fig. 6.39).

6.8.2.3 Decoration and Sculpture

Group 2 was tasked with designing, carving, and painting images on the components of the house such as the center pillar, cross beam, motif carving on the roof, etc. (Fig. 6.40). These images expressed the lifestyle of Cotu ethnic group and their religious beliefs, and depicted normal Cotu life, animals (buffalo, pig, chicken, etc.), images of flowers, leaves, trees, traditional dancing, etc. Upon comparing the Cotu Guol house to those Cotu in other areas, we can find that the sculpture in A Ka Guol house is simpler. The A Ka villagers only sculped the center pillar of the Guol house, while other components were painted without any sculpture. Table 6.11 shows the different topics used in the artwork. The decorative patterns are mainly distributed on the center pillar, cross beams, ridge beams, and round beams.

6.8.2.4 Guol Construction

Before the construction, the leader of group 4 organized a ceremony on May 20, 2018, for setting up the Guol house. The purpose of the ceremony was to pray for safety and advance. Besides, the villagers also performed other important ceremonies, including those for putting the ridge beam (May 22, 2018) and the inauguration ceremony (August 22, 2018). The actual construction of the Guol house in A Ka hamlet essentially followed nine steps (Fig. 6.41) and it was completed in four months (May 20, 2018–August 22, 2018) (Fig. 6.42).

- **Erecting main frame of Guol (pillars—floor beams—eaves beams—cross beams (steps 1, 2, 3, 4)**: After the ground is leveled,

Fig. 6.38 Application of body-based units in the reconstruction

Fig. 6.39 Implementing outwork

Fig. 6.40 Decorative painting on the center pillar

Table 6.11 Decoration on main components of Guol house in A Ka hamlet

No	Component	Decoration and Sculpture	Main color
1	Center pillar	Peacock, dragon, buffalo head, iguana, pigeon	Black, brown, ocean blue, red, yellow
		Wine jar, agate, stylized images (flower, leaf)	Black, white, red, brown, ocean blue, yellow
2	Cross beam	Tortoise, iguana, gold fish, crocodile, stylized images	Black, white, red, orange, ocean blue, green
3	Eaves beam	Normal life, ceremonies (buffalo sacrifice, hunting activity, dancing), elephant, buffalo, wild pig, tiger	Black, white, yellow, brown, ocean blue, green
	Upper wing beam	Normal life, ceremonies (buffalo sacrifice, hunting activity, dancing)	Black, white, yellow, brown, ocean blue
4	Lower ridge pole	Dragon	Ocean blue, white, red, yellow
5	Gable	Cock	Black, brown, red

13 cylindrical concrete blocks are placed exactly where the pillars are located. Next is the construction of the frame. Specifically, the two main pillars at the front and back of the central compartment and the lower half of the central pillar are joined by two big horizontal beams. This frame system is firmly fixed by a scaffolding system made of small wooden trunks. The next thing is to build the frame system of the remaining four pillars of the central compartment. These frames are assembled by eaves and cross beams,

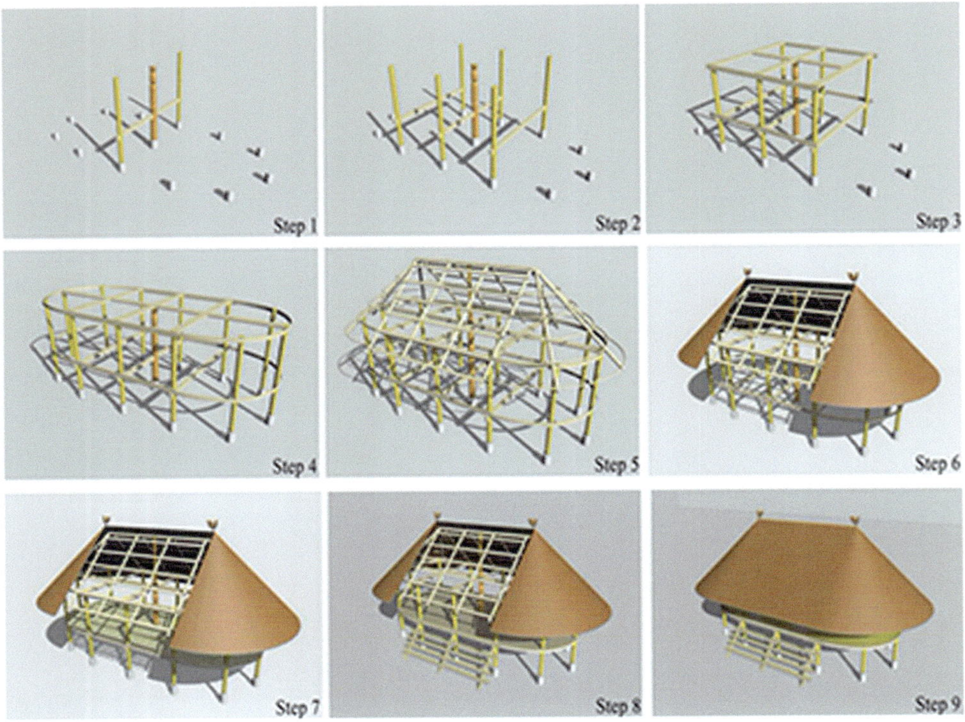

Fig. 6.41 Steps for constructing Guol house

forming a solid frame system. The next step is to erect the pillars of the two lean-tos. The three pillars of each lean-to are placed on cylindrical concrete blocks. Then they are connected by two horizontal round beams in position with the floor level at the bottom and the top of the column at the top, which have been curved in advance for three months.

- **Erecting roof frame (step 5)**: first, the upper central pillar (square section) is connected to the lower central pillar (circular cross-section) that was built earlier; After that, the ridge pole is placed on the center pillar at the midpoint position and supported by 16 roof beams. These roof beams are linked by 12 purlins in the central compartment and ten curved purlins made by the trunk of the rattan tree in the two lean-tos to create a solid roof frame that can withstand the impact of strong winds in the rainy season as well as having a suitable slope to prevent leakage.
- **Roofing (step 6)**: for roofing, the thatched roof sheets must be made in advance. The bamboo slats are woven together to create the sheet; After that, the conical leaves, after being fully collected and cut by the villagers, are attached to the bamboo frame. After the roofing sheets are done (size: 180 × 250 cm), the villagers conduct roofing. Roofing is carried out from top to bottom following the traditional and popular roofing method of the Cotu people.
- **Assembling floor beams and floor bamboo mats (step 7)**: at this stage, 21 wooden vertical floor beams are assembled parallel, spaced about 20 cm, and put on five main floor beams below. The floor mat is made after assembling the vertical floor beams. The assembly of these bars has been calculated so that after making the floor mat, it must be flat and horizontal. The floor mat is made from bamboo bars that are tied together by string and fastened to the floor beams.

Fig. 6.42 Guol house in A Ka hamlet in June 2021

- **Assembling stairs (step 8)**: first, three stair beams are placed at an angle suitable for the purpose of supporting the stairs. One end of the stair beam is attached to the mortise holes of the central compartment pillars. This connection ensures the necessary certainty. Then, four lower steps are attached to the slots that have been pre-punched with equal distance. Particularly, the top step is placed on the excess floor beams from the pillar with a larger distance from the lower step than the remaining steps to make it easier to sit on it.
- **Weaving and assembling bamboo walls (step 9)**: unlike bamboo used to make floor bamboo mats, the type used to make bamboo walls is Bop Bop (a type of small bamboo) due to its thin, soft, and flexible characteristics. To make the bamboo walls, the Bop Bop trunks are first split into flat pieces. The pieces are then knitted in an alternating inner-outer fashion to create an aesthetic effect. The bamboo wall sheets of different sizes are attached to the round lean-to beams at the top and pillars at the bottom. A total of 17 large and small sheets were installed, including eight sheets in two lean-tos, two sheets in the central compartment, two door sheets, and four sheets for the shelf at the back of the central compartment.

6.8.3 Significance and Experienced Lessons

The reconstruction project in A Ka hamlet exhibits the effort of the villagers with support from local government and consultant board. The completion of the reconstruction demonstrates that indigenous knowledge is still known by common villagers and can be applied to real construction. During the process of the reconstruction, certain modern techniques and tools derived from the King people were used for reducing the reconstruction time. This is suitable in the current modern context.

The new Guol house in A Ka hamlet may not be bigger and more beautiful than the other Guol houses in Quang Nam province, but it maintains the tangible and intangible values of the Cotu. The reconstruction was a change that allowed the young generation to learn and implement traditional construction knowledge and techniques by observing and participating in the process.

The Guol house will be a place to exhibit and store traditional tools and valuable items, as well as organize traditional ceremonies of the Cotu. Even more than two years after the inauguration, the Guol house has been preserved and it has been evaluated as the most beautiful Guol house in Nam Dong district. The villagers usually use it as a place for daily activities such as recreation and meetings, and to organize different events of the hamlet and commune. Finally, the Guol house in A Ka hamlet can be used as a reference for application for the reconstruction of other traditional community houses. This, in turn, can increase awareness of local villagers regarding preservation and maintenance of traditional, cultural values as well as passing traditional knowledge to the next generation.

6.9 Concluding Remark

Since 1986 when the government initiated a comprehensive reform, called Đổi Mới (renovation), Vietnam has enjoyed the rapid economic development, leading to urbanization. As described in Chap. 5, urbanization has caused several environmental impacts directly in cities, such as water pollution, air pollution, and poor waste management. It has also indirectly affected the livelihood and environment in rural areas, such as outmigration of young people from rural areas, change from subsistence to market-oriented crops, as discussed in Chap. 4. Of course, the government has taken initiatives to achieve SDGs (Chap. 1) and implemented several environmental policies (Chap. 3). Our activities of GSGES, Kyoto University (In this chapter) have also contributed to it. Considering that "No one will be left behind" is the final goal of SDGs, however, more studies and policy interventions should be warranted about the active participation of ethnic minorities and urban poor in the entire society. In addition, interconnectedness between urban and rural areas and among sectors such as agriculture, forestry, and manufacturing should be more aware in policy implementation and establishing research framework.

Column 9

"Safe Food" in Vietnam

As pesticides, chemical fertilizers, and genetically modified foods are introduced to our food production chains, an increasing number of people are getting interested in the safety of food—not only in developed countries but in developing countries as well. This tendency can also be seen in Vietnam, where certain projects for "safe food" have been launched in recent times.

After the Doi Moi (*Đổi Mới*) policy in 1986, agriculture in Vietnam was industrialized in order to increase production in an economically efficient manner. This led to an increase in the use of chemical substances in the food chain. Although regulations on the use of such chemicals have still not been provided, the national government did introduce the VietGAP (Vietnamese Good Agricultural Practices) certification system in 2008. This is in line with other GAP certifications such as Global GAP and EUREGAP. According to Vietnam Certification Center, VeitGAP includes 12 regulations for certifying vegetables and fruits, and currently, there are hundreds of organizations, units, and individuals whose products meet VietGAP standards.

For example, in Danang city, there are certain supermarkets and small shops that sell VietGAP certified vegetables and fruits (Fig. 6.43). These agricultural crops are grown in or near Danang city or in Dalat area. However, the prices of these certified products are usually higher than their non-certified counterparts sold in local markets—some even more than double the

Fig. 6.43 VietGAP certification system displayed in a small shop

price. According to staff working at one of these shops, their main customers include mothers raising kids, elderly people, and chefs at hotels targeting foreign tourists.

But safe food is not limited to certified products only. For example, in Tra Que (located in the Hoi An tourism area), traditional agriculture has been carried out for more than 400 years. Farmers here use seaweed that they can collect from brackish water around this area in order to make fertilizer (Fig. 6.44). Although the vegetables of this area are not VietGAP certified, local people recognize that the vegetables grown in Tra Que are safe. The farmers sell their products at almost the same price as other vegetables, although they sell them to the supermarket at 1.5 × the price.

Although the availability of safe food products is still limited, the above examples show that there are indeed customers who seek for products in Vietnam. Thus, safe food products, certified or otherwise, should be further promoted in future.

Column 10

Phytoremediation

"Phytoremediation" is an environmental purification technology that utilizes the ability of plants to absorb and decompose moisture and nutrients. Compared to other environmental restoration technologies, it is cost efficient and, at the same time, requires no energy, resulting in almost zero environmental impact, which makes it an attractive green option. The soil environment restoration business has already found a huge customer base in the United States and its future appears to be promising in Japan as well, where environmental regulations are becoming stricter.

Research papers have shown that vegetation plays an important role in bioremediation of highly toxic organic chemicals. At moderately contaminated sites with hazardous wastes, vegetation itself can effectively repair pollution and reduce maintenance costs. The use of plants for restoration is particularly suitable for soil

Fig. 6.44 Farming in Tra Que

contaminated with organic chemicals at a surface depth of less than two meters. It has been shown that the interaction between the roots of plants and the microbial groups of the rhizosphere promotes the decomposition of harmful organic chemicals in contaminated soil. The interaction between plants and harmful organic chemicals consists of accumulation and decomposition. The amount of contaminant taken up by the roots depends on the concentration of contaminants in the soil leachate, which is usually related to the separation mechanism of chemical substances on the surface of the roots as well as the permeation of chemical substances in plant conduits and hulls.

The plant itself indirectly helps in reducing the concentration of pollutants in the soil. The number and types of microorganisms that come into contact with the root are increasing. The reason for this is that there are substances that ooze from plants and substances such as shells of cell tissue, which can act as energy, carbon, nitrogen, or nutritional sources for growth for plants. Substances that ooze from plants include amino acids, carboxyl acids, hydrocarbons, nucleic acids and their leafy substances, nutrients, enzymes, and other related components.

Column 11

Sludge Reduction in Wastewater Treatment Plants in Vietnam

Vietnam is developing with each passing day. However, besides economic development, Vietnamese have been mindful of environment protection in general and wastewater treatment in particular. So far, we have built 35 urban wastewater treatment plants (WWTPs) in big cities/provinces. Based on the design data, we can treat a total of approximately 850,000 m^3/day. Besides that, we are designing and building 40 new WWTPs (total capacity of 1,600,000 m^3/day). These are based on different wastewater treatment technologies such as conventional activated sludge (CAS),

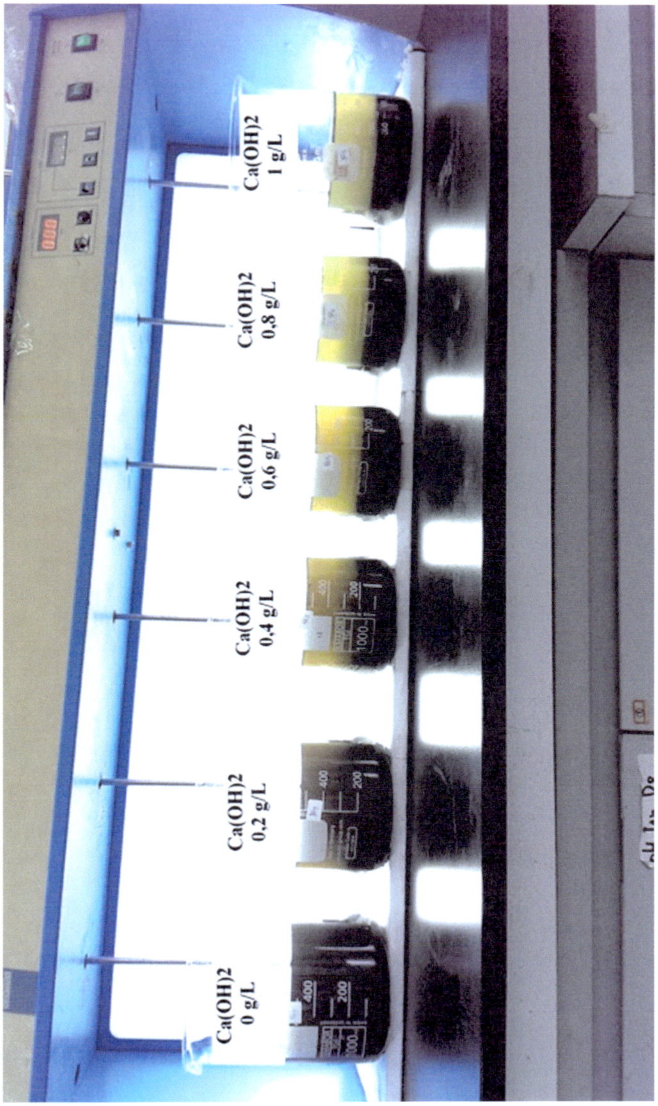

Fig. 6.45 Example of sludge pre-treated with different $Ca(OH)_2$ dosages

anaerobic-anoxic-aerobic (A_2O), oxidation ditch (OD), and sequencing batch reactor (SBR). During the operation of WWTPs, a lot of sludge is generated. This sludge, called sewage sludge, is collected from the wastewater treatment plant and then transported to a landfill. This is not only hard work but also expensive, because the sludge still contains a high portion of water and has a high volume. Therefore, there is a need for suitable methods to reduce sewage sludge. We tested several techniques for sludge reduction in the laboratory to find the best conditions for sludge reduction. It was found that the sludge volume and mass reduced significantly after pre-treatment with NaOH and $Ca(OH)_2$ (Fig. 6.45). In particular, we observed that the sludge further reduced when alkali dosage was increased. Moreover, this alkali sludge pre-treatment increased the soluble chemical oxygen demand (COD), which facilitates anaerobic digestion to produce biogas. Based on the obtained results, we tried to estimate the operating costs for the pre-treatment. In the case of NaOH used for sludge digestion at 1.2 g/L, we found that the chemical consumption was approximately 0.36 USD/kg TSS. Therefore, we recommend that alkali sludge pre-treatment could be considered as a potential method for sludge reduction in wastewater treatment plants in Vietnam.

References

Alshuwaikhat HM, Abubakar I (2008) An integrated approach to achieving campus sustainability: assessment of the current campus environmental management practices. J Clean Prod 16(16):1777–1785

Århem N (2009) In the sacred forest: landscape, livelihood and spirit beliefs among the Katu of Vietnam. Göteborgs Universitet

Bătinaş RH (2010) The methodology for assessing the potential attractiveness of waterfalls as tourist attractions. Studia Universitatis Babeş-Bolyai 2:205–212

Bh'Riu Liếc (2020) P'rá Cơtu Tiếng Cơtu B (Cotu language B). Hội Nhà Văn publishers

Biswanath S (2017) Role of local resources in rural development. Int J Bus Manag Invention 6(7):69–71

Boniface B, Cooper R, Cooper C (2016) Worldwide destinations: the geography of travel and tourism. Routledge

Brinkhurst M, Rose P, Maurice G, Ackerman JD (2011) Achieving campus sustainability: top-down, bottom-up, or neither? Int J Sustain High Educ 12(4):338–354. https://doi.org/10.1108/14676371111168269

Danang People Committee (2008) The project for danang environment city development. Danang People Committee

Dang NV, Chu TS, Luu H (2000) Ethnic minorities in Vietnam. The GIOI Publishers, Hanoi

Dlamini PN (2017) The use of information and communication technology tools in managing indigenous knowledge in the province of KwaZulu-Natal, South Africa [Doctoral dissertation, University of Zululand]. UNIZULU Institutional Repository

Gannon T, Singer J, McLellan B (2016) Collaborating for change. In: Singer J, Gannon T, Noguchi F, Mochizuki Y (eds) Educating for sustainability in Japan: fostering resilient communities after the triple disaster. Routledge, pp 119–142

Hồ TT, Le VA, Tanaka U (2010) Tiếp cận phát triển nông thôn dựa vào cộng đồng (Community-based rural development approaches). Agricultural Publishing House, Hanoi

Hồ TT, Nguyễn VC, Tanaka U (2016) Effects of supplementations of biocharcoal and wood vinegar in the diets on emissions of ammonia and hydrogen sulfide from pig manure. J Anim Husbandry Sci Tech Anim Husbandry Assoc Vietnam 203(2/2016):66–72 (in Vietnamese with English abstract)

Hoang TS, Le VA (2006) Creating opportunities for change: Strengthening social capital of women and the poor in upland communities in Hue, Viet Nam. In: Vernooy R (ed) Social and gender analysis in natural resource management: learning studies and lessons from Asia. Sage Publications, New Delhi, pp 155–179

Howard MC, Howard KB (2002) Textiles of the central highlands of Vietnam. Press, Bangkok, White L.O.T.U.S

Inskeep E (1991) Tourism planning: an integrated and sustainable development approach. Van Nostrand Reinhold

Kieu KT (2020) Report of building sustainable campus indicators for the Vietnamese higher education institutions. Project report, MOET

Kobayashi H, Nguyen NT (2013) Body-based units of measurement for building Cotu community houses in Central Vietnam. In: Proceedings of international conference on vernacular architecture CIAV2013 | 7°ATP | VerSus. Vernacular heritage and earthen architecture: contributions for sustainable development, pp 359–364

Kobayashi H et al (2008) Participatory construction of traditional community house in mountainous village of Central Vietnam. NXB Chính trị quốc gia, Hanoi

Lâm TQ (2013) Tổng quan về chính sách giáo dục Đại học Việt Nam và ý nghĩa đối với phát triển giáo dục đại học định hướng nghề nghiệp ứng dụng (Overview

of the Vietnamese higher education policies and their contribution to higher education development with job orientation towards applicability). Bộ Giáo Dục và Đào Tạo, Việt Nam (Ministry of Education and Training)
Le VA, Tanaka U, Kobayashi H (2013) Project report on local livelihood diversification for vulnerable people in natural disaster prone areas. Agricultural Publishing House, Hanoi
Le UV (2018) The sustainability efforts of Ton Duc Thang University in the South of Vietnam. E3S Web Conf 48:3–4. https://doi.org/10.1051/e3sconf/20184804008
Li Y (2004) Exploring community tourism in China: the case of Nanshan cultural tourism zone. J Sustain Tour 12(3):175–193. https://doi.org/10.1080/09669580408667232
Lozano R, Lukman R, Lozano FJ, Huisingh D, Lambrechts W (2013) Declarations for sustainability in higher education: Becoming better leaders, through addressing the university system. J Clean Prod 48:10–19. https://doi.org/10.1016/j.jclepro.2011.10.006
Lưu H (2007) A contribution to Cotu ethnography. Thế Giới publishers, Hanoi
Manhas PS, Manrai LA, Manrai AK (2016) Role of tourist destination development in building its brand image: a conceptual model. J Econ Financ Adm Sci 21(40):25–29. https://doi.org/10.1016/j.jefas.2016.01.001
MOET (2006) Program of higher education reform 2006 to 2020
Ngo LM, Trinh TA (2016) A university-city complex, a model for sustainable development: a case study in Vietnam. Procedia Eng 142:92–99. https://doi.org/10.1016/j.proeng.2016.02.018
Nguyen TC, Jintrawet A, Promburom P (2015) Impacts of seasonal climate variability on rice production in the central highlands of Vietnam. Agric Agric Sci Procedia 5:83–88. https://doi.org/10.1016/j.aaspro.2015.08.012
Nguyen TTP, Masuda M, Iwanaga S (2016) The effect of forestland allocation to the livelihoods of local people in the North Central Coast of Vietnam: a case in Nam Dong district. Tropics 24:169–180. https://doi.org/10.3759/tropics.24.169
Nguyen NT, Kobayashi H, Truong HP, Yoshizumi M, Le AT, Tran DS (2019) Reconstruction process of traditional community house of Cotu ethnic minority—case study of Aka hamlet in Nam Dong district, Thua Thien Hue province, Vietnam. In: Yong S et al (eds) Vernacular and earthen architecture towards local development. Tongji University Press, Shanghai, pp 518–524
Nguyễn NT, Kobayashi H, Võ NĐ (2014) Architecture of Guol house and construction method of body modules of Co tu ethnic group. J Sci Technol 1(2):203–212 (in Vietnamese with English abstract)
Nguyễn HT et al (2004) Katu kẻ sống đầu ngọn nước (Katu - the people living at the water-head). Thuận Hóa publishers
Pham TLP (2012) The renovation of higher education governance in Vietnam and its impact on the teaching quality at universities. Tert Educ Manag 18(4):289–308. https://doi.org/10.1080/13583883.2012.675350
Phạm VT (2017) Nghiên cứu mô hình Trường đại học Xanh - Kinh nghiệm quốc tế và đề xuất áp dụng cho Đại học Quốc gia Hà Nội, Báo cáo tổng kết đề tài cấp Đại học Quốc gia (Research on green university model from international experiences and recommendations for the Hanoi National University, Report)
Phan AT (2020) Xây dựng mô hình tự quản sống xanh trong các khu kí túc xá đại học, Báo cáo tổng kết Nhiệm vụ môi trường (Building green dormitory model at higher education institutions, Research report)
Priskin J (2001) Assessment of natural resources for nature-based tourism: the case of the Central Coast Region of Western Australia. Tour Manag 22(6):637–648. https://doi.org/10.1016/S0261-5177(01)00039-5
Rahayuningsih T, Muntasib EH, Prasetyo LB (2016) Nature based tourism resources assessment using geographic information system (GIS): case study in Bogor. Procedia Environ Sci 33:365–375. https://doi.org/10.1016/j.proenv.2016.03.087
Sandermann H Jr (1992) Plant metabolism of xenobiotics. Trends Biochem Sci 17(2):82–84. https://doi.org/10.1016/0968-0004(92)90507-6
Sasana H (2017) The strategy development of the region in support Borobudur tourism cluster competitiveness regions in Indonesia. J Environ Manag Tourism 8:1517–1529. https://doi.org/10.14505/jemt.v8.8(24).07
Savelyeva T, McKenna JR (2011) Campus sustainability: emerging curricula models in higher education. Int J Sustain High Educ 12(1):55–66
Singer J, Kieu KT, Pravitasari AE (2019) Solid waste management in tourist destinations in developing nations: case studies in Hoi An, Vietnam, and Puncak, Indonesia. In: So WWM, Chow CF, Lee JCK (eds) Environmental sustainability and education for waste management. Springer, Singapore, pp 189–206
Thua Thien Hue Statistical Office (2007) Statistical Year Book 2006. Thua Thien Hue Statistical Office, Hue
Tran TA (2014) Renovation of higher education and missions of the lecturers. Teach Learn Today 6:5–8
Tran CDTT, Villano RA (2017) An empirical analysis of the performance of Vietnamese higher education institutions. J Furth High Educ 41(4):530–544. https://doi.org/10.1080/0309877X.2015.1135886
Tran TD, Tanaka U, Kobayashi H (2012) Living with typhoon and flood disasters: a case study in Huong Phong commune, Tam Giang lagoon area, central Vietnam. SANSAI 6:85–96
Trần TV (2009) The Cotu in Vietnam. Vietnam News Agency Publishing House
Truong HP, Kobayashi H, Nguyen NT (2013) Typological research on traditional community house of the Cotu ethnic minority in Vietnam. In: Proceedings of international conference on vernacular

architecture CIAV2013 | 7ºATP | VerSus. Vernacular Heritage and Earthen Architecture: Contribution for Sustainable Development, pp 343–349
Uchiyama Y, Kohsaka R (2016) Cognitive value of tourism resources and their relationship with accessibility: a case of Noto region, Japan. Tourism Manag Perspect 19:61–68. https://doi.org/10.1016/j.tmp.2016.03.006
Waller V, Malherbe F, Tuc HD (2014) Green transformations at a university campus in Vietnam [Conference paper]. In: 14th international Australasian ACTS conference. Hobart, Australia

Correction to: Livelihood and Its Surrounding Environments in Rural Areas

Izuru Saizen, Shinya Funakawa, Yuki Okamoto, Tin Cong Hoang, Binh Huu Ngo, Ni Ngoc Khanh Tran, Hoa Thi Thai Hoang, Lam Ho Nguyen, Minh Tuan Vu, Hitoshi Shinjo, Giang Huong Pham, Ryo Sakaguchi, Masataka Kuroda, Hidenori Harada, Shigeo Fujii, Minori Tokito, Jane Singer, Hai Hoang, Huu Ty Pham and Kinh Thi Kieu

Correction to:
Chapter 4 in: S. Fujii et al. (eds.), *Livelihood and the Environment in Vietnam*, Sustainable Development Goals Series, https://doi.org/10.1007/978-981-97-2289-1_4

In the original version of the book, the following belated correction has been incorporated: The author's name "Thuc Dinh Do" has been removed in the Front Matter, and in Chapter 4. The book has been updated with the changes.

The updated version of this chapter can be found at
https://doi.org/10.1007/978-981-97-2289-1_4

S. Fujii et al. (eds.), *Livelihood and the Environment in Vietnam*, Sustainable Development Goals Series,
https://doi.org/10.1007/978-981-97-2289-1_7

Index

S. Fujii et al. (eds.), Livelihood and the Environment in Vietnam, Sustainable Development Goals Series,
https://doi.org/10.1007/978-981-97-2289-1